Springer Proceedings in Physics

Volume 302

Indexed by Scopus

The series Springer Proceedings in Physics, founded in 1984, is devoted to timely reports of state-of-the-art developments in physics and related sciences. Typically based on material presented at conferences, workshops and similar scientific meetings, volumes published in this series will constitute a comprehensive up to date source of reference on a field or subfield of relevance in contemporary physics. Proposals must include the following:

- Name, place and date of the scientific meeting
- A link to the committees (local organization, international advisors etc.)
- Scientific description of the meeting
- List of invited/plenary speakers
- An estimate of the planned proceedings book parameters (number of pages/articles, requested number of bulk copies, submission deadline).

Please contact:

For Americas and Europe: Dr. Zachary Evenson; zachary.evenson@springer.com
For Asia, Australia and New Zealand: Dr. Loyola DSilva; loyola.dsilva@springer.com

Nicolae Herisanu · Vasile Marinca
Editors

Acoustics and Vibration of Mechanical Structures—AVMS-2023

Proceedings of the 17th AVMS, Timişoara, Romania, May 26–27, 2023

Springer

Editors
Nicolae Herisanu
Department of Mechanics and Strength of Materials
University Politehnica Timisoara
Timişoara, Romania

Vasile Marinca
Department of Mechanics and Strength of Materials
University Politehnica Timisoara
Timişoara, Romania

ISSN 0930-8989 ISSN 1867-4941 (electronic)
Springer Proceedings in Physics
ISBN 978-3-031-48086-7 ISBN 978-3-031-48087-4 (eBook)
https://doi.org/10.1007/978-3-031-48087-4

This Springer imprint is published by the registered company Springer Nature Switzerland AG
The registered company address is: Gewerbestrasse 11, 6330 Cham, Switzerland

Paper in this product is recyclable.

Organization

Organized By

University Politehnica Timişoara—Acoustics and Vibration Laboratory
University of Nis—Noise and Vibration Laboratory
Romanian Academy—Centre for Fundamental and Advanced Technical Research
Romanian Acoustical Society

Organizing Committee

Nicolae Herisanu—Chairman, University Politehnica Timişoara, Romania
Momir Praščevič—Co-chairman, University of Nis, Serbia

Members

Florin Dragan—Rector—University Politehnica Timişoara, Romania
Vasile Marinca—University Politehnica Timişoara, Romania
Darko Mihajlov—University of Nis, Serbia
Ramona Nagy—University Politehnica Timişoara, Romania
Karoly Menyhardt—University Politehnica Timişoara, Romania
Cosmina Vigaru—University Politehnica Timişoara, Romania
Marko Ličanin—University of Nis, Serbia

Program Committee Chairs

Nicolae Herisanu, University Politehnica Timişoara, Timişoara, Romania
Vasile Marinca, University Politehnica Timişoara.

Program Committee Members

Jan Awrejcewicz, Lodz University of Technology, Poland
Vasile Bacria, University Politehnica Timişoara
Mahmoud Bayat, University of South Carolina, U.S.A.
Polidor Bratu, ICECON Bucharest, Romania
Livija Cveticanin, University of Novi Sad, Serbia
Dumitru Caruntu, University of Texas Rio Grande Valley, Mechanical Engineering, Edinburg, United States
Maciej Dutkiewicz, Bydgoszcz University of Science and Technology, Faculty of Civil, Environmental Engineering and Architecture, Bydgoszcz, Polska
Gillich Gilbert-Rainer, Babeş-Bolyai University, Cluj-Napoca, Romania
Nicolae Herisanu, University Politehnica Timişoara, Timişoara, Romania
Emil Manoach, Bulgarian Academy of Sciences, Bulgaria
Dan Marghitu, Auburn University, USA
Vasile Marinca, University Politehnica Timişoara
Darko Mihajlov, University of Nis, Serbia
Sotirios Natsiavas, Aristotle University of Thessaloniki, Greece
Carla Pinto, University of Porto, Portugal
Momir Praščevič, University of Nis, Faculty of Occupational Safety, Department of Preventive Engineering, Nis, Serbia
Sorin Vlase, Transilvania University Brasov, Romania.

Preface

This conference proceedings volume contains selected papers presented at the 17th International Conference on Acoustics and Vibration of Mechanical Structures (AVMS-2023) during 26–27 May 2023. This conference hosted as in the previous editions by the University Politehnica of Timişoara and the Romanian Academy—branch of Timişoara, was organized on five sections covering a broad range of topics related to acoustics and vibration problems:

1. Analytical Approaches to Vibration Problems
2. Environmental and Occupational Noise
3. Biomechanics and Bioacoustics
4. Vibration Problems in Industrial Processes
5. Structural Vibration.

Recent developments in analytical, numerical, and experimental techniques for linear and nonlinear noise and vibration problems were presented and discussed during the conference, which was an excellent opportunity for participants from USA, Brazil, and many European countries to present and discuss their newest achievements in the respective research areas.

All the papers went through a rigorous review process performed by the members of the International Scientific Committee and external reviewers, and the accepted papers are included in this volume.

We would like to thank all participants for their contributions to the conference proceedings. We also express our gratitude and appreciation to the members of the International Scientific Committee and Organizing Committee for their contribution to the success of this conference.

Finally, special thanks are given to Springer for producing this volume.

Timişoara, Romania

Nicolae Herisanu
Vasile Marinca

Contents

Contributors

Alexandra-Teodora Aman Doctoral School of Engineering, Babes-Bolyai University, Cluj-Napoca, Romania

Gina Diana Musca Anghelache Research Center for Mechanics of Machines and Technological Equipments, Engineering and Agronomy Faculty of Braila, Engineering Sciences and Management, Dunarea de Jos University of Galati, Braila, Romania

Tiberiu Baldogi Department of Mechanical Engineering, Technical University of Cluj-Napoca, Cluj-Napoca, Romania;
Qualification Laboratory Sibiu, Continental Sibiu, Sibiu, Romania

Milan Banić Faculty of Mechanical Engineering, University of Niš, Niš, Serbia

Ştefania Bîrsan Department of Mechanical Engineering, Transilvania University of Braşov, Brasov, România

Mišo Bjelić Faculty of Mechanical and Civil Engineering in Kraljevo, University of Kragujevac, Kraljevo, Serbia

Alexandru Bolcu Faculty of Mechanics, University of Craiova, Craiova, Romania

Dumitru Bolcu Faculty of Mechanics, University of Craiova, Craiova, Romania

Paul Nicolae Borza Department of Mechanical Engineering, Transilvania University of Braşov, Braşov, Romania

Polidor Bratu Romanian Academy of Technical Sciences, Bucharest, Romania;
The Institute of Solid Mechanics of the Romanian Academy, Bucharest, Romania

Daniela Georgiana Burtea Doctoral School of Engineering, Babes-Bolyai University, Cluj-Napoca, Romania

Gigel-Florin Capatana Faculty of Engineering and Agronomy in Brăila, Research Center for Mechanics of Machines and Technological Equipments, "Dunărea de Jos" University of Galaţi, Galaţi, Romania;

Faculty of Engineering and Agronomy in Braila, "Dunarea de Jos" University of Galati, Galati, Romania;
Research Center for Mechanics of Machines and Technological Equipments, Engineering and Agronomy Faculty of Braila, Engineering Sciences and Management, Dunarea de Jos University of Galati, Braila, Romania

Dumitru I. Caruntu University of Texas Rio Grande Valley, Edinburg, TX, USA

Cristina Chilibaru-Opritescu University Politehnica Timisoara, Timisoara, Romania

Eliza Chircan Department of Mechanical Engineering, Transilvania University of Brașov, Brașov, Romania

Mihaela Crisan-Vida Politehnica University of Timisoara, Timisoara, Romania

Lavinia Culda University Aurel Vlaicu of Arad, Arad, Romania

Jefferson da Silva Coelho Department of Mechanical Engineering, University of Brasilia, Brasil, Brazil

Carmen-Nicoleta Debeleac Faculty of Engineering and Agronomy in Braila, "Dunarea de Jos" University of Galati, Galati, Romania

Cornelia Dobrescu "Dunărea de Jos" University of Galati, Galati, Romania

Nicusor Dragan Research Center for Mechanics of Machines and Technological Equipments, Engineering and Agronomy Faculty of Braila, Engineering Sciences and Management, Dunarea de Jos University of Galati, Braila, Romania

Corneliu Drugă Transilvania University of Brașov, Brașov, Romania

Mircea-Cristian Dudescu Department of Mechanical Engineering, Technical University of Cluj-Napoca, Cluj-Napoca, Romania

Maciej Dutkiewicz Faculty of Civil, Architecture and Environmental Engineering and Architecture, Bydgoszcz University of Science and Technology, Bydgoszcz, Poland

Rita Fernandes Polytechnic Institute of Setubal, Setúbal, Portugal

Angela Găman National Institute for Research and Development in Mine Safety and Protection to Explosion—INSEMEX Petroşani, Petroşani, Romania

Aleksandar Gajicki Faculty of Occupational Safety, University of Niš, Niš, Serbia; Institute of Transportation CIP, Belgrade, Serbia

Norbert Gal-Nadasan Politehnica University of Timisoara, Timisoara, Romania

Mihaela Galescu-Blidariu Politehnica University Timisoara, Timişoara, Jud. Timiş, România

Attila Gere Hungarian University of Agriculture and Life Sciences, Gödöllő, Hungary

Attila Gerőcs University Aurel Vlaicu of Arad, Arad, Romania;
Babeș-Bolyai University, Cluj-Napoca, Romania;
University of Szeged, Szeged, Hungary

Gilbert-Rainer Gillich Doctoral School of Engineering, Babes-Bolyai University, Cluj-Napoca, Romania;
Department of Engineering Science, Babes-Bolyai University, Cluj-Napoca, Romania

Nicoleta Gillich Doctoral School of Engineering, Babes-Bolyai University, Cluj-Napoca, Romania

Vasile Ghiorghe Gliga Departament of Mechanical Engineering, Transilvania University of Brașov, Brasov, Romania;
S.C. Gliga Instrumente Muzicale S.A. Reghin, Mures, Romania

Adrian Mihai Goanta Research Center for Mechanics of Machines and Technological Equipments, Engineering and Agronomy Faculty of Braila, Engineering Sciences and Management, Dunarea de Jos University of Galati, Braila, Romania

Constantin-Adelin Grigorescu Transilvania University, Brașov, Romania

Maria-Violeta Guiman Faculty of Mechanical Engineering, "Transilvania" University of Brasov, Brasov, Romania;
Department of Mechanical Engineering, Transilvania University of Brașov, Brasov, România

Codruta Oana Hamat Department of Engineering Science, Babes-Bolyai University, Cluj-Napoca, Romania

Cornel Hațiegan Department of Engineering Science, Babes-Bolyai University, Cluj-Napoca, Romania

Nicolae Herisanu University Politehnica Timisoara, Timisoara, Romania;
Romanian Academy, Timisoara Branch, Timisoara, Romania

Brian Horsak St. Polten University of Applied Sciences, St. Pölten, Austria

Benjamin M. Huerta University of Texas Rio Grande Valley, Edinburg, TX, USA

Calin Itu Department of Mechanical Engineering, Transilvania University of Brașov, Brașov, Romania

Dragan S. Jovanović Faculty of Mechanical Engineering, University of Niš, Niš, Serbia

Laura Jurca University Politehnica Timisoara, Timisoara, Romania

Andrei Komjaty University Aurel Vlaicu of Arad, Arad, Romania;
University of Szeged, Szeged, Hungary

Izabella Kovacs National Institute for Research and Development in Mine Safety and Protection to Explosion—INSEMEX Petroşani, Petroşani, Romania

Marius Kovacs National Institute for Research and Development in Mine Safety and Protection to Explosion—INSEMEX Petroşani, Petroşani, Romania

Linda Lancere Vidzeme University of Applied Sciences, Valmiera, Latvia

Marcela R. Machado Faculty of Civil, Environmental Engineering and Architecture, Bydgoszcz University of Science and Technology, Bydgoszcz, Poland;
Faculdade de Tecnologia Asa Norte, University of Brasília, Campus Universitário Darcy Ribeiro, Brasília, DF, Brazil

Marko Mančić Faculty of Mechanical Engineering, University of Niš, Niš, Serbia

Milena Mančić Faculty of Occupational Safety, University of Niš, Niš, Serbia

Tiberiu Manescu Doctoral School of Engineering, Babes-Bolyai University, Cluj-Napoca, Romania;
Department of Industrial Engineering, Aurel Vlaicu University, Arad, Romania

Dan B. Marghitu Department of Mechanical Engineering, Auburn University, Auburn, AL, USA

Vasile Marinca University Politehnica Timisoara, Timisoara, Romania;
Romanian Academy, Timisoara Branch, Timisoara, Romania

Remus Stefan Maruta Politehnica University Timisoara, Timisoara Timis, Romania

Karoly Menyhardt Politehnica University Timisoara, Timisoara Timis, Romania

Manuela Micu Departament of Mechanical Engineering, Transilvania University of Brașov, Brasov, Romania;
Siemens Industry Software Romania, Brasov, Brașov, Romania

Darko Mihajlov Institute of Transportation CIP, Belgrade, Serbia

Mircea Mihalcica Department of Mechanical Engineering, Transilvania University of Brașov, Brasov, România

Goran Miodragović Academy of Applied Study Šumadija, Department Trstenik, Trstenik, Serbia

Tanja Miodragović Faculty of Mechanical and Civil Engineering in Kraljevo, University of Kragujevac, Kraljevo, Serbia

Cosmin Mihai Miriţoiu Faculty of Mechanics, University of Craiova, Craiova, Romania

Diogo Moco Polytechnic Institute of Setubal, Setúbal, Portugal

Elena Muncut University Aurel Vlaicu of Arad, Arad, Romania;
Babeș-Bolyai University, Cluj-Napoca, Romania

Gina Diana Muscă Faculty of Engineering and Agronomy in Braila, Research Centre for Machines and Technological Equipment's Mechanics, "Dunarea de Jos" University of Galati, Galati, Romania

Ramona Nagy Politehnica University Timisoara, Timisoara Timis, Romania

Silviu-Marian Nastac Departament of Mechanical Engineering, Transilvania University of Brașov, Brasov, Romania;
Faculty of Engineering and Agronomy in Braila, "Dunarea de Jos" University of Galati, Galati, Romania;
Faculty of Mechanical Engineering, "Transilvania" University of Brasov, Brasov, Romania

Alina Maria Nauncef Faculty of Music, Transilvania University of Brașov, Brașov, Romania

Petronela Nechita Faculty of Engineering and Agronomy in Braila, "Dunarea de Jos" University of Galati, Galati, Romania

Claudiu Nicolicescu Faculty of Mechanics, University of Craiova, Drobeta-Turnu Severin, Romania

Flavius Nițu Transilvania University of Brașov, Brașov, Romania

Marinela Cristina Nițu Institute of Solid Mechanics of the Romanian Academy, Bucharest, Romania

Sebastian Parlac University of Pitesti, Pitesti, Romania

Rusalin-Lucian Paun Doctoral School of Engineering, Babes-Bolyai University, Cluj-Napoca, Romania

Dan Alexandru Pîrşan Doctoral School of Engineering, Babes-Bolyai University, Cluj-Napoca, Romania

Constantin Picu "Dunarea de Jos", University of Galati, Galaţi, Romania

Mihaela Picu "Dunarea de Jos", University of Galati, Galaţi, Romania

Claudia-Mari Popa University of Pitesti, Pitesti, Romania

Dinel Popa University of Pitesti, Pitesti, Romania

Dragos-Mihai Popa University of Pitesti, Pitesti, Romania

Maria Aurora Potirniche University Dunarea de Jos, Galati, Romania;
Research Center for Mechanics of Machines and Technological Equipments, Engineering and Agronomy Faculty of Braila, Engineering Sciences and Management, Dunarea de Jos University of Galati, Braila, Romania;
Faculty of Engineering and Agronomy in Brăila, Research Center for Mechanics of Machines and Technological Equipments, "Dunărea de Jos" University of Galați, Galați, Romania

Zeno-Iosif Praisach Doctoral School of Engineering, Babes-Bolyai University, Cluj-Napoca, Romania;
Department of Engineering Science, Babes-Bolyai University, Cluj-Napoca, Romania

Momir Praščevič Faculty of Occupational Safety, University of Niš, Niš, Serbia

Milan Protić Faculty of Occupational Safety, University of Niš, Niš, Serbia

Alexandru Ioan Rădoi Faculty of Mechanics, University of Craiova, Craiova, Romania

Branko Radičević Faculty of Mechanical and Civil Engineering in Kraljevo, University of Kragujevac, Kraljevo, Serbia

Miomir Raos Faculty of Occupational Safety, University of Niš, Niš, Serbia

Andreea Rohan Transilvania University of Brașov, Brașov, Romania

Ioan Calin Rosca Departament of Mechanical Engineering, Transilvania University of Brașov, Brasov, Romania

Cristina Sandu Transilvania University of Brașov, Brașov, Romania

Mihai Seciureanu Faculty of Engineering and Agronomy in Braila, "Dunarea de Jos" University of Galati, Galati, Romania

Ionel Șerban Transilvania University of Brașov, Brașov, Romania

Sorin Simion National Institute for Research and Development in Mine Safety and Protection to Explosion—INSEMEX Petroşani, Petroşani, Romania

Mark Simonlehner St. Polten University of Applied Sciences, St. Pölten, Austria

Marius Marinel Stănescu Faculty of Mechanics, University of Craiova, Craiova, Romania

Mariana Domnica Stanciu Departament of Mechanical Engineering, Transilvania University of Brașov, Brasov, Romania

Yasmine Sterckx UC Leuven-Limburg University of Applied Sciences, Leuven, Belgium

Andreea Stoia Politehnica University of Timisoara, Timisoara, Romania

Dan Ioan Stoia Politehnica University of Timisoara, Timisoara, Romania

Daniela Tarnita Faculty of Mechanics, University of Craiova, Craiova, Romania

Oana Tonciu Technical University of Civil Engineering of Bucharest, Bucharest, Romania

Lorand Toth National Institute for Research and Development in Mine Safety and Protection to Explosion—INSEMEX Petroşani, Petroşani, Romania

Alin Totorean Politehnica University of Timisoara, Timisoara, Romania

Andrei Tudor The Institute of Solid Mechanics of the Romanian Academy, Bucharest, Romania

Irina-Virginia Tudor University of Pitesti, Pitesti, Romania

Cristian Tufisi Department of Engineering Science, Babes-Bolyai University, Cluj-Napoca, Romania

Alexandru-Constantin Tulică Transilvania University, Brașov, Romania

Ovidiu Vasile National University of Sciences and Technology POLITEHNICA Bucharest, Bucharest, Romania

Cosmina Vigaru University Politehnica Timisoara, Timisoara, Romania

Sorin Vlase Department of Mechanical Engineering, Transilvania University of Brașov, Brașov, Romania;
Romanian Academy of Technical Sciences, Bucharest, Calea Victoriei, Romania

Cornelia Zaharia Transilvania University of Brașov, Brașov, Romania

Abdul Zulkarnain Hungarian University of Agriculture and Life Sciences, Gödöllő, Hungary

Chapter 1
Parametric Resonance of Electrostatically Actuated MEMS Angled Uniform Cantilever Resonators: Amplitude-Frequency Response

Dumitru I. Caruntu and Benjamin M. Huerta

Abstract This paper deals with amplitude-frequency response of parametric resonance of electrostatically actuated microelectromechanical system (MEMS) uniform cantilever resonators. The model consists of an elastic cantilever beam positioned at an angle relative to a ground plate. They are not parallel. There is an initial gap distance between the beam and the plate. Alternating current (AC) at frequency near natural frequency of the MEMS resonator produces an electrostatic force that leads to parametric resonance. The electrostatic force, which includes the fringe effect, is nonlinear, while the viscous damping force is linear. The amplitude-frequency response consists of two bifurcations, namely subcritical and supercritical. The effect of the angle between the beam and the ground plate, and the effect of the ratio of the beam length to the gap length, on the amplitude-frequency response of the system are investigated. A comparison, amplitude-frequency response of parallel beam versus amplitude-frequency response of angled beam, is reported. Two methods, the Method of Multiple Scales (MMS) and Reduced Order Model (ROM), are used in this work. For small amplitudes, the predictions of the two methods are in agreement. However, MMS is only accurate for small amplitudes, while ROM, given a sufficient number of terms, is accurate for all amplitudes. The results show that as the angle between the MEMS cantilever beam and ground plate increases, the amplitude-frequency response is shifted to higher frequencies and amplitudes. A sufficiently large angle results in the disappearance of the subcritical and supercritical bifurcation points, and the appearance of a saddle-node bifurcation. A larger length-to-gap ratio amplifies the effect of the angle on the response.

D. I. Caruntu (✉) · B. M. Huerta
University of Texas Rio Grande Valley, Edinburg, TX 78539, USA
e-mail: dumitru.caruntu@utrgv.edu

N. Herisanu and V. Marinca (eds.), *Acoustics and Vibration of Mechanical Structures—AVMS-2023*, Springer Proceedings in Physics 302,
https://doi.org/10.1007/978-3-031-48087-4_1

1.1 Introduction

Microelectromechanical systems (MEMS) are used in a variety of modern devices. Their small size, low cost, and low energy consumption make them excellent candidates for applications such as energy harvesters [1], sensors [2], switches [3], and filters [4]. They have been utilized extensively in the automotive [5], medical [6], and smartphone [7] industries. The diversity of MEMS applications has made them the subject of intense research in recent years.

There are several methods by which MEMS devices can be actuated, including piezoelectric, electrostatic, electromagnetic, and pneumatic [8]. Among the actuation methods, electrostatic actuation is most common. This method of actuation offers the simplest design with accurate control [9]. Electrostatic actuation typically utilizes an elastic beam positioned parallel to a rigid ground plate with a gap distance between them. An AC and/or DC voltage is applied between the beam and plate, causing the beam to deform. If the applied voltage is DC, the beam deflects statically, whereas if the applied voltage is AC, the beam vibrates [9]. A phenomenon associated with this configuration is pull-in, in which the beam collapses onto the ground plate due to the electrostatic force [10]. This occurs once the applied voltage reaches a certain critical value. Pull-in is detrimental to the functionality of certain MEMS devices, particularly resonators. Several designs and modifications have been proposed to eliminate or reduce pull-in [11, 12]. In addition to the electrostatic force, there is an additional force present due to the fringe effect. This fringe effect force acts on the beam due to the electric field that is outside the volume between the beam and the ground plate, further contributing to its deflection towards the ground plate [13]. The fringe effect is sensitive to gap distance between the beam and the ground plate, and width of the beam [13]. In addition, viscous damping, due to air resistance on the beam, influences the amplitude-frequency response. At the microscale, fringe and damping affect the stability of the system, as well as pull-in behavior [13].

This work proposes a design in which the cantilever beam is positioned at an angle relative to the ground plate. The advantage of the angled design is the larger maximum deflection at the tip of the cantilever before contacting the ground plate. In addition, it is desired to understand the effect of this angle on the response of the structure. This work investigates the amplitude-frequency response of parametric resonance of angled MEMS cantilever beams under electrostatic actuation. In the model, the electrostatic force, fringe effect, and viscous damping are considered. Two methods, the Method of Multiple Scales (MMS) and Reduced Order Model (ROM) with two modes of vibration, are used to solve the equation of motion that describes the deflection of the beam.

1.2 Equation of Motion

The model considered in this study is shown in Fig. 1.1. This model consists of an elastic cantilever beam positioned above a ground plate at an initial gap distance g. The beam is positioned at an angle θ relative to the ground plate. An electrostatic force between the beam and plate is generated by an applied AC voltage. The partial differential equation describing beam's deflection along with boundary conditions are [14].

$$\frac{\partial^2 u}{\partial \tau^2} + b^* \frac{\partial u}{\partial \tau} + \frac{\partial^4 u}{\partial z^4} = \frac{\delta V^2}{(1 + Pz \tan\theta - u)^2} + \frac{f \delta V^2}{(1 + Pz \tan\theta - u)} \tag{1.1}$$

$$u(0) = \frac{\partial u}{\partial z}(0) = \frac{\partial^2 u}{\partial z^2}(1) = \frac{\partial^3 u}{\partial z^3}(1) = 0 \tag{1.2}$$

The dimensionless and dimensional variables, beam geometry and material properties, as well as dimensionless parameters are given in Table 1.1.

It should be noted that the present model is valid only for slender beams. This refers to beams with a length-to-thickness ratio l/h greater than or equal to 100. So, the length-to-gap ratio $P = l/g$ satisfies

$$Pg \geq 100h \tag{1.3}$$

The dimensionless variables and parameters are given by (also see Table 1.1)

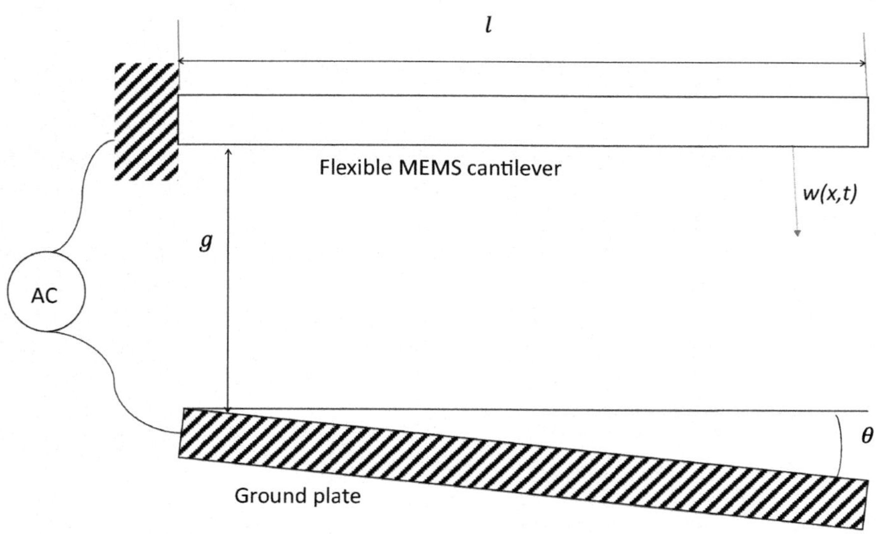

Fig. 1.1 Angled MEMS resonator schematic

Table 1.1 Variables/parameters

Symbol	Variable/parameter	Symbol	Variable/parameter
u	Dimensionless deflection	w	Dimensional deflection
τ	Dimensionless time	t	Dimensional time
z	Dimensionless longitudinal coordinate	x	Dimensional long coordinate
f	Dimensionless fringe	l	Beam length
δ	Dimensionless voltage	W	Beam width
b^*	Dimensionless damping	g	Initial gap length
P	Length to gap ratio	A_0	Ref. cross-sectional area
σ	Dimensionless detuning frequency	I_0	Ref. moment of inertia
Ω	Dimensionless AC frequency	ρ	Density of material
θ	Beam to plate angle	E	Young's modulus
V	Voltage	b	Damping coefficient

$$u = \frac{w}{g}, z = \frac{x}{l}, \tau = \frac{t}{l^2}\sqrt{\frac{EI_0}{\rho A_0}}, \delta = \frac{\epsilon_0 W l^4}{2g^3 EI_0}, f = \frac{0.65g}{W}, b^* = \frac{bl^2}{\sqrt{\rho A_0 EI_0}} \tag{1.4}$$

1.3 Parametric Resonance

The dimensionless AC frequency Ω is near first natural frequency of the beam ω_1

$$V(\tau) = \cos\Omega\tau \;\; \Omega = \omega_1 + \varepsilon\sigma \tag{1.5}$$

where σ is detuning frequency and ε a small bookkeeping parameter used in MMS. The detuning frequency is the difference between the natural frequency of the beam and the applied AC frequency. A negative detuning frequency corresponds to an applied frequency that is slightly less than the natural frequency of the beam. The dimensionless AC frequency and the natural frequency are given by $\Omega = \overline{\Omega} l^2 \sqrt{\rho A_0/EI_0}$ and $\omega_1 = \overline{\omega}_1 l^2 \sqrt{\rho A_0/EI_0}$, respectively, where $\overline{\Omega}$ and $\overline{\omega}_1$ are the corresponding dimensional frequencies. The electrostatic force is proportional to the voltage squared, Eq. (1.1). Squaring the voltage in Eq. 1.5, one can notice that the electrostatic force has a frequency near twice the first natural frequency of the beam, resulting in parametric resonance.

1.4 Method of Multiple Scales

The Method of Multiple Scales (MMS) is a perturbation method. While it provides accurate results at small amplitudes, it tends to underestimate the softening effect, which is the leftward bending of the branches as the amplitude increases, (see Fig. 1.2) [15]. The small bookkeeping parameter ε is utilized to mark small terms. The fringe, excitation, and viscous damping parameters in this work are considered to be small, therefore they are multiplied by ε. Expanding in Taylor series the right side of Eq. (1.1), and retaining the terms up to the cubic power, it results

$$\begin{aligned} \ddot{u} + \epsilon b^* \dot{u} + u^{(4)} &= \epsilon \delta V^2 \left(\frac{1}{H^2} + \frac{2u}{H^3} + \frac{3u^2}{H^4} + \frac{4u^3}{H^5} \right) \\ &\quad + \epsilon \delta f V^2 \left(\frac{1}{H} + \frac{u}{H^2} + \frac{u^2}{H^3} + \frac{u^3}{H^4} \right) \end{aligned} \tag{1.6}$$

$$H(z) = 1 + Pz \tan\theta \tag{1.7}$$

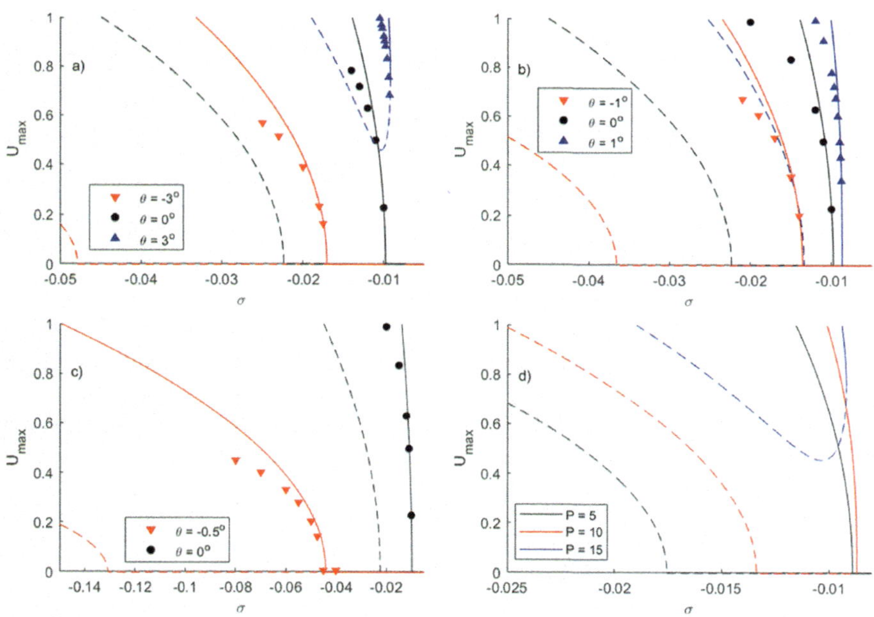

Fig. 1.2 For all figures, $f = 0.26$, $\delta = 0.1$, $b^* = 0.01$. . Solid lines show the MMS results stable branch, whereas dashed lines are the unstable branch. Markers display 2-term ROM. **a** The effect of θ is considered when $P = 5$. **b** The effect of θ is considered when $P = 10$. **c** The effect of θ is considered when $P = 60$. **d** The effect of P is considered when $\theta = 1°$

Using in MMs the fast time scale $T_0 = t$ and the slow time scale $T_1 = \varepsilon t$, the time derivatives become

$$\dot{\square} = \frac{\partial}{\partial \tau} = D_0 + \epsilon D_1 \quad \ddot{\square} = \frac{\partial^2}{\partial \tau^2} = D_0^2 + 2\epsilon\, D_0 D_1 + \epsilon^2 D_1^2 \tag{1.8}$$

where D_0 represents the derivative with respect to T_0, and D_1 is the derivative with respect to T_1. A first-order expansion of the dimensionless displacement u is as follows

$$u = u_0 + \epsilon u_1 \tag{1.9}$$

Equations (1.8) and (1.9) are substituted into Eq. (1.6). All terms containing the bookkeeping parameter at any power higher than or equal to two are considered to be very small and are neglected. Collecting the terms of like powers of ε and setting their sum equal to zero, two problems, namely zeroth-order and first-order, result as follows

$$D_0^2 u_0 + u_0^{(4)} = 0 \tag{1.10}$$

$$\begin{aligned} D_0^2 u_1 + 2D_0 D_1 u_0 + u_1^{(4)} = & - b^* D_0 u_1 + \left(\frac{\delta V^2}{H^2} + \frac{f\delta V^2}{H}\right) + \left(\frac{2\delta V^2}{H^3} + \frac{f\delta V^2}{H^2}\right) u_0 \\ & + \left(\frac{3\delta V^2}{H^4} + \frac{f\delta V^2}{H^3}\right) u_0^2 + \left(\frac{4\delta V^2}{H^5} + \frac{f\delta V^2}{H^4}\right) u_0^3 \end{aligned} \tag{1.11}$$

The solution to Eq. (1.10) is given by

$$u_0 = \phi_1(z)\left[A(T_1)e^{i\omega_1 T_0} + \overline{A}(T_1)e^{-i\omega_1 T_0}\right] \tag{1.12}$$

where ϕ_1 is the first mode shape and ω_1 is the first natural frequency. Equation (1.12) is substituted into Eq. (1.11). Then, the sum of all secular terms (terms containing $e^{i\omega_1 T_0}$) is set equal to zero, resulting the secular terms equation which is then multiplied by the mode shape ϕ_1 and integrated from 0 to 1 resulting the following g_{mn} coefficients

$$g_{mn} = \int_0^1 \frac{\phi_1^m(z)}{H^n(z)} dz, \quad m = 2, 4, \quad n = 0, 2, 3, 4, 5 \tag{1.13}$$

Then, the following substitutions for the complex amplitude A and its conjugate

$$A = \frac{1}{2} a e^{i\beta} \quad \overline{A} = \frac{1}{2} a e^{-i\beta} \tag{1.14}$$

are made into the resulting equation. Also, the following substitution used

$$\gamma = \sigma T_1 - \beta \; \gamma' = \sigma - \beta' \tag{1.15}$$

The steady state solutions (amplitude-frequency response) result from setting the derivatives with respect to slow time scale T_1 of a and γ equal to zero. The trivial solution, i.e. $a = 0$, is a steady state solution. The non-trivial case where amplitude is not equal to zero results in Eqs. (1.16–1.17) as follows

$$a^2 = \frac{8b^*\omega_1 g_{20} - 2\delta(2g_{23} + fg_{22})\sin(2\gamma)}{\delta(4g_{45} + fg_{44})\sin(2\gamma)} \tag{1.16}$$

$$\sigma = -\frac{\delta}{16\omega_1 g_2}[8g_{23} + 4g_{22}f + 12g_{45}a^2 + 3g_{44}fa^2 + 4g_{23}\cos(2\gamma) + 2g_{22}f\cos(2\gamma) + 8g_{45}a^2\cos(2\gamma) + 2g_{44}fa^2\cos(2\gamma)]. \tag{1.17}$$

1.5 Reduced Order Model

A second method of investigation is Reduced Order Model (ROM) using two modes of vibration in which the partial differential equation of motion is transformed into a system of two second-order ordinary differential equations that is numerically integrated. This method is accurate for both small and large amplitudes, as well small or large nonlinearities, if a sufficient number of terms is used. The solution is assumed as

$$u(z, \tau) = \sum_{i=1}^{N} u_i(\tau)\phi_i(z) \tag{1.18}$$

where N is the number of terms (number of modes of vibration) considered, $u_i(\tau)$ are time dependent functions to be determined, and $\phi_i(z)$ are the corresponding mode shapes of the beam. Mode shapes $\phi_i(z)$ are given by

$$\phi_i(z) = -\{\cos(\sqrt{\omega_k}z) - \cosh(\sqrt{\omega_k}z) + c_k[\sin(\sqrt{\omega_k}z) - \sinh(\sqrt{\omega_k}z)]\} \tag{1.19}$$

The first five mode shape coefficients c_k and dimensionless natural frequencies ω_k are given in Table 1.2.

Substituting Eq. (1.18) into Eq. (1.1), multiplying the resulting equation by the mode shape ϕ_i, $i = 1, 2$, and integrating from $z = 0$ to 1, this results in a system of $N = 2$ second-order differential equations that are solved numerically.

Table 1.2 First five mode shape coefficients and natural frequencies for a cantilever beam

	$k = 1$	$k = 2$	$k = 3$	$k = 4$	$k = 5$
ω_k	3.51562	22.0336	61.70102	120.91202	199.85929
c_k	– 0.734	– 1.0185	– 0.9992	– 1.00003	– 1.00000

1.6 Results and Conclusions

Figure 1.2 shows the amplitude-frequency response as predicted by MMS and ROM in four different cases. Figure 1.3 shows an example of a time response obtained using ROM.

In Fig. 1.2a–d the horizontal axis is the detuning frequency σ, and the vertical line is the amplitude of the tip of the MEMS cantilever resonator $U_{\max}$. Solid lines show the MMS stable solutions, dash lines show the MMS unstable solutions, whereas the markers show the time responses obtained using ROM. The effect of varying the length to gap ratio P on the amplitude-frequency response of the system can be seen in Fig. 1.2a–c. If the case of θ=0°, Fig. 1.2a, the response consists of two bifurcations, (1) a subcritical bifurcation with a subcritical bifurcation point at the intersection between the unstable (dash line) branch and σ-axis, and (2) a supercritical bifurcation with the subcritical bifurcation point at the intersection between the stable (solid line) branch and σ-axis. The zero solution (σ-axis) branch of steady states is stable except between the subcritical and supercritical bifurcation points.

Same behavior occurs in the case of θ=−3°. In the case of θ=3°, the subcritical and supercritical points vanished, and a saddle-node bifurcation (common point of the stable (solid line) and unstable (dash line)) branches in higher amplitudes is born. The zero solution (σ-axis) branch is stable for all values of σ. Results show that as the angle θ increases, the response of the system is shifted towards higher frequencies, Fig. 1.2a–c, while subcritical and supercritical bifurcation points move

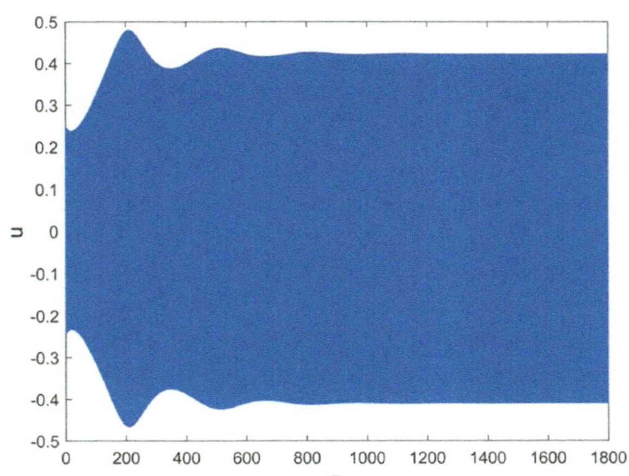

Fig. 1.3 The position of the tip of the beam is plotted in the time response, obtained using 2-term ROM. Amplitude at steady state is considered. $f = 0.26$, $\delta = 0.1$, $b^* = 0.01$, $\sigma = -0.015$. Initial displacement $u_0 = 0.25$. Angle $\theta = 0°$. Length-to-gap ratio $P = 10$

closer together, Fig. 1.2a–c, until they coalesce and then vanish, Fig. 1.2a. Figure 1.2d shows the effect of length to gap ratio P for θ=1°. A larger ratio P results in a greater sensitivity to changes in the angle θ. Given a sufficiently large angle θ and ratio P, the two branches, stable and unstable, coalesce. This can be also seen in Fig. 1.2a for the case of θ=3°. The bifurcation points can be controlled by adjusting the angle θ and ratio P. An example of the time responses obtained using ROM are shown in Fig. 1.3. The horizontal axis is dimensionless time, and the vertical axis is the oscillation of the free end of the MEMS cantilever resonator. Limitation of this work are (1) it only predicts the behavior of slender MEMS Cantilever resonators and (2) it does not include quantum dynamics effects, i.e. Casimir and van der Waals forces, for NEMS structures [16, 17].

References

1. S. Azizi, A. Ghodsi, H. Jafari, M.R. Ghazavi, A conceptual study on the dynamics of a piezoelectric MEMS energy harvester. Energy **96**, 495–506 (2016)
2. M.H. Hasan, F.M. Alsaleem, H.M. Ouakad, Novel threshold pressure sensors based on nonlinear dynamics of MEMS resonators. J. Micromech. Microeng. **28**(6), 65007–65019 (2018)
3. I.E. Lysenko, A.V. Tkachenko, O.A. Ezhova, S.I. Shafrostova, Design and simulation of RF MEMS switch for use in the millimetre wavelength range and SPDT RF MEMS switch. J. Phys. **1695**(1), 12162–12170 (2020)
4. Z. Shi, L. Fang, L. Zhong, MEMS-based filter integrating tunable Fabry–Perot cavity and grating. Optics Commun. **402**, 472–477 (2017)
5. Z. Pentek, T. Hiller, A. Czmerk, Algorithmic enhancement of automotive MEMS gyroscopes with consumer-type redundancy. IEEE Sens. J. **21**(2), 2092–2103 (2021)
6. Q. Zhen, W. Piyawattanametha, MEMS-based medical endomicroscopes. IEEE J. Sel. Top. Quantum Electron. **21**(4), 376–391 (2015)
7. A. Elhattab, N. Uddin, E. O'Brien, Extraction of bridge fundamental frequencies utilizing a smartphone MEMS accelerometer. Sensors **19**(14), 3143–3163 (2019)
8. D.I. Caruntu, R. Oyervides, Frequency response reduced order model of primary resonance of electrostatically actuated MEMS circular plate resonators. Commun. Nonlinear Sci. Numer. Simul. **43**, 261–270 (2017)
9. D.I. Caruntu, R. Oyervides, Amplitude–frequency response of parametric resonance of electrostatically actuated MEMS clamped circular plate. Int. J. Non-Linear Mech. **149**, 104310–104319 (2023)
10. D.I. Caruntu, M.A. Botello, C.A. Reyes, J. Beatriz, Frequency–amplitude response of superharmonic resonance of second order of electrostatically actuated MEMS cantilever resonators. Int. J. Non-Linear Mech. **133,** 103719 (9 pages) (2021)
11. M. Pallay, M. Daeichin, S. Towfighian, Dynamic behavior of an electrostatic MEMS resonator with repulsive actuation. Nonlinear Dyn. **89**(2), 1525–1538 (2017)
12. X. Yang, P. Kästner, E. Käkel, M. Smolarczyk, S. Liu, Q. Li, H. Hillmer, Study of dynamics in metallic MEMS cantilevers—Pull-in voltage and actuation speed. Appl. Sci. **13**(2), 1118–1140 (2023)
13. D.I. Caruntu, M.A. Botello, C.A. Reyes, J. Beatriz, Voltage–amplitude response of superharmonic resonance of second order of electrostatically actuated MEMS cantilever resonators. J. Comput. Nonlinear Dyn. **14**(3), 031005 (8 pages) (2019)
14. D.I. Caruntu, M. Knecht, On nonlinear response near half natural frequency of electrostatically actuated microresonators. Int. J. Struct. Stab. Dyn. **11**(4), 641–672 (2011)

15. D.I. Caruntu, I. Martinez, Reduced order model of parametric resonance of electrostatically actuated MEMS cantilever resonators. Int. J. Non-Linear Mech. **66**, 28–32 (2014)
16. D.I. Caruntu, C.A. Reyes, Casimir effect on amplitude-frequency response of parametric resonance of electrostatically actuated NEMS cantilever resonators. In: Developments Novel Approaches Biomech. Metamaterials. Adv. Struct. Mat. **132,** 267–289, (Springer, Cham, 2020) (Chapter 15)
17. D.I. Caruntu, J.S. Beatriz, Quantum Dynamics Effects on Amplitude-Frequency Response of Superharmonic Resonance of Second-Order of Electrostatically Actuated NEMS Circular Plates. In: Theoretical Analyses, Computations, and Experiments of Multiscale Materials, Series: Advanced Structured Materials, vol. 175, (Springer, Cham, 2022), pp. 69–104 (Chapter 4)

Chapter 2
The Approximate Entropy of Human Knee with Osteoarthritis

Dan B. Marghitu and Daniela Tarnita

Abstract We analyze the knee joint to estimate the flexion and extension motion dynamics for humans with osteoarthritis. The motion is aperiodic. Knee joint angle time series were analyzed and reconstructed with lag and embedding dimensions. The Lorenz maps and the approximate entropy were calculated from experimental data. The osteoarthritis patients have higher values of approximate entropy compared with healthy subjects.

2.1 Introduction

Shannon was one of the first researchers to describe entropy as the information content [1]. The regularity and complexity of a time series are evaluated by entropy. Pincus introduced the approximate entropy [2]. Next, the sample entropy was studied by Richman and Moorman [3], and permutation entropy and multi-scale entropy were analyzed in [4, 5]. For physiological data Gan and Learmonth measured the sample entropy [6].

Yentes et al., examine the resilience of approximate entropy and sample entropy for sets of short biological data [7]. They explore the length of the data segment, the similarity criterion, and the length of the data. Their results show sensitivity for very short data sets, and extreme attention is required to select experimental data parameters. The study's conclusion shows that young adults walk has more regular patterns than older adults.

The nonlinear dynamics of the human knee with osteoarthritis were studied with the largest Lyapunov exponents from experimental data in [8]. The largest Lyapunov exponents of osteoarthritis patients have higher values than healthy humans. The calculation of Lyapunov exponents may depend on filtering the experimental time

D. B. Marghitu
Department of Mechanical Engineering, Auburn University, 36349 Auburn, AL, USA
e-mail: marghitu@auburn.edu

D. Tarnita (✉)
Faculty of Mechanics, University of Craiova, Craiova, Romania
e-mail: daniela.tarnita@edu.ucv.ro

N. Herisanu and V. Marinca (eds.), *Acoustics and Vibration of Mechanical Structures—AVMS-2023*, Springer Proceedings in Physics 302,
https://doi.org/10.1007/978-3-031-48087-4_2

series. The investigation of filtering was studied in [9]. They show that the level of smoothing of the data affects the calculation of the exponents.

The objective of the study reported here was to develop from kinematic data an approach to study the locomotion of human knee with osteoarthritis using approximate entropy.

2.2 Approximate Entropy

The statistical entropy defines the probability of information in a time series [1]. The approximate entropy is calculated for a uniform series. Approximate entropy quantifies the regularity and unpredictability of changes in a signal.

We reconstruct the phase space of a time series θ, the knee angle. First, the delay in phase space reconstruction is specified. The vector $\boldsymbol{\theta}$ is reconstructed using the delay or lag. Random noise is introduced in the signal if the lag is too small. The reconstructed system does not characterize the dynamics if the lag is too large. The phase space reconstruction is also specified by embedding dimension. The knee data is sampled at a regular rate T_s. The knee angle signal is made at discrete times $\theta(t) = \theta(t_0 + n\,T_s)$, where t_0 is the starting time and $n = 1, 2, ...N$ and a set $\theta(n)$ can be created. With the delay coordinates vectors method, the state space is reconstructed.

$$\boldsymbol{\theta} = \{\theta(n), \theta(n+T), ..., \theta(n+(d-1)T)\}, \tag{2.1}$$

where T is time delay and d is called the embedding dimension. The Lorenz attractor has three dimensions, and the estimated embedding dimension is 3. For more information on lag and embedding dimension, see [10, 11].

Approximate entropy quantifies the amount of regularity and unpredictability of fluctuations over a time series. The approximate entropy function first generates a delayed reconstruction $\theta(1, 2, ..N)$ for N data points with embedding dimension d and lag T. The approximate entropy is calculated with [12]

$$\text{ApEn} = [\phi^m - \phi^{m+1}], \tag{2.2}$$

where

$$\phi^m = \frac{1}{N-m+1} \sum_{i=1}^{N-m+1} \log C_i, \tag{2.3}$$

$$C_i = \sum_{i=1, i\neq k}^{N} \mathbf{1}\left(||\theta_i - \theta_k||_\infty \leq r\right). \tag{2.4}$$

An indicator function is denoted with **1**, and r is the tolerance. Pincus developed approximate entropy to measure the rate of information generation of the system. He estimated that this parameter would give us a good estimation of the change in the complexity of data.

2.3 Lorenz Map

For the Lorenz map, we focus on θ_m, the mth local maximum of $\theta(t)$. Lorenz's idea is that θ_m should predict $\theta_{(m+1)}$, [13, 14]. We plot $\theta_{(m+1)}$ versus θ_m. The function $\theta_{(m+1)} = f(\theta_m)$ is now called the Lorenz Map. The Lorenz map characterized the trajectory by only one number.

2.4 Experimental Data

We analyzed three men healthy subjects and five men elderly patients with knee osteoarthritis [8]. The healthy subjects were all men between 27 and 33 years old. The participants walked on a treadmill. We collect data for walking on the treadmill at a speed of 3.6 km/h. The healthy subjects were pain-free, with no surgery on the legs. The duration of the experiments for healthy subjects was three minutes. The osteoarthritis patients with knee osteoarthritis were assessed in the first or second stage of osteoarthritis. The osteoarthritis patients walked on the treadmill for one minute at a speed of 3.6 km/h. The experimental method is based on electro-goniometers and a data acquisition system. For all participants in the experimental tests, patients and healthy subjects, the acquisition system consists of two electro goniometers, one for each knee of the two legs, so it is possible to read data simultaneously during walking on the treadmill. The flexion-extension process was performed in a 2D space in the sagittal plane. Data series for human knee joints during flexion-extension repetitive cycles are obtained. The resulting data series were analyzed, and angular displacements of knee joints were obtained. The participants were given the test protocol, and rehearsals were managed before the final procedure. The participants expressed their agreement regarding participation in the experimental tests. The experimental protocol was approved by the Ethics Committee of the University of Craiova.

Results

Time series for the flexion-extension knee angles were developed from experimental data.

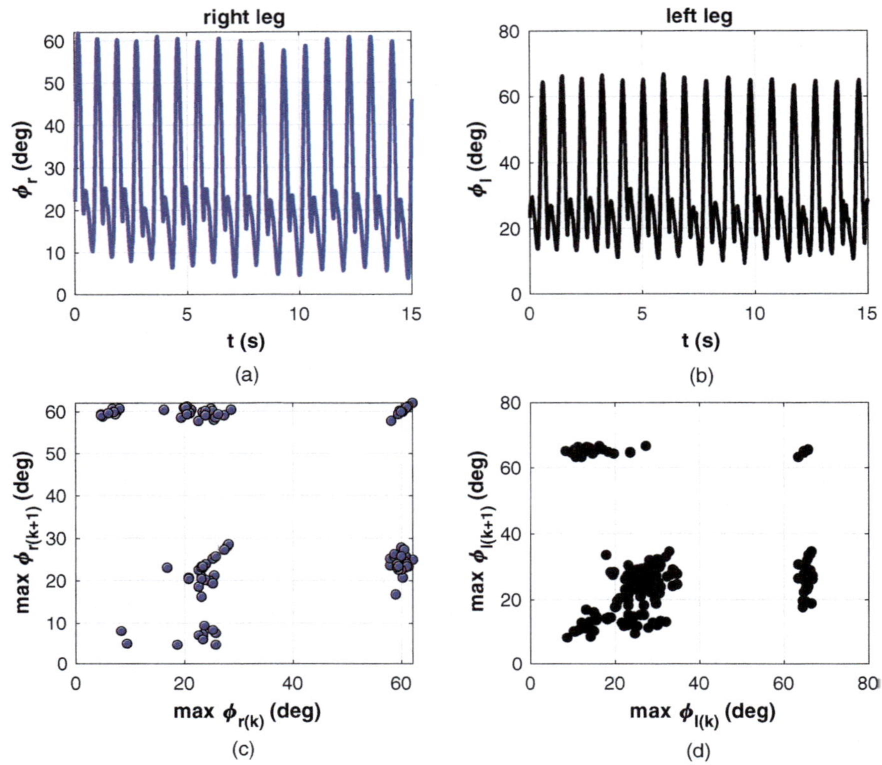

Fig. 2.1 Osteoarthritis patient: **a** consecutive walking cycles of the right knee; **b** consecutive walking cycles of the osteoarthritis (left) knee; **c** Lorentz map of right knee variation angle; **d** Lorenz map of left knee variation angle

The amplitudes ϕ of the knee angles for healthy subjects were larger than for the osteoarthritis patients. The osteoarthritis patients had less range of motion during the gait cycle than the healthy subjects.

Figure 2.1a shows the variation of the right knee angle, ϕ_r and Fig. 2.1b depicts the left knee angle, ϕ_l for osteoarthritis patient number 2. Figure 2.2a represents the right knee angle, ϕ_r, for a healthy subject. For the osteoarthritis patient, the Lorenz map for the maximum right knee angle is in Fig. 2.1c, and for the maximum left knee angle is in Fig. 2.1d. Figure 2.2b shows the Lorenz map for a healthy subject. No structure is evident in the plot for healthy human subjects and osteoarthritis patients. The motion is not periodic or quasiperiodic. For healthy human subjects, there is a more concentration of the data. For osteoarthritis patients, the data is distributed over a larger area. For the osteoarthritis patients shown in Fig. 2.1b, c, we can see a higher dispersion for the left leg on the Lorenz map. The left leg looks more affected, confirming the clinical data, with the left knee being the one affected by osteoarthritis. For the healthy human subjects shown in Fig. 2.1b, the data on the Lorenz map looks more clustered. Lorenz maps can characterize the behavior of the system.

Fig. 2.2 Healthy subject: **a** consecutive walking cycles of the right knee; **b** Lorenz map of right knee variation angle

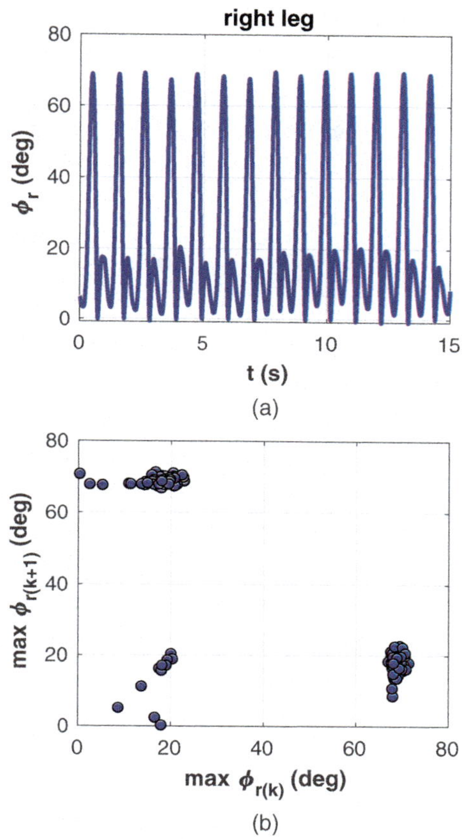

It was difficult to distinguish from the previous figures the evolution of the dynamic system That is why we use approximate entropy and must first find the embedding dimension computed by using the false nearest neighbor method [15]. The embedding dimension was 4 and 5 for osteoarthritis patients, while for each of the healthy subjects, the embedding dimension was three, as seen in Fig. 2.3.

Figure 2.4 shows the average value of approximate entropy for the sample of healthy subjects and the values of each osteoarthritis patient. A lower entropy means that the gait of the concerning subject is less chaotic and more periodic. The average approximate entropy of healthy subjects is lower than osteoarthritis patients. The healthy subjects have a more regular gait. For osteoarthritis, patient number 2, the left leg was weaker. The approximate entropy of the left leg was ApEn = 0.302, and for the right leg was ApEn = 0.261. The left leg was more affected than the right leg. The motion of the left leg was more non-periodic. This remark is valid for osteoarthritis patients 3, 4, and 5. The approximate entropy is lower for humans

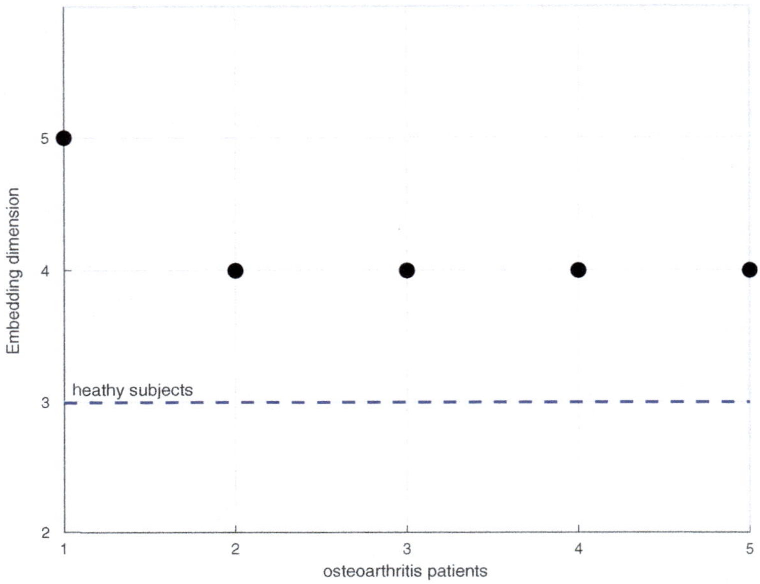

Fig. 2.3 Embedding dimension

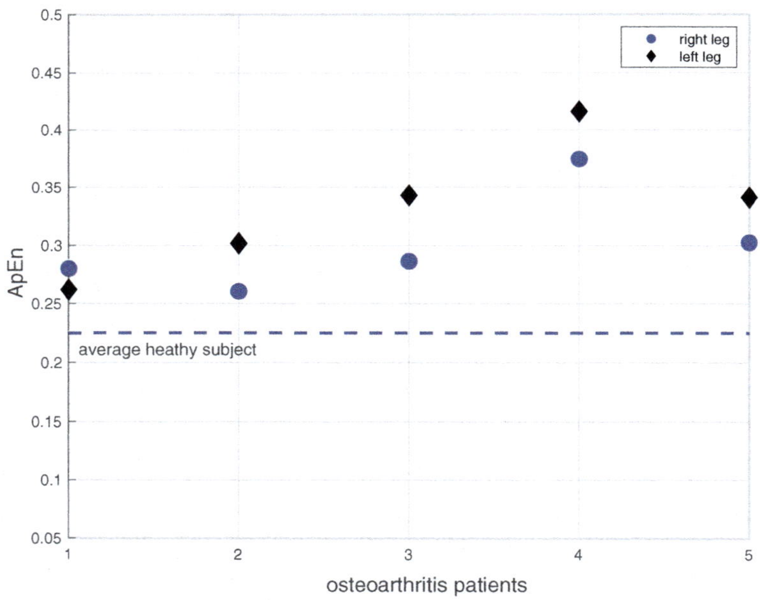

Fig. 2.4 Approximate entropy

with lower embedding dimensions (healthy subjects). The healthy subjects can be described with fewer degrees of freedom. The approximate entropy increases with the embedding dimensions for osteoarthritis patients.

2.5 Conclusion

The purpose of this study was to determine the applicability of approximate entropy to osteoarthritis patients and healthy subjects. The osteoarthritis patients have higher values of approximate entropy and more irregular gait than the healthy subject. The osteoarthritis patients also have a more dispersed distribution on Lorenz maps. The dispersion of the distribution can represent an indicator of the degree of gait regularity, with a big dispersion indicating an affected and uneven gait, while a lower distribution indicates a normal gait.

References

1. C. Shannon, A mathematical theory of communication. Bell Syst. Techn. J. **27**(3), 379–423 (1948)
2. S.M. Pincus, Approximate entropy as a measure of system complexity. Proc. Natl. Acad. Sci. USA **88**, 2297–2301 (1991)
3. J.S. Richman, J.R. Moorman, Physiological time-series analysis using approximate entropy and sample entropy. Am. J. Physiol. Heart Circ. Physiol. (2000)
4. C. Bandt, B. Pomp, Permutation entropy: a natural complexity measure for time series. Phys. Rev. Lett. **88**, 174102 (2002)
5. M. Costa, A.L. Goldberger, C.K. Peng, Multiscale entropy analysis of complex physiologic time series. Phys Rev Lett. (2002)
6. C.C. Gan, G. Learmonth, *Comparing Entropy with Tests for Randomness as a Measure of Complexity in Time Series*. 1512.00725, arXiv, stat.ME (2015)
7. J.M. Yentes, N. Hunt, K.K. Schmid, J.P. Kaipust, D. McGrath, N. Stergiou, The appropriate use of approximate entropy and sample entropy with short data sets. Ann. Biomed. Eng. **41**(2), 349–65 (2013)
8. D. Tarnita, D.B. Marghitu, Nonlinear Dynamics of human knee with osteoarthritis. Proc. Romanian Acad. Ser. A Math. Phys. Techn. Sci. Inf. Sci. **18**(4), 353–360 (2017)
9. P.C. Raffalt, B. Senderling, N. Stergiou, Filtering affects the calculation of the largest Lyapunov exponent. Comput. Biol. Med. **122**. https://doi.org/10.1016/j.compbiomed.2020.103786 (2020)
10. J.L. Semmlow, B. Griffel, *Biosignal and Medical Image Processing* (CRC Press, Taylor & Francis Group, Boca Raton, 2014)
11. J.L. Semmlow, *Circuits, Signals, and Systems for Bioengineers* (Academic Press, London, 2018)
12. https://www.mathworks.com/
13. E.N. Lorenz, Deterministic nonperiodic flow. J. Atmosph. Sci. **20**(2), 130–141 (1963)
14. S.H. Strogatz, *Nonlinear Dynamics and Chaos* (Addison-Wesley, MA, 1994)
15. M.B. Kennel, R. Brown, H.D.I. Abarbanel, Determining minimum embedding dimension using a geometrical construction. Phys. Rev. A **45**, 3403–3411 (1992)

Chapter 3
Effect of Mechanical Impact and Electromagnetic Actuation on the Nonlinear Vibration of a Beam

Nicolae Herisanu and Vasile Marinca

Abstract This work is devoted to an investigation of the nonlinear vibration response of a simply-supported flexible beam under a mechanical impact in the presence of an electromagnetic actuator. The governing differential equation and the boundary conditions are modeled within the framework of a Euler-Bernoulli beam and using so-called Hamiltonian principle and Galerkin-Bubnov procedure. An explicit analytical solution is proposed for a complex problem by means of some auxiliary functions derived from the governing equation. The proposed procedure is proved to be very accurate, simple and easy to implement.

3.1 Introduction

The flexural vibration of a beam under mechanical impact, electromagnetic actuation, moving body or nonlinear foundation is interesting as the basic research on vibration problems related to practical bridges.

Some of the previous important studies have been the works of Emans et al. [1]. The periodic motion and co-existence of attractors was found at modest oscillation levels during investigations with realistic parameters. Finite element method and Neimark's integration is used by Abiala [2] to obtain dynamic response of beams under uniformly distributed moving loads. Three types of conventional boundary conditions are investigated by Ding et al. [3] via the fourth-order Runge-Kutta method. Nbendjo and Woafo [4] took into account the geometrical nonlinearities and showed that the simple mode dynamic of the beam can be described by a potential with various configurations. Poorjamshidian et al. [5] analyzed the nonlinear vibration for a simply-supported beam with a constant velocity carrying moving mass. Dynamic stability in the Hamiltonian formulation is studied in [6]. Vibration

N. Herisanu (✉) · V. Marinca
University Politehnica Timisoara, Piața Victoriei No. 2, 300006 Timisoara, Romania
e-mail: nicolae.herisanu@upt.ro

Romanian Academy, Timisoara Branch, Bd. M. Viteazu 24, 300223 Timisoara, Romania

N. Herisanu and V. Marinca (eds.), *Acoustics and Vibration of Mechanical Structures—AVMS-2023*, Springer Proceedings in Physics 302,
https://doi.org/10.1007/978-3-031-48087-4_3

of viscoelastic axially moving Rayleigh and Euler-Bernoulli beam are explored by Shariati et al. [7].

In our study, after deriving the equation governing the vibrational motion of a simply-supported beam under mechanical impact and electromagnetic actuation, the time response of the beam has been obtained using the Optimal Homotopy Asymptotic Method (OHAM). The results show that the analytical procedure has a very good correspondence with numerical method. Also, an analytical expression for the nonlinear natural frequency is obtained.

3.2 Formulation of the Problem

In Fig. 3.1 is presented the physical model of a simply-supported beam of length L subjected to a mechanical impact by means of the force F_0 and an electromagnetic load.

The transversal and longitudinal displacements are W(x,t) and U(x,t) respectively. The Young's modulus E, a mass density ρ and a cross-sectional area A of the beam are supposed to be constants.

Considering an element of length dx of the beam at rest, the corresponding deflection can be written as

$$ds = \left[\left(1 + \frac{\partial u}{\partial x}\right)^2 + \left(\frac{\partial W}{\partial x}\right)^2\right]^{1/2} dx \tag{3.1}$$

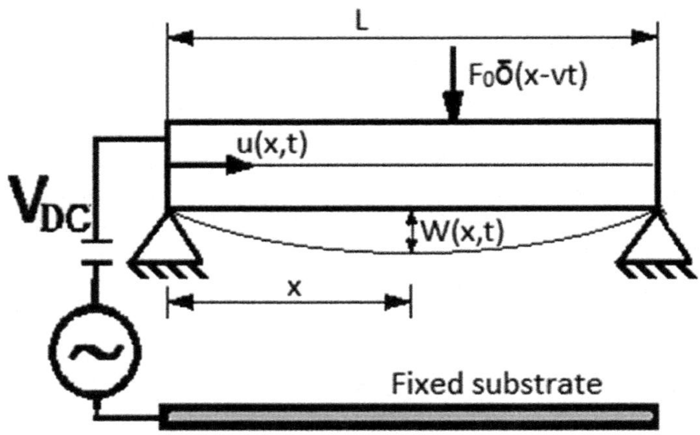

Fig. 3.1 The physical model of the simply-supported beam subjected to impact force and electromagnetic actuation

The unit vector parallel to the defined element is of the form

$$\overline{k} = \left[\left(1 + \frac{\partial u}{\partial x}\right)\overline{i} + \frac{\partial W}{\partial x}\overline{j}\right]\frac{dx}{ds} \tag{3.2}$$

The length of the beam is variable during the motion, such that the tension becomes

$$T = -EAe, \text{ where } e = \frac{dx - ds}{dx} = 1 - \frac{ds}{dx} \tag{3.3}$$

The dynamics of the beam is governed by the equation [4]:

$$\rho A\frac{\partial^2 W}{\partial t^2} + EI\frac{\partial^4 W}{\partial x^4} = -\frac{\partial}{\partial x}(T\overline{k})\overline{j} + F_{mi} + F_{em} \tag{3.4}$$

where I is the moment of inertia of the beam cross section, F_{mi} is the impact force and F_{em} is the electromagnetic force related to the electromagnetic actuation.

Neglecting the terms of the forms $U_x^2, U_x W_x^2, U_x^3, U_x^2 W_x^2, U_x^4$, where $U_x = \partial U/\partial x$, from Eq. (3.1) one can get

$$\frac{dx}{ds} = \left[(1 + U_x)^2 + W_x^2\right]^{-1/2} \approx 1 - U_x - \frac{1}{2}W_x^2 + \frac{3}{8}W_x^4 \tag{3.5}$$

From Eqs. (3.3) and (3.5) we have that

$$e = 1 - \frac{dx}{ds} = U_x + \frac{1}{2}W_x^2 - \frac{3}{8}W_x^4 \tag{3.6}$$

such that taking into account Eqs. (3.3) and (3.6), the Eq. (3.4) can be rewritten as

$$\rho A\frac{\partial^2 W}{\partial t^2} + EI\frac{\partial^4 W}{\partial x^4} = -EA\frac{\partial}{\partial x}\left(e\frac{\partial W}{\partial x}\right) + F_{mi} + F_{em} \tag{3.7}$$

In general, the material term in Eq. (3.3) is small such that $\frac{d}{dx}(e) = 0$. It follows that

$$U_x + \frac{1}{2}W_x^2 - \frac{3}{8}W_x^4 = e = C \tag{3.8}$$

where C is a constant which can be determined by integration of Eq. (3.8):

$$U(x) - U(0) = Cx - \frac{1}{2}\int_0^L\left(\frac{\partial W}{\partial x}\right)^2 dx + \frac{3}{8}\int_0^L\left(\frac{\partial W}{\partial x}\right)^4 dx \tag{3.9}$$

If we use the following boundary conditions for the longitudinal displacement, it holds that U(0,t) = U(L,t) = 0. From Eq. (3.9) it is clear that

$$e = C = \frac{1}{2L}\int_0^L\left(\frac{\partial W}{\partial x}\right)^2 dx - \frac{3}{8L}\int_0^L\left(\frac{\partial W}{\partial x}\right)^4 dx \tag{3.10}$$

Substituting Eq. (3.10) into Eq. (3.7) we obtain

$$\rho A\frac{\partial^2 W}{\partial t^2} + EI\frac{\partial^4 W}{\partial x^4} = \frac{EA}{2L}\frac{\partial^2 W}{\partial x^2}\left[\int_0^L\left(\frac{\partial W}{\partial x}\right)^2 dx - \frac{3}{4}\int_0^L\left(\frac{\partial W}{\partial x}\right)^4 dx\right] + F_{mi} + F_{em} \tag{3.11}$$

The nonlinear term in aforementioned equation is produced to a high amplitude and midplane stretching in the beam. The term F_{mi} represents the mechanical impact of load F_0:

$$F_{mi} = F_0\delta(x - vt) \tag{3.12}$$

where v is the speed of the load and δ is the Dirac delta function. The vibration of the beam depends on the action of DC voltage source, such that electromagnetic actuation can be considered as [8]:

$$F_{em} = \frac{1}{2}\frac{C_0 V_{DC}^2}{(d - W(x,t))^2} - \frac{1}{2}\frac{C_0 V_{DC}^2}{(d + W(x,t))^2} \tag{3.13}$$

in which C_0 is the capacitance of the actuator, d is the gap width and V_{DC} is the voltage. More details on the two terms of (3.13) can be found in the mentioned paper. Moreover, it is to remark the difference between the electromagnetic actuation (3.13) and the electrostatic actuation [9].

The expression (3.13) can be simplified as

$$\frac{1}{2}\frac{C_0 V_{DC}^2}{(d - W)^2} - \frac{1}{2}\frac{C_0 V_{DC}^2}{(d + W)^2} = \frac{C_0 V_{DC}^2}{2d^2}\left[1/\left(1 - \frac{W}{d}\right)^2 - 1/\left(1 + \frac{W}{d}\right)^2\right] \approx \frac{C_0 V_{DC}^2}{2d^2}\left[4\left(\frac{W}{d}\right) + 8\left(\frac{W}{d}\right)^3 + 12.37\left(\frac{W}{d}\right)^5\right] \tag{3.14}$$

Inserting Eqs. (3.12), (3.13) and (3.14) into Eq. (3.11), one retrieves

$$\rho A\frac{\partial^2 W}{\partial t^2} + EI\frac{\partial^4 W}{\partial x^4} - \frac{EA}{2L}\frac{\partial^2 W}{\partial x^2}\left[\int_0^L\left(\frac{\partial W}{\partial x}\right)^2 dx - \frac{3}{4}\int_0^L\left(\frac{\partial W}{\partial x}\right)^4 dx\right]$$

$$- F_0\delta(x - vt) - \frac{C_0 V_{DC}^2}{2d^2}\left[4\left(\frac{W}{d}\right) + 8\left(\frac{W}{d}\right)^3 + 12.37\left(\frac{W}{d}\right)^5\right] = 0 \quad (3.15)$$

The equation is obtained for a simply-supported beam, such that the boundary conditions are

$$W(0,t) = W(L,t) = \frac{\partial^2 W(0,t)}{\partial x^2} = \frac{\partial^2 W(L,t)}{\partial x^2} = 0 \quad (3.16)$$

In order to express the governing Eqs. (3.15) and (3.16) in nondimensional form, the following parameters are defined:

$$\overline{W} = \frac{W}{d},\ \overline{x} = \frac{x}{L},\ \overline{t} = \sqrt{\frac{EI}{\rho A}}\frac{t}{L^2},\ \overline{v} = vL\sqrt{\frac{\rho A}{EI}},\ \overline{F}_0 = \frac{F_0 L^5}{dEI},$$
$$\overline{\alpha} = \frac{d^2 E}{2\rho},\ \overline{\beta} = \frac{3d^2}{4L^2},\ \overline{\gamma} = \frac{2C_0 L^4 V_{DC}^2}{d^3 EI} \quad (3.17)$$

Omitting the bar, Eq. (3.15) can be rewritten in nondimensional form as

$$\frac{\partial^2 W}{\partial t^2} + \frac{\partial^4 W}{\partial x^4} - \alpha\frac{\partial^2 W}{\partial x^2}\left[\int_0^1\left(\frac{\partial W}{\partial x}\right)^2 dx - \beta\int_0^1\left(\frac{\partial W}{\partial x}\right)^4 dx\right]$$
$$- \gamma\left(W + 2W^3 + 3.0925W^5\right) = F_0\delta(x - vt) \quad (3.18)$$

The solution of Eq. (3.18) can be assumed to be of the form

$$W(x,t) = X(x)T(t) \quad (3.19)$$

where, taking into account the conditions (3.16), X(x) can be expressed as

$$X(x) = \sqrt{2}\sin\pi x \quad (3.20)$$

In accordance with Galerkin-Bubnov procedure, Eq. (3.15) will be written as follows:

$$\ddot{T} + \omega^2 T + aT^3 + bT^5 = F_0 \sin\pi vt \quad (3.21)$$

where the dot denotes derivative with respect to time and

$$\omega^2 = \int_0^1 \frac{d^2X}{dx^2} X dx - \gamma \int_0^1 X^2(x) dx$$

$$a = -\alpha \left(\int_0^1 \left(\frac{dX}{dx} \right)^2 dx \right) \left(\int_0^1 \frac{d^2X}{dx^2} X dx - 2\gamma \int_0^1 X^4 dx \right); \tag{3.22}$$

$$b = \alpha\beta \left[\int_0^1 \left(\frac{dX}{dx} \right)^4 dx \Big/ \int_0^1 \frac{d^2X}{dx^2} X dx \right] - 3.0925\gamma \int_0^1 X^6 dx\,;\ f = \sqrt{2} F_0$$

3.3 The Optimal Homotopy Asymptotic Method

We apply OHAM [10–12] to the following nonlinear differential equation

$$L(T(t)) + N(T(t)) = 0 \tag{3.23}$$

with the boundary/initial conditions

$$B\left(T(t), \frac{dT(t)}{dt} \right) = 0 \tag{3.24}$$

If $\overline{T}(t)$ is the unknown function [an approximate solution of Eq. (3.23)], such as

$$\overline{T}(t) = T_0(t) + T_1(t) \tag{3.25}$$

where the initial approximation $T_0(t)$ can be determined from the linear equation

$$L(T_0(t)) = 0\,,\ B\left(T_0(t), \frac{dT_0(t)}{dt} \right) = 0 \tag{3.26}$$

The first approximation $T_1(t)$ is determined from the linear equation:

$$L(T_1(t)) = H(\tau, C_1, C_2, \ldots, C_n) N[T_0(t)]\,,\ B\left(T_1(t), \frac{dT_1(t)}{dt} \right) \tag{3.27}$$

where $H(t, C_1, C_2, \ldots, C_n)$ is an arbitrary auxiliary function. This auxiliary function is determined such that the product $H\ N[T_0(t)]$ and $N[T_0(t)]$ be of the same shape. The parameters $C_1, C_2, \ldots, C_n$ which appear in the first-order approximate solution given by Eq. (3.25) can be determined in many ways: using the least square method, the Galerkin method, the collocation method, the Ritz method, by minimizing the residual error.

With these parameters known, (namely convergence-control parameters), the approximate solution (3.25) is well-determined.

3.4 Application of OHAM to Nonlinear Vibration of the Beam

In what follows we apply our technique to obtain an approximate solution of Eq. (3.21):

$$\ddot{T} + \omega^2 T + aT^3 + bT^5 = f \sin \pi \upsilon t \tag{3.28}$$

with the initial conditions

$$T(0) = A, \ \dot{T}(0) = 0 \tag{3.29}$$

Making the transformations

$$\tau = \Omega t, \ T(t) = A\theta(\tau) \tag{3.30}$$

where Ω is unknown natural frequency of the beam, the Eq. (3.28) becomes

$$\theta'' + \frac{\omega^2}{\Omega^2}\theta + \frac{aA^2}{\Omega^2}\theta^3 + \frac{bA^4}{\Omega^2}\theta^5 - \frac{f}{A\Omega^2}\sin \pi\frac{\upsilon\tau}{\Omega}, \ \theta(0) = 1 \ , \ \theta'(0) = 0 \tag{3.31}$$

with the prime denoting derivative with respect to τ.

The linear and nonlinear operators are respectively

$$L[\theta(\tau)] = \theta'' + \theta; \ N[\theta(\tau)] = \left(\frac{\omega^2}{\Omega^2} - 1\right)\theta + \frac{aA^2}{\Omega^2}\theta^3 + \frac{bA^4}{\Omega^2}\theta^5 - \frac{f}{A\Omega^2}\sin \pi\frac{\upsilon\tau}{\Omega} \tag{3.32}$$

The initial approximation is determined from Eq. (3.26), which can be rewritten as

$$\theta_0'' + \theta_0 = 0, \ \theta_0(0) = 1, \theta_1\prime(0) = 0 \tag{3.33}$$

Its solution is

$$\theta_0(\tau) = \cos \tau \tag{3.34}$$

The first approximation $\theta_1(\tau)$ is obtained from Eq. (3.27)

$$\theta_1'' + \theta_1 = H(\tau, C_1, C_2, ..., C_n)N[\theta_0(\tau)], \ \theta_1(0) = \theta_1\prime(0) = 0 \tag{3.35}$$

where N $[\theta_0(\tau)]$ is determined by inserting Eq. (3.34) into the nonlinear operator (3.32).

$$N[\theta_0(\tau)] = A_1 \cos\tau + A_3 \cos 3\tau + A_5 \cos 5\tau + B \sin \frac{\pi v \tau}{\Omega} \tag{3.36}$$

The constants from the above equation are

$$A_1 = \frac{\omega^2}{\Omega^2} - 1 + \frac{3aA^2}{4\Omega^2} + \frac{5bA^4}{8\Omega^2},\ A_3 = \frac{aA^2}{4\Omega^2} + \frac{5bA^4}{16\Omega^2},\ A_5 = \frac{bA^4}{16\Omega^2},\ B = -\frac{f}{A\Omega^2} \tag{3.37}$$

The auxiliary function H from Eq. (3.35) is chosen such that the product H N[θ_0(t)] and N[θ_0(t)] be of the same form. The auxiliary function H and the natural number n are not unique. For example, we can choose these auxiliary functions in the following expressions

$$H_1(t, C_1, C_2, C_3, C_4) = C_1 + 2C_2 \cos 2\tau + 2C_3 \cos 4\tau + 2C_4 \cos 6\tau \tag{3.38}$$

$$H_2(t, C_1, C_2, C_3) = C_1 + 2C_2 \cos 2\tau + 2C_3 \cos 4\tau \tag{3.39}$$

$$\begin{aligned} H_3(t, C_1, C_2, C_3, C_4, C_5) = {} & C_1 + 2C_2 \cos 2\tau + 2C_3 \cos 4\tau \\ & + 2C_4 \cos 6\tau + 2C_5 \cos 8\tau \end{aligned} \tag{3.40}$$

$$H_4(t, C_1, C_2, C_3) = C_1 + 2C_2 \cos 4\tau \tag{3.41}$$

Using only the expression (3.38), the Eq. (3.35) can be written after simple manipulations as follows:

$$\begin{aligned} \theta_1'' + \theta_1 = {} & [A_1(C_1 + C_2) + A_3(C_2 + C_3) + A_5(C_3 + C_4)]\cos\tau + [A_3(C_1 + C_4) \\ & + A_1(C_2 + C_3) + A_5C_2]\cos 3\tau + [A_5C_1 + A_3C_2 + A_1(C_3 + C_4)]\cos 5\tau \\ & + (A_5C_2 + A_3C_3 + A_1C_4)\cos 7\tau + (A_5C_3 + A_3C_4)\cos 9\tau \\ & + A_5C_4 \cos 11\tau,\ \theta_1(0) = \theta_1{}'(0) = 0 \end{aligned} \tag{3.42}$$

No secular terms in Eq. (3.42) requires that the coefficient of cos τ be zero. Form this condition one can put the natural frequency:

$$\Omega^2 = \omega^2 + \frac{3aA^2}{4} + \frac{5bA^4}{8} + \left(\frac{aA^2}{4} + \frac{5bA^4}{16}\right)\frac{C_2 + C_3}{C_1 + C_2} + \frac{bA^4}{16}\frac{C_3 + C_4}{C_1 + C_2} \tag{3.43}$$

The solution of the linear Eq. (3.42) is

$$\begin{aligned} \theta_1(\tau) = {} & \frac{A_3(C_1 + C_4) + A_1(C_2 + C_3) + A_5C_2}{8}(\cos\tau - \cos 3\tau) \\ & + \frac{A_5C_1 + A_3C_2 + A_1(C_3 + C_4)}{24}(\cos\tau - \cos 5\tau) \end{aligned}$$

$$+\frac{A_5C_2+A_3C_3+A_1C_4}{48}(\cos\tau-\cos 7\tau)$$
$$+\frac{A_5C_3+A_3C_4}{80}(\cos\tau-\cos 9\tau)$$
$$+\frac{A_5C_4}{120}(\cos\tau-\cos 11\tau) \tag{3.44}$$

The approximate solution of Eq. (3.28) is obtained from Eqs. (3.25), (3.30), (3.34) and (3.44):

$$\overline{T}(t)=A\cos\Omega t+\frac{A[A_3(C_1+C_4)+A_1(C_2+C_3)+A_5C_2]}{8}(\cos\Omega t-\cos 3\Omega t)$$
$$+\frac{A[A_5C_1+A_3C_2+A_1(C_3+C_4)]}{24}(\cos\Omega t-\cos 5\Omega t)$$
$$+\frac{AA_5C_4}{120}(\cos\Omega t-\cos 11\Omega t)$$
$$+\frac{A[A_5C_2+A_3C_3+A_1C_4]}{48}(\cos\Omega t-\cos 7\Omega t)$$
$$+\frac{A[A_5C_3+A_3C_4]}{80}(\cos\Omega t-\cos 9\Omega t) \tag{3.45}$$

where Ω is given by Eq. (3.43).

3.5 Numerical Results

To illustrate the accuracy of OHAM, we consider the parameters $\omega = 1.1$, $a = 0.023$, $b = 0.03$, $f = 0.03$, $v = 0.95$, for which applying the OHAM procedure and minimizing the residual of the initial equation, it is obtained the optimal values of the convergence-control parameters $C_1 = 0.430116872481$; $C_2 = 0.000287558193$; $C_3 = 0.006874588249$; $C_4 = 0.049982573619$, and $\Omega = 1.116$.

In Fig. 3.2 we present the solution (3.45) in comparison with numerical integration result. One can be seen that the error between the approximate analytical solution and numerical solution is very good, the solutions being in an excellent agreement.

3.6 Conclusions

In the present study we propose the Optimal Homotopy Asymptotic Method (OHAM) to obtain an approximate analytical solution for nonlinear vibration of a beam subjected to mechanical impact and electromagnetic actuation. The validity of our technique was demonstrated by adequately choosing the linear operator and auxiliary functions. An example is given and a very good agreement was found between

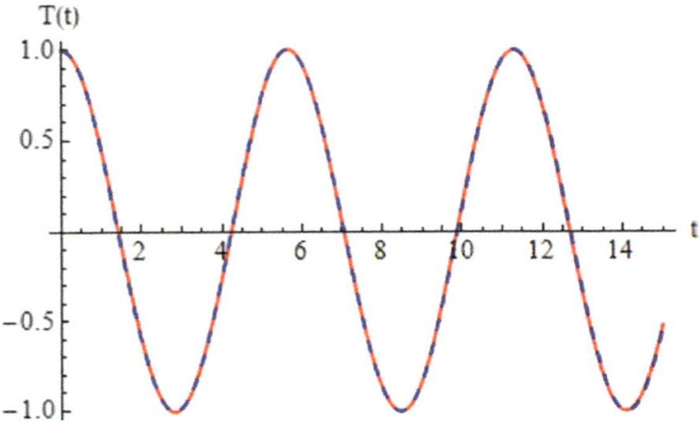

Fig. 3.2 Comparison between analytical OHAM solution and numerical integration results: redsoild line numerical blue dashed line analytical

the approximate analytical solution and numerical simulation results. The proposed approach is valid even if the nonlinear differential equation does not contain any small or large parameters.

References

1. J. Emans, M. Wiercigroch, A.M. Krivtsov, Cumulative effect of structural nonlinearities: chaotic dynamics of cantilever beam system with impacts. Chaos Solitons Fractals **23**, 1661–1670 (2003)
2. O. Abiala, Finite element evaluation of the dynamic response of beams under uniformly distributed moving loads. J. Natural Sci. Eng. Technol. **8**, 95–105 (2009)
3. H. Ding, L.Q. Chen, S.P. Yang, Convergence of Galerkin truncation for dynamic response of finite beams on nonlinear foundations under a moving load. J. Sound Vib. **331**, 2426–2442 (2012)
4. B.R.N. Nbendjo, P. Woafo, Modelling of the dynamic of Euler's beam by ϕ^6 potential. Mech. Res. Commun. **38**, 542–545 (2011)
5. M. Poorjamshidian, J. Sheikhi, S.M. Moghadas, M. Nakhaie, Nonlinear vibration analysis of the beam carrying a moving mass using modified homotopy. J. Solid Mech. **6**(4), 389–396 (2014)
6. M. Pirmoradian, H. Karimpour, Parametric resonance and jump analysis of a beam subjected to a periodic mass transition. Nonlinear Dyn. **89**, 214–254 (2017)
7. A. Shariati, D.W. Jung, H.M. Sedighi, K.K. Zur, M. Habibi, M. Safa, Stability and dynamics of viscoelastic moving Rayleigh beams with an asymmetrical distribution of material parameters. Symmetry **12**, 586 (2020)
8. M. Belhaq, A. Bichri, J. Der Hogabian, J.M. Mahfoud, Effect of electromagnetic actuations of the dynamics of a harmonically excited cantilever beam. Int. J. Nonlinear Mech. **40**, 828–833 (2011)
9. J. Duan, L. Zongxue, L. Jinyuan, Pull-in instability analyses for NEMS actuators with quartic shape approximation. Appl. Math. Mech. **37**, 303–314 (2016)
10. V. Marinca, N. Herisanu, The optimal homotopy asymptotic method. Engineering Application. (Springer, Cham, 2015)
11. N. Herisanu, V. Marinca, G. Madescu, Nonlinear dynamics of a wind turbine permanent magnet generator system in different wind profile conditions. AIP Conf. Proc. **1863**, 460002 (2017)
12. N. Herisanu, V. Marinca, Optimal homotopy asymptotic method to large post-buckling deformation of MEMS. MATEC Web of Conf. **148**, 13003 (2018)

Chapter 4
Three-Dimensional Model of Wind Turbine NREL-Monopile 5 MW for Vibration Characterisation

Marcela R. Machado and Maciej Dutkiewicz

Abstract The development of wind energy is a key solution in the field of environmental protection by reducing greenhouse gas emissions and limiting changes taking place in ecosystems. In this context, the development of onshore and offshore wind farms is important. In the article the 5 MW wind turbine developed by National Renewable Energy Lab (NREL) is the subject of the research. A 3D-Timoshenko beam element coupled to a lumped mass is used in the wind turbine model. The results investigate various modal properties and mode-coupling instability in the WT vibrations, essential information for the turbine's monitoring and control. The receptance frequency response function (FRF) comparison between 2D Spectral Element Method (SEM) and 3D Finite Element Model (FEM) with excitement at the wind turbine top is investigated.

4.1 Introduction

One of the renewable energy sources that is rising in development and use globally is wind energy. In the past ten years, worldwide countries have installed a remarkable amount of wind farms. It continues to be a reliable source of energy for the nation. Wind energy production depends on steady wind occurrences of the proper strength and free of hidden variations in speed or direction. In 2021, new wind

---

The original version of the chapter has been revised: Acknowledgement section has been added. A correction to this chapter can be found at https://doi.org/10.1007/978-3-031-48087-4_39

M. R. Machado · M. Dutkiewicz (✉)
Faculty of Civil, Environmental Engineering and Architecture, Bydgoszcz University of Science and Technology, Al. Prof. Sylwestra Kaliskiego 7, Bydgoszcz, Poland
e-mail: macdut@pbs.edu.pl

M. R. Machado
e-mail: marcelam@unb.br

M. R. Machado
Faculdade de Tecnologia Asa Norte, University of Brasília, Campus Universitário Darcy Ribeiro, Brasília, DF 70910-900, Brazil

N. Herisanu and V. Marinca (eds.), *Acoustics and Vibration of Mechanical Structures—AVMS-2023*, Springer Proceedings in Physics 302,
https://doi.org/10.1007/978-3-031-48087-4_4

turbine (WT) installation reached 93.6 GW, with 72.5 GW generated onshore and 21.1 GW generated offshore. The global WT market expects to grow US$ 102.4 billion by 2030. Within the next five years, the Compound Annual Growth Rate (CAGR) predicts global output of 466 GW from new onshore WT installations and 90 GW from offshore WT capacity [1]. In May 2022, the European Union (EU) Commission proposed in the REPowerEU plan to increase renewable energy share and consumption to 45% to reduce reliance on fossil fuels further [2]. The 5 MW wind turbine (WT) is the subject of this article. This turbine was developed by the National Renewable Energy Laboratory (NREL) [3]. Considering the environment, the people living in it, the comfort of life, the vibrations of wind turbines should be constantly monitored [4]. In the article the 3D finite element model and the spectral element method (SEM) were used in [5, 6]. The application of SEM is very wide: beams [7], plates, cables [8–10] and composite structures [11, 12], wind turbine [13], and smart structures [14–16]. This paper explores the dynamic characteristic of the 5 MW WT monopile modelled by finite element 3D-Thimoshenko attached to a lumped mass. The dynamic response is compared to other NREL commercial software models and the 2D-spectral model presented in [13]. The study of the modal properties and the coupling mode instability in the WT vibration characteristics is essential to control designs and integrity monitoring.

4.2 NREL 5MW Wind Turbine

Offshore wind turbines-three-bladed typically drive a horizontally mounted generator, and their main components include the tower, foundation, nacelle, rotor blades, and hub[17]. The nacelle hosts the generator and the gearbox, and the tower structure supports the weight of the nacelle and rotor (RNA) and the wind loads. The foundation engineer considers seabed conditions and environmental regulations. The dynamic simplified WT models consist of the tower as a beam structural element and the RNA as a lumped mass with or without rotational inertia. The foundation is often idealised as translational and rotational springs at the base. Figure 4.1 idealised the monopile offshore wind turbine, where m_{RNA} is the mass of the rotor-nacelle assembly, L_T is the tower height, D_t is the tower top diameter, D_b is the tower bottom diameter, t_T is the average tower wall thickness, E_T is the tower Young's modulus, m_T is the tower mass, L_S is the platform height above mudline, D_p is the pile diameter, L_p is pile depth, E_p is the pile Young's modulus, K_L is the lateral stiffness of foundation, K_V is the vertical stiffness of foundation, K_{LR} is the cross-stiffness of foundation, K_R is the rocking stiffness of foundation.

In this work, the simplified mathematical model of the WT is employed. Therefore, the tower is considered a tubular steel element, and the lumped mass represents the rotor-nacelle-blade of the WT. The tower is modelled with a 3D-element Timoshenko beam with two nodes and 6°C of freedom (dof) per node comprising three translational displacements u, v, w and three rotational degrees of freedom θ_u, θ_v, θ_w (w.r.t. x-, y- and z-axis, respectively) as shown in Fig. 4.2. The element has a length

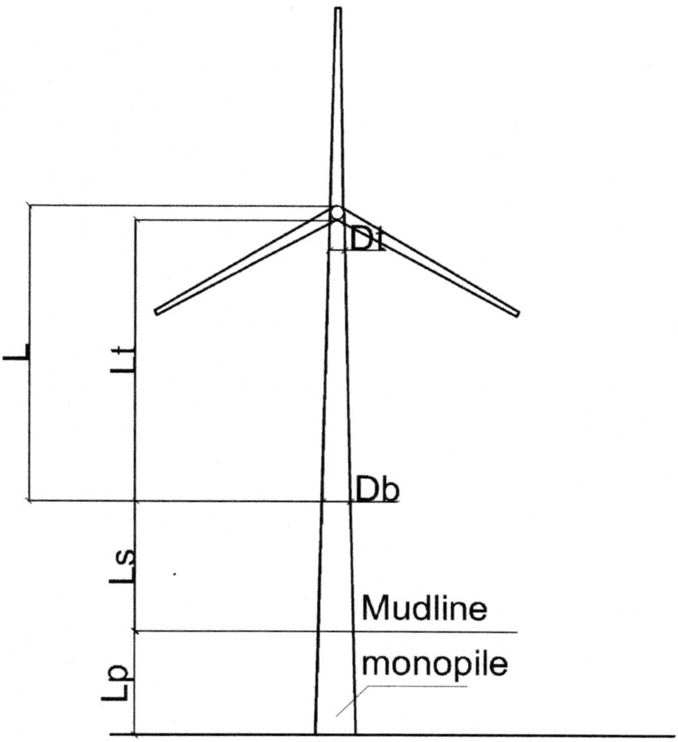

Fig. 4.1 Monopile offshore wind turbine representation

of le, thickness t, and tubular cross-section $Dt - 2 * t$, where the Dt is the external diameter of the tube and area A. For an isotropic, linear elastic material density is ρ, Young's modulus E, Poisson's ratio ν, and its shear modulus by $G = E/2(\nu + 1)$. The element's second moments of area are the I_y and I_z, and the polar moment of inertia is I_p. For the element's torsion, the constant is I_t. The element displacement is defined as

$$\mathbf{u} = \{u_1, v_1, w_1, \theta_{u1}, \theta_{v1}, \theta_{w1}, u_2, v_2, w_2, \theta_{u2}, \theta_{v2}, \theta_{w2}\}^T \tag{4.1}$$

The local mass and stiffness matrices for the three-dimensional finite Timoshenko beam element are presented in [18, 19], and defined as

$$\mathbf{M}_e = \begin{bmatrix} \mathbf{M}_{11}^e & \mathbf{M}_{12}^e \\ \mathbf{M}_{21}^e & \mathbf{M}_{22}^e \end{bmatrix}, \quad \mathbf{K}_e = \begin{bmatrix} \mathbf{K}_{11}^e & \mathbf{K}_{12}^e \\ \mathbf{K}_{21}^e & \mathbf{K}_{22}^e \end{bmatrix} \tag{4.2}$$

where $()_{11}^e$ and $()_{22}^e$ describe the interdependency of the dofs of the first and second node, respectively, $()_{12}^e$ and $()_{21}^e$ is the interconnection between both diagonal variable.

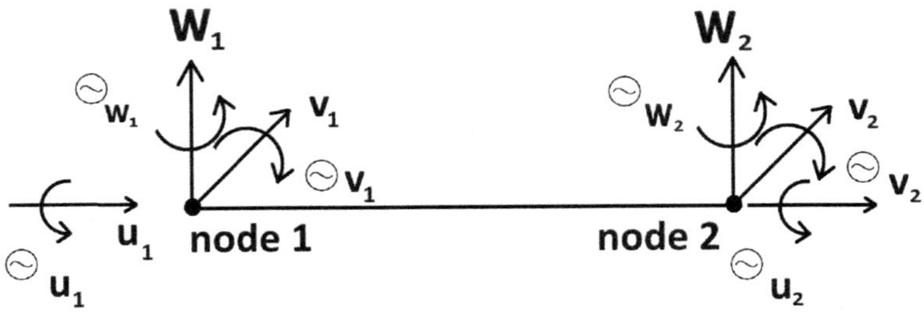

Fig. 4.2 Three-dimensional Timoshenko element

M_{22}^e and K_{22}^e are similar to M_{11}^e and K_{11}^e respectively, except for the sign of the off-diagonal entries. The mass element matrix is given

$$\mathbf{M}_{11}^e = \begin{bmatrix} \frac{1}{3} & 0 & 0 & 0 & 0 & 0 \\ 0 & \frac{13}{35}+\frac{6I_z}{5Ale^2} & 0 & 0 & 0 & \frac{11le}{210}+\frac{I_z}{10Ale} \\ 0 & 0 & \frac{13}{35}+\frac{6I_y}{5Ale^2} & 0 & -\frac{11le}{210}+\frac{I_y}{10Ale} & 0 \\ 0 & 0 & 0 & \frac{I_p}{3A} & 0 & 0 \\ 0 & 0 & -\frac{11le}{210}+\frac{I_y}{10Ale} & 0 & \frac{le^2}{105}+\frac{2I_y}{15A} & 0 \\ 0 & -\frac{11le}{210}+\frac{I_z}{10Ale} & 0 & 0 & 0 & \frac{le^2}{105}+\frac{2I_z}{15A} \end{bmatrix} \tag{4.3}$$

$$\mathbf{M}_{21}^e = \mathbf{M}_{21}^{eT} = \begin{bmatrix} \frac{1}{6} & 0 & 0 & 0 & 0 & 0 \\ 0 & \frac{9}{70}-\frac{6I_z}{5Ale^2} & 0 & 0 & 0 & -\frac{13le}{420}+\frac{I_y}{10Ale} \\ 0 & 0 & \frac{9}{70}-\frac{6I_y}{5Ale^2} & 0 & -\frac{13le}{410}+\frac{I_y}{10Ale} & 0 \\ 0 & 0 & 0 & \frac{I_p}{6A} & 0 & 0 \\ 0 & 0 & \frac{13le}{420}-\frac{I_y}{10Ale} & 0 & -\frac{le^2}{140}-\frac{2I_y}{30A} & 0 \\ 0 & -\frac{13le}{420}+\frac{I_z}{10Ale} & 0 & 0 & 0 & -\frac{le^2}{140}-\frac{2I_z}{30A} \end{bmatrix} \tag{4.4}$$

$$\mathbf{K}_{11}^e = \frac{1}{le^3} \begin{bmatrix} Ale^2 & 0 & 0 & 0 & 0 & 0 \\ 0 & \frac{12EI_z}{1+P_y} & 0 & 0 & 0 & \frac{6EI_zle}{1+P_y} \\ 0 & 0 & \frac{12EI_y}{1+P_z} & 0 & -\frac{6EI_yle}{1+P_z} & 0 \\ 0 & 0 & 0 & GI_tle^2 & 0 & 0 \\ 0 & 0 & -\frac{6EI_yle}{1+P_z} & 0 & -\frac{EI_yle^2(4+P_z)}{1+P_z} & 0 \\ 0 & -\frac{6EI_zle}{1+P_z} & 0 & 0 & 0 & \frac{EI_zle^2(4+P_z)}{1+P_{yz}} \end{bmatrix} \tag{4.5}$$

$$\mathbf{M}_{21}^{e} = \mathbf{M}_{21}^{eT} = \begin{bmatrix} -EAle^2 & 0 & 0 & 0 & 0 & 0 \\ 0 & -\frac{12EI_z}{1+P_y} & 0 & 0 & 0 & -\frac{6EI_zle}{1+P_y} \\ 0 & 0 & -\frac{12EI_y}{1+P_z} & 0 & \frac{6EI_yle}{1+P_z} & 0 \\ 0 & 0 & 0 & -GI_tle^2 & 0 & 0 \\ 0 & 0 & -\frac{6EI_yle}{1+P_z} & 0 & \frac{EI_yle^2(2-P_z)}{1+P_z} & 0 \\ 0 & \frac{6EI_zle}{1+P_y} & 0 & 0 & 0 & \frac{EI_zle^2(2+P_y)}{1+P_y} \end{bmatrix} \tag{4.6}$$

The lumped mass matrix includes the RNA mass $m_{RNA}$and the hub inertia, J_{hub}, and the nacelle inertia about the yaw axis, J_{nac},(see Fig. 4.1a for axis orientation). Thus lumped mass matrix yields

$$\mathbf{m}_{lumped} = \begin{bmatrix} m_{RNA} & 0 & 0 & 0 & 0 & 0 \\ 0 & m_{RNA} & 0 & 0 & 0 & 0 \\ 0 & 0 & m_{RNA} & 0 & 0 & 0 \\ 0 & 0 & 0 & J_{nac} & 0 & 0 \\ 0 & 0 & 0 & 0 & J_{hub} & 0 \\ 0 & 0 & 0 & 0 & 0 & 0 \end{bmatrix} \tag{4.7}$$

4.3 Spectral Wind Turbine Model

SEM is a mesh method similar to the finite element method with its shape function established from the exact solution of the wave equation formulated in the frequency domain [6]. The spectral model of the WT proposed in [13]. The equation of motion that represents the wind turbine is

$$EI\frac{d^4\hat{v}}{dz^4} + \omega^2\left[\rho A + \tilde{m}_n\delta(z-L)\right]\hat{v} = \hat{q}$$

where $\tilde{m}_n = m_n EI\beta^3$, $\delta(z-L)$ is Dirac delta function at $z = L$. The global wind turbine spectral matrix is

$$\mathbf{S}_{\text{OWT}}(\omega) = \mathbf{S}_b(\omega) + S_{m_n}(z = L, \omega)$$
$$= \begin{bmatrix} S_b(1,1) & S_b(1,2) & S_b(1,3) & S_b(1,4) \\ S_b(2,1) & S_b(2,2) & S_b(2,3) & S_b(2,4) \\ S_b(3,1) & S_b(3,2) & S_b(3,3) + S_{m_n} & S_b(3,4) \\ S_b(4,1) & S_b(4,2) & S_b(4,3) & S_b(4,4) + S_{m_n} \end{bmatrix} \tag{4.8}$$

where $S_{m_n} = -\omega^2\tilde{m}_n$.

4.4 Numerical Results

The tower model is shown in Fig. 4.3. The mass of the rotor-nacelle assembly has 350 tons, a tower height of 107.6 m, an average tower diameter of 6.6 m, a tower wall thickness of 0.6 m, Young's modulus of 210 GPa, and a density of 8500 kg/m^3. It assumed a fixed base neglecting the soil-structure interaction. In the FEM model, 10 elements were assumed in the mesh and 2 elements in the SEM. A dynamic analysis was conducted to explore an optimised model for this study. The analysis compared the wind turbine (WT) response calculated using two different models: a two-dimensional spectral element model (SEM-2D) and a three-dimensional finite element model (FEM-3D). The models were excited by a unitary force, and the vertical and rotational responses were estimated at the top of the tower. The validation of the 3D model was performed using the receptance response, employing the spectral element model proposed in [13].

Figure 4.4 compares the vertical and rotational receptance frequency response functions (FRFs) for the continuous tower modelled by SEM and FEM. In the vertical receptance FRFs, the first mode shape exhibits the highest amplitudes. As we move to higher mode shapes, the amplitudes of the resonance frequencies for vertical and rotational modes become equal. This alignment of resonant frequencies indicates a

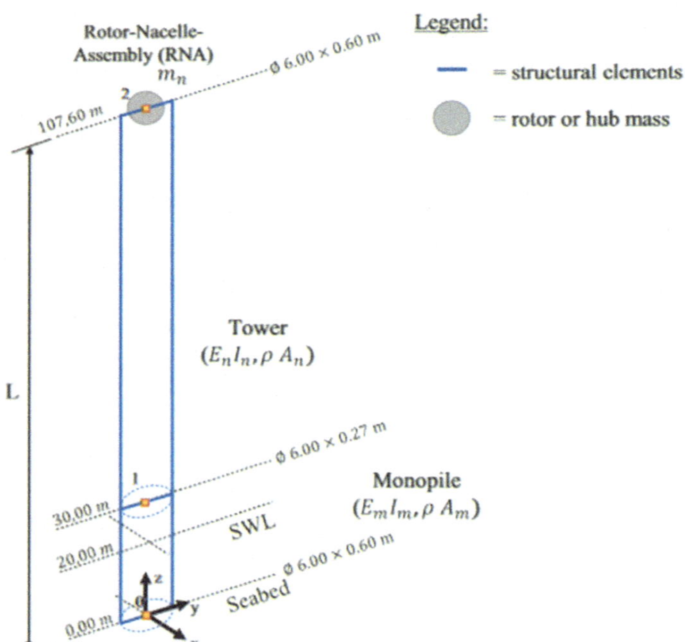

Fig. 4.3 Graphic representation of continuous monopile offshore wind turbine model

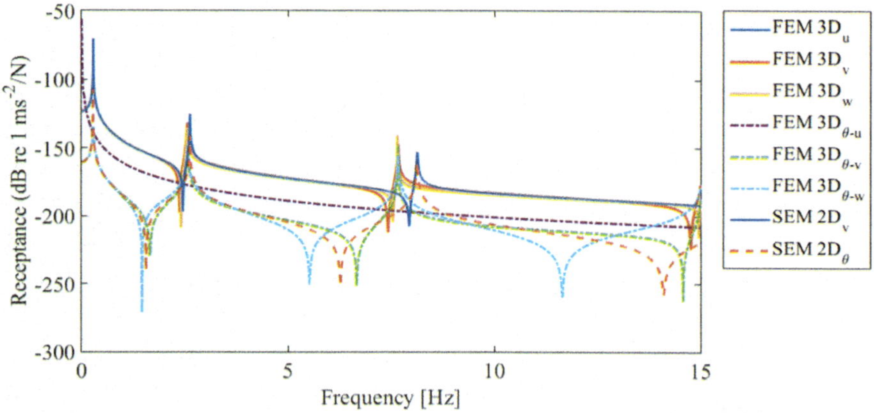

Fig. 4.4 Vertical and rotational receptance FRF comparison between SEM and FEM model with excitement at the WT top

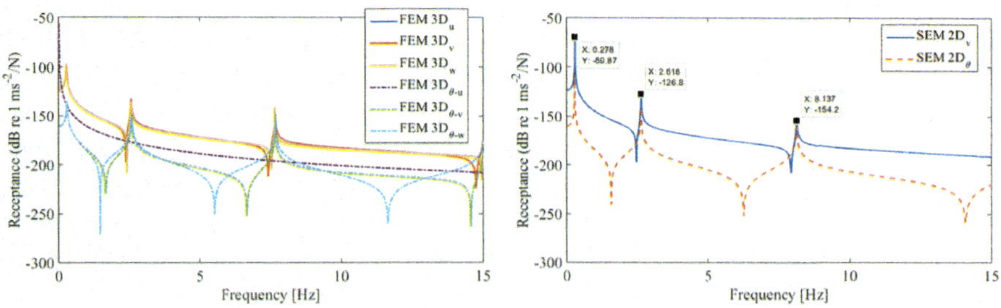

Fig. 4.5 Vertical and rotational receptance FRF modelled by (LHS) FEM and (RHS) SEM obtained on the WT top

correspondence between the vertical and rotational modes. The dynamic response depicted in Fig. 4.5 is similar for both the SEM-2D and FEM-3D models.

The flexural frequency response function (FRF) obtained at the zv-plane of the wind turbine (WT) using the spectral element model (SEM) is comparable to the flexural FRF obtained from the finite element model (FEM) in the same plane (see Fig. 4.1a). This similarity is evident in the low-frequency range, as depicted in Fig. 4.6 (LHS) for flexural displacement and in Fig. 4.6 (RLHS) for rotational displacement. Consequently, the SEM 2D model accurately captures the dynamic behavior of the wind turbine.On the other hand, the 3D model offers a more detailed analysis by considering the modal coupling between the vertical and rotational mode shapes. This coupling becomes more pronounced in the second and third mode shapes, as illustrated in Fig. 4.5a.

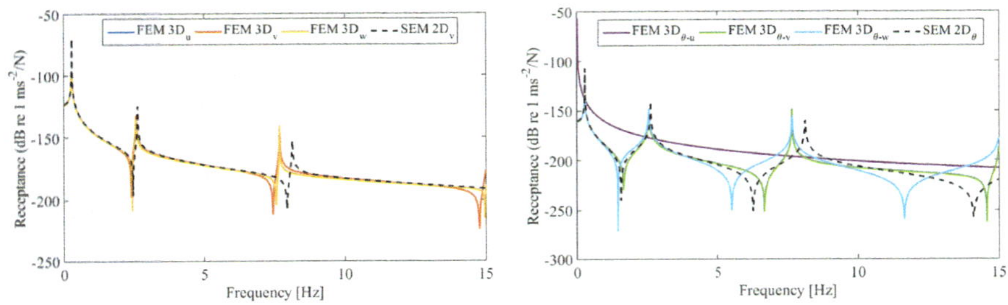

Fig. 4.6 Vertical (LHS) and rotational (RHS) receptance FRF modelled by FEM and SEM obtained on the WT top

4.5 Conclusions

This paper explored the vibration characteristics of a selected 5 MW wind turbine. SEM-2D and FEM-3D models were used to model a tower with a continuous average diameter. The research shows the modal properties and mode-coupling instability in the wind turbine vibrations. In the vertical FRFs, the first mode shape exhibits the highest amplitudes and for the higher mode shapes, the amplitudes of the resonance frequencies for vertical and rotational modes become equal. The SEM 2D model accurately captures the dynamic behavior of the wind turbine. The 3D model gives a more detailed analysis by considering the modal coupling between the vertical and rotational mode shapes. The next step of this study is the implementation of the WT in 3-dimensional using SEM and adding the soil interaction in both models.

Acknowledgements This research is part of the project No. 2022/45/P/ST8/02123 co-funded by the National Science Centre and the European Union Framework Programme for Research and Innovation Horizon 2020 under the Marie Skłodowska-Curie grant agreement no. 945339.

References

1. J. Lee, F. Zhao, Global wind report. Tech. Rep., Global Wind Energy Council, 2022. Available online: https://gwec.net/wp-content/uploads/2022/03/GWEC-GLOBAL-WIND-REPORT-2022.pdf (2022). Accessed on 25 May 2022
2. Energy, In focus: Employment in EU's renewable energy sector. Available online: https://ec.europa.eu/info/news/focus-employment-eus-renewable-energy-sector-2022-may-16-en (2022). Accessed on 25 May 2022
3. J. Jonkman, S. Butterfield, W. Musial, G.T. Scott, *Definition of a 5-MW Reference Wind Turbine for Offshore System Development* (National Renewable Energy Laboratory NREL, Colorado, 2009)
4. A. Awada, R. Younes, A.S. Ilinca, Review of vibration control methods for wind turbines. Energies **14**(11), 3058 (2021)
5. J.F. Doyle, *Wave Propagation in Structures*, 2nd edn. (Springer, New York, 2012)
6. U. Lee, *Special Element Method in Structural Dynamics* (John Wiley & Sons, Singapore, 2009)

7. M.R. Machado, S. Adhikari, J.M.C. Dos Santos, Spectral element-based method for a one-dimensional damaged structure with distributed random properties. J. Brazil. Soc. Mech. Sci. Eng. **40**(9), 415 (2018)
8. M. Machado, M. Dutkiewicz, C. Matt, D. Castello, Spectral model and experimental validation of hysteretic and aerodynamic damping in dynamic analysis of overhead transmission conductor. Mech. Syst. Signal Process **136**, 106483 (2020)
9. M. Dutkiewicz, M. Machado, Spectral approach in vibrations of overhead transmission lines. IOP Conf. Ser.: Mater. Sci. Eng. **471**(5), 052029 (2019)
10. M. Dutkiewicz, M. Machado, Dynamic response of overhead transmission line in turbulent wind flow with application of the spectral element method. IOP Conf. Ser. Mater. Sci. Eng. **471**(5), 052031 (2019)
11. Y. Song, T. Kim, U. Lee, Vibration of a beam subjected to a moving force: frequency-domain spectral element modeling and analysis. Int. J. Mech. Sci. **113**, 162–174 (2016)
12. M.R. Machado, J.M.C. Dos Santos, Effect and identification of parametric distributed uncertainties in longitudinal wave propagation. Appl. Math. Modelling **98**, 498–517 (2021)
13. G.B. Colherinhas, M.V. Girao de Morais, M.R. Machado, Spectral model of offshore wind turbines and vibration control by pendulum tuned mass dampers. Int. J. Struct. Stab. Dyn. **22**(5), 2250053 (2022)
14. M.R. Machado, B.B. Moura, S. Dey, T. Mukhopadhyay, Bandgap manipulation of single and multi-frequency smart metastructures with random impedance disorder. Smart Mater. Struct. **31**(10), 105020 (2022)
15. B.B. Moura, M.R. Machado, T. Mukopadhyay, S. Dey, Dynamic and wave propagation analysis of periodic smart beams coupled with resonant shunt circuits: passive property modulation. Euro. Phys. J. Spec. Top. **231**, 1415–1431 (2022)
16. Y.M.F. Soares, M.R. Machado, M. Dutkiewicz, The spectral approach of love and Mindlin-Herrmann theory in the dynamical simulations of the tower-cable interactions under the wind and rain loads. Energies **15**(20), 7725 (2022)
17. R. Varghese, V. Pakrashi, S. Bhattacharya, A compendium of formulae for natural frequencies of offshore wind turbine structures. Energies **15**(8), 2967 (2022)
18. H. Peuscher, J. Hubele, R. Eid, B. Lohmann, Generating a Parametric Finite Element Model of a 3D Cantilever Timoshenko Beam Using MATLAB. Technical Reports on Automatic Control, TRAC-4 (2009)
19. J.S. Przemieniecki, *Theory of Matrix Structural Analysis* (McGraw-Hill Book Company, New York, 1968)

Chapter 5
Nonlinear Forced Vibration of an Axially Functionally Graded Beam Under Combined Effects of an Electromagnetic Actuator and Mechanical Impact

Vasile Marinca, Cristina Chilibaru-Opritescu, and Nicolae Herisanu

Abstract The present paper deals with the nonlinear forced vibration of an axially functionally graded beam (FGB) subjected to a mechanical impact and an electromagnetic actuation. Material properties are expressed according to a power law function. The governing equation of motion is derived via Hamilton's principle and Galerkin-Bubnov approach is implemented to reduce the nonlinear partial differential equation of the system to a nonlinear ordinary differential equation. The Optimal Homotopy Asymptotic Method (OHAM) is applied to obtain approximate analytical expressions for the nonlinear dynamic response and for the nonlinear frequency. This study shows a high accuracy of OHAM proved by comparison with numerical results.

5.1 Introduction

Functionally graded materials are usually known as materials with nonhomogeneous properties which are designed to be varied continuously. Functional graded structural dynamics or statics have been studied extensively in the last years. The free vibration characteristics of functionally graded (FG) nanobeams based on third-order shear deformation beam theory are proposed by Ebrahimi and Barati [1], presenting a Navier-type solution. Sedighi et al. [2] employed the dynamic instability of FG nanobridges considering Casimir attraction and electric filed actuation. Ghayes [3] derived a coupled nonlinear longitudinal-transverse-rotational set of equations governing motion of a nonlinear vibration of axially FG shear deformable tapered

V. Marinca · C. Chilibaru-Opritescu · N. Herisanu (✉)
University Politehnica Timisoara, Piața Victoriei No. 2, 300006 Timisoara, Romania
e-mail: nicolae.herisanu@upt.ro

V. Marinca · N. Herisanu
Romanian Academy, Timisoara Branch, Bd. M. Viteazu 24, 300223 Timisoara, Romania

N. Herisanu and V. Marinca (eds.), *Acoustics and Vibration of Mechanical Structures—AVMS-2023*, Springer Proceedings in Physics 302,
https://doi.org/10.1007/978-3-031-48087-4_5

beams subjected to external harmonic excitations, which are solved by a continuation scheme. Alimorazadeh et al. [4] investigated nonlinear vibration of a simply supported axially FGB subjected to a moving harmonic load as an Euler-Bernoulli beam utilizing Green's strain tensor. Other studies about FGM are presented in [5–10].

In this paper the nonlinear forced vibration of a shear-deformable tapered functionally graded Euler-Bernoulli beam under the influence of an electromechanical actuator and mechanical impact is investigated by means of the Optimal Homotopy Asymptotic Method (OHAM) and an accurate explicit analytical solution is obtained.

5.2 Governing Equation

Consider a shear-deformable tapered FGB of length L in x-direction and thickness h in z-direction clamped at both ends (Fig. 5.1).

The width b in y-direction varies along the length of the beam. The material of the Euler-Bernoulli axially functionally graded beam (AFGB) such as Young's modulus E, Poisson's ratio ν, the shear modulus μ and the density ρ and geometric properties such as cross-sectional area A and the second moment of inertia vary continuously through the length of the beam according to the power-law as follows:

$$\begin{aligned}
&E(x) = E_R + \left(1 - \frac{x}{L}\right)^m (E_L - E_R)), \ \nu_x = \nu_R + \left(1 - \tfrac{x}{L}\right)^m (\nu_L - \nu_R)) \\
&\mu_x = \frac{E(x)}{2[1 + \nu(x)]}, \ \rho(x) = \rho_R + \left(1 - \tfrac{x}{L}\right)^m (\rho_L - \rho_R)), \ h_R = h_L = h \\
&b(x) = b_R + \left(1 - \frac{x}{L}\right)^m (b_L - b_R)), \ A(x) = \int_A dA, \ I(x) = \int_A z^2 dA
\end{aligned} \tag{5.1}$$

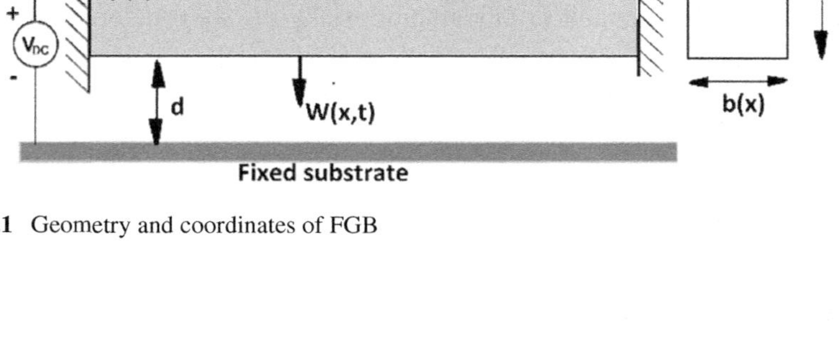

Fig. 5.1 Geometry and coordinates of FGB

where m is the volume fraction and exponent and subscripts L and R denote "left" and "right" ends of the beam.

According to the Euler-Bernoulli beam theory, the components of the displacement field along the x, y and z directions can be expressed as [3, 4, 8]:

$$u_x(x,z,t) = u(x,t) - z\frac{\partial W(x,t)}{\partial x}\ ,\ u_y(x,z,t) = 0\ ,\ u_z(x,z,t) = W(x,t) \quad (5.2)$$

where u(x,t) and W(x,t) are the displacement components in the mid-plane along x and z directions respectively, and $\partial W(x,t)/\partial x$ is the rotation angle about y-axis.

The non-zero nonlinear strain displacement is

$$\varepsilon_{xx} = \frac{\partial u}{\partial x} + \frac{1}{2}\left(\frac{\partial W}{\partial x}\right)^2 - z\frac{\partial^2 W}{\partial x^2} \quad (5.3)$$

The elastic potential energy of the shear deformable tapered AFGB yields to

$$U = \frac{1}{2}\int_0^L \left\{ E(x)A(x)\left[\frac{\partial u}{\partial x} + \frac{1}{2}\left(\frac{\partial W}{\partial x}\right)^2\right]^2 + E(x)I(x)\left(\frac{\partial^2 W}{\partial x^2}\right)^2 \right\} dx \quad (5.4)$$

The kinetic energy of the AFGB shear deformable tapered beam can be written as

$$T = \frac{1}{2}\int_0^L \rho(x)\left[A(x)\left(\frac{\partial u}{\partial t}\right)^2 + I(x)\left(\frac{\partial^2 W}{\partial x \partial t}\right)^2 + A(x)\left(\frac{\partial W}{\partial t}\right)^2\right] dx \quad (5.5)$$

The external work done by the impact force F and the electromagnetic actuator is

$$V = \int_0^L \left[F\delta(x - vt)W(x,t) + \frac{C_0 V_{DC}^2}{2}\left(\frac{1}{d+W} - \frac{1}{d-W}\right)\right] dx \quad (5.6)$$

where δ(x-vt) is the Dirac delta function used to deal mechanical impact, C_0 is the capacitance of the actuator, V_{DC} is the voltage and g is the gap width.

Substituting Eqs. (5.4), (5.5) and (5.6) into the expression of Hamilton's principle:

$$\delta \int_{t_1}^{t_2} [T - (U + V)] dt = 0 \quad (5.7)$$

after some simple mathematical manipulations and integrating by parts, then setting the coefficients of δu and δW equal to zero, one can obtain the following partial differential equations

$$\rho(x)A(x)\frac{\partial^2 u}{\partial t^2} - \frac{\partial}{\partial x}\left\{E(x)A(x)\left[\frac{\partial u}{\partial x} + \frac{1}{2}\left(\frac{\partial W}{\partial x}\right)^2\right]\right\} = 0 \quad (5.8)$$

$$\rho(x)A(x)\frac{\partial^2 W}{\partial t^2} - \frac{\partial}{\partial x}\left\{E(x)A(x)\left[\frac{\partial u}{\partial x} + \frac{1}{2}\left(\frac{\partial W}{\partial x}\right)^2\right]\left(\frac{\partial W}{\partial x}\right)\right\}$$
$$- \frac{\partial^2}{\partial x^2}\left[E(x)I(x)\frac{\partial^2 W}{\partial x^2}\right] - F\delta(x - vt)$$
$$- \frac{C_0 V_{DC}^2}{2}\left(\frac{1}{(d-W)^2} - \frac{1}{(d+W)^2}\right) = 0 \tag{5.9}$$

The clamped-clamped boundary conditions at both ends implies that

$$W(L,t) = W(0,t) = 0,\ \frac{\partial W(L,t)}{\partial x} = \frac{\partial W(0,t)}{\partial x} = 0 \tag{5.10}$$

In Eq. (5.8) the axial inertial in general is assumed to be negligible, so Eq. (5.8) gives

$$\frac{\partial}{\partial x}\left\{E(x)A(x)\left[\frac{\partial u}{\partial x} + \frac{1}{2}\left(\frac{\partial W}{\partial x}\right)^2\right]\right\} = 0 \Leftrightarrow E(x)A(x)\left[\frac{\partial u}{\partial x} + \frac{1}{2}\left(\frac{\partial W}{\partial x}\right)^2\right] = C \tag{5.11}$$

where C is the integration constant to be calculated with respect to the boundary conditions. Since the considered end supports of the beam are immovable, i.e. u(L,t) = u(0,t) = 0, integration of Eq. (5.11) yields:

$$C = \frac{1}{2}\left[\int_0^L \frac{dx}{E(x)A(x)}\right]^{-1} \int_0^L \left(\frac{\partial W}{\partial x}\right)^2 dx \tag{5.12}$$

The component corresponding to electromagnetic actuation from Eq. (5.9) can be simplified in the form:

$$\frac{C_0 V_{DC}^2}{2}\left(\frac{1}{(d-W)^2} - \frac{1}{(d+W)^2}\right) \approx \frac{C_0 V_{DC}^2}{2d^2}\left[4\left(\frac{W}{d}\right) + 8\left(\frac{W}{d}\right)^3 + 12.37\left(\frac{W}{d}\right)^5\right] \tag{5.13}$$

Inserting Eqs. (5.11–5.13) into Eq. (5.9) on can get:

$$\rho(x)A(x)\frac{\partial^2 W}{\partial t^2} - \frac{1}{2}\left[\int_0^L \frac{dx}{E(x)A(x)}\right]^{-1}\left[\int_0^L \left(\frac{\partial W}{\partial x}\right)^2 dx\right]$$
$$\left\{E(x)A(x)\frac{\partial^2 W}{\partial x^2} + \frac{\partial W}{\partial x}\frac{d}{dx}[E(x)A(x)] - \frac{\partial^2}{\partial x^2}\left[E(x)A(x)\frac{\partial^2 W}{\partial x^2}\right]\right\}$$
$$- \frac{2C_0 V_{DC}^2}{d^2}\left[\frac{W}{d} + 2\left(\frac{W}{d}\right)^3 + 3.095\left(\frac{W}{d}\right)^5\right] = F\delta(x - vt) \tag{5.14}$$

The following nondimensional quantities are introduced:

$$\overline{W} = \frac{W}{d}, \ \overline{x} = \frac{x}{L}, \ \overline{E} = \frac{E(x)}{E_R}, \ \overline{\rho} = \frac{\rho(x)}{\rho_R}, \ \overline{A} = \frac{A(x)}{A_0}, \ \overline{I} = \frac{I(x)}{I_R},$$
$$\overline{t} = \frac{t}{L^2}\sqrt{\frac{E_R I_R}{\rho_R A_R}}, \ \overline{\alpha} = \frac{C_0 V_{DC}^2 L^2}{E_R d^3 I_R}, \ \overline{v} = vL^3\sqrt{\frac{\rho_R A_R}{E_R I_R}}, \ \overline{f} = \frac{FL^3}{E_R I_R d^3} \tag{5.15}$$

These parameters are inserted into Eq. (5.14) and the bar is dropped for brevity, so that one can obtain the following nondimensional equation:

$$\begin{aligned}\rho(x)A(x)\frac{\partial^2 W}{\partial t^2} - \frac{1}{2}\left[\int_0^1 \frac{dx}{E(x)A(x)}\right]^{-1}\left[\int_0^1 \left(\frac{\partial W}{\partial x}\right)^2 dx\right] \\ \left\{E(x)A(x)\frac{\partial^2 W}{\partial x^2} + \frac{d}{dx}[E(x)A(x)]\right\} \\ - \frac{\partial^2}{\partial x^2}\left[E(x)A(x)\frac{\partial^2 W}{\partial x^2}\right] \\ - \alpha\left(W + 2W^3 + 3.095W^5\right) = f\delta(x - vt)\end{aligned} \tag{5.16}$$

In order to apply the Galerkin-Bubnov method, the displacement function W(x,t) can be expressed in the form

$$W(x,t) = X(x)T(t) \tag{5.17}$$

where X(x) is the basis function which must satisfy the boundary conditions (10), which can be rewritten as

$$X(0) = X(1) = X'(1) = X'(0) = 0 \tag{5.18}$$

Substituting Eq. (5.17) into Eq. (5.16), then multiplying both sides of resulting equation with X(x), integrating over the domain [0,1], yields to

$$\ddot{T} + \omega^2 T + aT^3 + bT^5 = fX(vt) \tag{5.19}$$

where $\ddot{T}$ is the second derivative of T(t) with respect to time. The coefficients which appear in the last equation can be expressed as

$$\omega^2 = -\frac{\int_0^1 EIX''Xdx + \int_0^1 X''X\left[E''X + 2E'I' + EI''\right]dx}{\int_0^1 \rho(x)AXdx}$$
$$b = -\frac{3.095\alpha \int_0^1 X^6(x)dx}{\int_0^1 \rho(x)AXdx} \tag{5.20}$$
$$a = -\frac{2\alpha \int_0^1 X^4 dx + \frac{1}{2}\left[\int_0^1 \frac{dx}{EA}\right]^{-1}\left[\int_0^1 X'^2 dx\right] \left\{\int_0^1 EAX''Xdx + \int_0^1 X'X\left[E'A + EA'\right]dx\right\}}{\int_0^1 \rho(x)AXdx}$$

where $X' = dX(x)/dx$. The shear deformation tapered beam AFGB is subject to the following initial conditions:

$$T(0) = A\,,\ \dot{T}(0) = 0 \tag{5.21}$$

The exact solution of nonlinear differential equation is very limited and very time consuming, and for this reason an approximate method is inevitable to be applied to solve Eqs. (5.19) and (5.20).

An approximate analytical solution for these equations can be obtained by the Optimal Homotopy Asymptotic Method (OHAM).

5.3 The Optimal Homotopy Asymptotic Method

The aim of this Section is to obtain an approximate solution by means of a combination of the second and third alternative of the OHAM [11] to the general nonlinear differential equation expressed using the linear and nonlinear operators L and N:

$$L[T(t)] + N[T(t)] = 0 \tag{5.22}$$

in the presence of boundary/initial conditions

$$B\left[T(t), \frac{dT(t)}{dt}\right] = 0 \tag{5.23}$$

The approximate solution of Eqs. (5.22) and (5.23) is supposed to be the sum of only two components:

$$\overline{T}(t) = T_0(t) + T_1(t) \tag{5.24}$$

The initial component T_0 can be obtained from the following linear equation:

$$L[T_0(t)] = 0\,,\ B\left[T_0(t), \frac{dT_0(t)}{dt}\right] = 0 \tag{5.25}$$

The second component T_1 can be found from the equation

$$L_1[T_1(t)] = H\big(t, C_1, C_2, \ldots, C_p\big)N[T_0(t)],\ B\Big[T_1(t), \tfrac{dT_1(t)}{dt}\Big] = 0 \tag{5.26}$$

in which is involved the so-called auxiliary function depending on p arbitrary parameters unknown at this moment. It should be emphasized that the functions H are determined from the condition that the product H $N[T_0(t)]$ to be of the same shape. It follows that the auxiliary function H is not unique. The parameters C_i can be optimally determined by means of various procedures, such as the least square method, the collocations method, the Ritz method, the Galerkin method, by minimizing the residual error. In this way, the approximate solution (24) of Eqs. (5.22) and (5.23) is well determined.

5.4 Approximate Solution of the Considered Problem

Using the transformations

$$\tau = \omega_s t\ ,\ T(t) = A\phi(\tau) \tag{5.27}$$

in which ω_s is the natural frequency of the system, Eq. (5.19) can be rewritten as

$$\ddot{\phi} + \frac{\omega^2}{\omega_s^2} + \frac{aA^2}{\omega_s^2}\phi^3 + \frac{bA^4}{\omega_s^2}\phi^5 = \frac{fX(vt)}{A\omega_s^2} \tag{5.28}$$

The linear and nonlinear operators for Eq. (5.38) are respectively:

$$L[\phi(\tau)] = \ddot{\phi} + \phi\ ,\ N[\phi(\tau)] = \left(\tfrac{\omega^2}{\omega_s^2} - 1\right)\phi + \tfrac{aA^2}{\omega_s^2}\phi^3 + \tfrac{bA^4}{\omega_s^2}\phi^5 - \tfrac{fX(vt)}{A\omega_s^2} \tag{5.29}$$

The approximate solution of Eq. (5.28) is

$$\overline{\phi}(\tau) = \phi_0(\tau) + \phi_1(\tau) \tag{5.30}$$

The initial component φ_0 is determined from the linear equation:

$$\ddot{\phi}_0 + \phi_0 = 0,\ \ \phi_0(0) = 1,\ \ \dot{\phi}(0) = 0 \tag{5.31}$$

such that

$$\phi_0(\tau) = \cos\tau \tag{5.32}$$

Inserting Eq. (5.32) into Eq. (5.29) one retrieves

$$N[\phi_0(\tau)] = B_1 \cos\tau + B_3 \cos 3\tau + B_5 \cos 5\tau - \frac{fX(vt)}{A\omega_s^2} \tag{5.33}$$

where

$$B_1 = \left(\frac{\omega^2}{\omega_s^2} - 1\right) + \frac{3aA^2}{4\omega_s^2} + \frac{5bA^4}{8\omega_s^2},\ B_3 = \frac{aA^2}{4\omega_s^2} + \frac{5bA^4}{16\omega_s^2},\ B_5 = \frac{bA^5}{16\omega_s^2} \tag{5.34}$$

and for the clamped–clamped beam, in the conditions (10), the function X becomes

$$X(x) = \sqrt{\frac{2}{3}}(1 - \cos\pi x) \tag{5.35}$$

The first component $\varphi_1(\tau)$, using $H(\tau,C_i) = C_1 + 2C_2\cos2\tau + 2C_3\cos4\tau + 2C_4\cos6\tau$, can be determined from the linear Eq. (5.26):

$$\begin{aligned}\ddot{\phi}_1 + \phi_1 &= (C_1 + 2C_2 \cos 2\tau + 2C_3 \cos 4\tau + 2C_4 \cos 6\tau)(B_1 \cos\tau \\ &\quad + B_3 \cos 3\tau + B_5 \cos 5\tau),\ \phi_1(0) = \dot{\phi}_1(0) = 0\end{aligned} \tag{5.36}$$

The Eq. (5.36) can be rewritten as

$$\begin{aligned}\ddot{\phi}_1 + \phi_1 &= [B_1(C_1 + C_2) + B_3(C_1 + C_3) + B_5(C_3 + C_4)] \cos\tau \\ &\quad + [B_1(C_2 + C_3) + B_3(C_1 + C_4) + B_5C_2]\cos 3\tau + [B_5C_1 + B_3C_2 \\ &\quad + B_1(C_3 + C_4)] \cos 5\tau + (B_5C_1 + B_3C_3 + B_1C_4) \cos 7\tau \\ &\quad + (B_5C_3 + B_3C_4) \cos 9\tau + B_5C_4 \cos 11\tau\end{aligned} \tag{5.37}$$

Avoiding secular terms, we can obtain the natural frequency of the system:

$$\overline{\omega}^2 = \omega^2 + \frac{3aA^2}{4} + \frac{5bA^4}{8} + \left(\frac{aA^2}{4} + \frac{5bA^4}{16}\right)\frac{C_2 + C_3}{C_1 + C_2} + \frac{bA^4}{16}\frac{C_3 + C_4}{C_1 + C_2} \tag{5.38}$$

The solution of Eq. (5.37) becomes

$$\begin{aligned}\phi_1(x) &= \frac{B_3(C_1 + C_4) + B_1(C_2 + C_3) + B_5C_2}{8}(\cos\tau - \cos 3\tau) \\ &\quad + \frac{B_5C_1 + B_3C_2 + B_1(C_3 + C_4)}{24}(\cos\tau - \cos 5\tau) \\ &\quad + \frac{B_5C_2 + B_3C_3 + B_1C_4}{48}(\cos\tau - \cos 7\tau) \\ &\quad + \frac{B_5C_3 + B_3C_4}{80}(\cos\tau - \cos 9\tau) \\ &\quad + \frac{B_5C_4}{120}(\cos\tau - \cos 11\tau)\end{aligned} \tag{5.39}$$

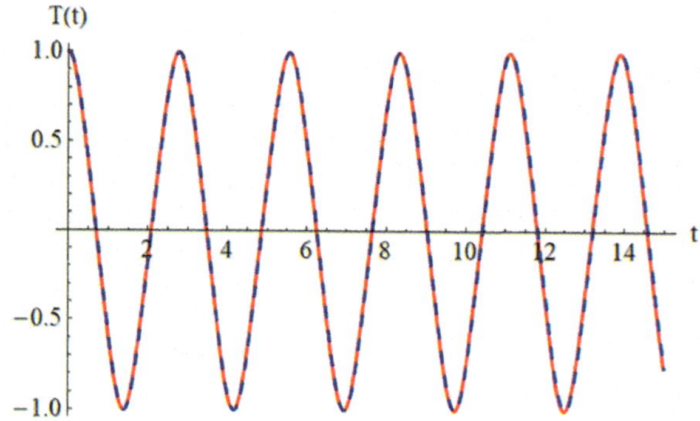

Fig. 5.2 Comparison between analytical OHAM solution and numerical integration results: red soild line numerical blue dashed line analytical

The approximate solution of Eqs. (5.19) and (5.21) is obtained from (5.30), (5.32) and (5.39):

$$\begin{aligned}\overline{T}(t) = A\cos\overline{\omega}t &+ \frac{A[B_3(C_1 + C_4) + B_1(C_2 + C_3) + B_5C_2]}{8}(\cos\overline{\omega}t - \cos 3\overline{\omega}t) \\ &+ \frac{A[B_5C_1 + B_3C_2 + B_1(C_3 + C_4)]}{24}(\cos\overline{\omega}t - \cos 5\overline{\omega}t) \\ &+ \frac{A[B_5C_2 + B_3C_3 + B_1C_4]}{48}(\cos\overline{\omega}t - \cos\overline{7\omega}t) \\ &+ \frac{A[B_5C_3 + B_3C_4]}{80}(\cos\overline{\omega}t - \cos 9\overline{\omega}t) \\ &+ \frac{AB_5C_4}{120}(\cos\overline{\omega}t - \cos 11\overline{\omega}t) \end{aligned} \tag{5.40}$$

To illustrate the accuracy of OHAM solution, we consider the parameters: $\omega = 1.5$, $f = 0.001$, $a = 0.96$, $b = 3.67$.

The obtained optimal values of the convergence-control parameters in this case are $C_1 = -0.71630664819$, $C_2 = 0.0098375896$, $C_3 = 0.09795853555$, $C_4 = -0.27220137053$.

In Fig. 5.2, the present solution (5.40) is compared with numerical integration results, emphasizing the accuracy of the analytical results which overlap the numerical integration ones.

5.5 Conclusions

In this research is analyzed the nonlinear forced vibration of a shear-deformable tapered functionally graded Euler-Bernoulli beam under the influence of an electromechanical actuator and mechanical impact. The beam under study is tapered, i.e. Young's modulus, Poisson ratio, the shear modulus, the density, with area and moment of inertia vary along the length. The kinetic and potential energy and the potential work are derived and implemented in Hamilton's principle. The nonlinear partial differential equation is discretized making use of the Galerkin-Bubnov procedure and solved with the aid of OHAM.

The main novelty of our technique consists in the presence of auxiliary functions some optimal convergence-control parameters, and the construction of the two iterations. An example is given and a very good accuracy is found in comparison with numerical integration results.

References

1. F. Ebrahimi, M.R. Barati, A nonlocal higher-order shear deformation beam theory for vibration analysis of size-dependent functionally graded nanobeams. Arab. J. Sci. Eng. **41**, 1679–1690 (2016)
2. H.M. Sedighi, F. Daneshmand, M. Abadyan, Dynamic instability analysis of electrostatic functionally graded doubly-clamped nano-actuators. Compos. Struct. **124**, 55–64 (2015)
3. M. Ghayesh, Nonlinear vibration analysis of axially functionally graded shear-deformable tapered beams. Appl. Math. Model. **59**, 583–596 (2018)
4. M. Alimoradzadeh, M. Salehi, S.M. Esfarjani, Nonlinear dynamic response of an axially functionally graded beam resting on nonlinear elastic foundation subjected to moving load. Nonlinear Eng. **8**, 250–260 (2019)
5. N. Shafiei, A.R. Setoodeh, An analytical study on the nonlinear forced vibration of functionally graded carbon nanotubes-reinforced composite beams on nonlinear viscoelastic foundation. Arch. Mech. **72**(2), 81–107 (2020)
6. M.H. Yas, S. Rahimi, Thermal vibration of functionally graded porous nanocomposite beams reinforced by graphene platelets. Appl. Math. Mech. English Edition **41**(8), 1209–1226 (2020)
7. Y. Chen, M. Zhang, Y. Su, Z. Zhou, Coupling analysis of flexoelectric effect of functionally graded piezoelectric cantilever nanobeams. Micromachines **12**(6), 595 (2021)
8. J.N. Reddy, E. Ruocco, J.A. Loya, A.M.A. Neves, Theories and analysis of functionally graded beams. Appl. Sci. **11**(15), 7159 (2021)
9. H.V. Long, V.L. Nguyen, M.T. Tran, D.K. Thai, Exact solution for nonlinear static behaviors of functionally graded beams with porosities resting on elastic foundation using neutral surface concept. Proc. Inst. Mech. Eng. C J. Mech. Eng. Sci. Eng. C J. Mech. Eng. Sci. **236**(1), 481–495 (2022)
10. J. Wu, L. Chen, R. Wu, X. Chen, Nonlinear forced vibration of bidirectional functionally graded material beam. Shock. Vib. **2021**, 6675125 (2021)
11. V. Marinca, N. Herisanu, in *The Optimal Homotopy Asymptotic Method*. Engineering Applications. (Springer, Cham, 2015)

Chapter 6
Dynamic Behavior of Vibratory System to Discrete Variation of Viscous Damping in Linear Viscoelastic Material Processing Technologies

Polidor Bratu, Ovidiu Vasile, Cornelia Dobrescu, Marinela Cristina Nițu, and Oana Tonciu

Abstract The experimental research carried out under forced vibrations, by applying a harmonic excitation inertial force, highlighted the fact that the processed viscoelastic materials, when put into operation (fresh concrete, soil, asphalt mixes, filling mastics, mortar with additives, bitumens), can be characterized by a relevant dynamic response that corresponds to a certain type of rheological model. It was found that the product family behaves differently, from a rheological point of view, depending on the weight of the physico-chemical composition components. Thus, the schematization of the calculation model must correspond to the dynamic characteristic, in response to the monotonous variation of the excitation pulse, by ensuring a continuous variation both numerically and experimentally.

P. Bratu
Romanian Academy of Technical Sciences, 030167 Bucharest, Romania

O. Vasile (✉)
National University of Sciences and Technology POLITEHNICA Bucharest, 060042 Bucharest, Romania
e-mail: ovidiu_vasile2002@yahoo.co.uk

C. Dobrescu
"Dunărea de Jos" University of Galati, 800008 Galati, Romania

M. C. Nițu
Institute of Solid Mechanics of the Romanian Academy, 030167 Bucharest, Romania

O. Tonciu
Technical University of Civil Engineering of Bucharest, 020396 Bucharest, Romania

N. Herisanu and V. Marinca (eds.), *Acoustics and Vibration of Mechanical Structures—AVMS-2023*, Springer Proceedings in Physics 302,
https://doi.org/10.1007/978-3-031-48087-4_6

6.1 Introduction

The paper presents for the first time in the field of vibration compaction processes the difference in the rheological behavior of the land, depending on the geotechnical structure, through the experimental "in situ" establishment of the two Voigt-Kelvin or Maxwell models. The novelty of the paper consists in the fact that it highlights the "in situ" dynamic method by which the rheological model of the Voigt-Kelvin or Maxwell material can be established at natural scale and the real dynamic quantities in resonance.

The results of laboratory tests on materials from the same product family, but with different compositions and gravimetric weights, highlighted the fact that, in the vibratory processing regime, the discrete variation of viscous damping leads to different rheological behavior. Thus, in the Voigt-Kelvin modeled materials, the amplitude-frequency characteristic is influenced by the fact that the discrete size of the damping, on successive technological stages of processing, causes the amplitude to tend to very small values at resonance, leading to the disappearance of the resonance state [1–5].

For viscoelastic materials from the same product family but in different compositions and proportions, the Maxwell model is used for which the resonance amplitude becomes more important the higher the damping [6–9]. In the studies carried out both in the laboratory and "in situ" within the Research Institute for Construction Equipment and Technologies ICECON Bucharest, Romania, the amplitude-frequency characteristics of the discrete variation of the viscous damping could be highlighted numerically and experimentally. On this basis, the nature of the antagonistic behavior of the two Voigt-Kelvin and Maxwell rheological models is displayed [10–15].

Certainly, both in the laboratory and "in situ" or based on analytical calculation, depending on the dynamic response of the material for two distinct values of viscous damping $c_1 \neq c_2$ the Voigt-Kelvin or Maxwell rheological behavior can be appreciated [16–19]. In this case, the rheological schematization can be established with the stiffness and discrete damping parameters c_j, where $j = 1, 2, 3, 4, 5$ [20, 21].

6.2 Dynamic Response Parameterized as a Function of Discrete Variation of Viscous Damping

To define the rheological models, Voigt–Kelvin or Maxwell schematizations are used, in the linear domain, where the damper is represented by the discrete values c_j, with $j = 1, 2, 3, 4, 5$, and the stiffness k is constant. The disturbing force is of the inertial type of the form $F(t) = m_0 r \omega^2 \sin \omega t$, where $m_0 r$ is the static moment of the dynamic imbalance bodies and ω is the excitation pulsation, the same as the angular speed of rotation of the eccentric bodies with mass m_0 and eccentricity r.

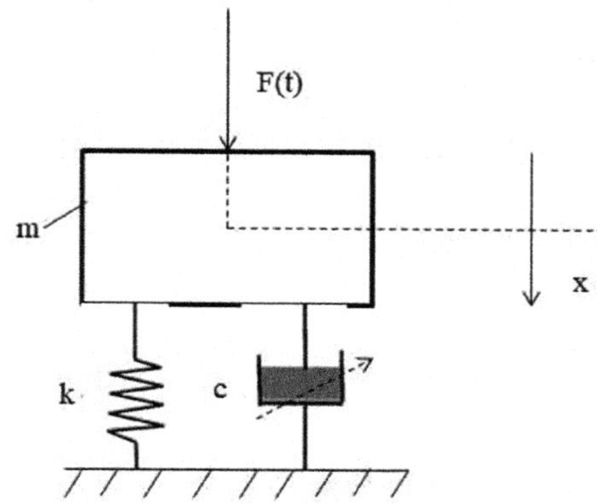

Fig. 6.1 Voigt–Kelvin model

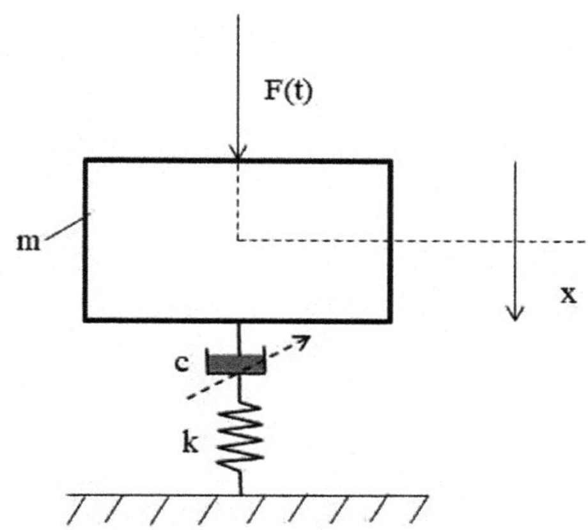

Fig. 6.2 Maxwell model

6.2.1 Composite Rheological Models

The composite rheological model *E/V* represented by the stiffness *k* and the discrete variable viscous damping coefficient c_j with $j = 1, 2, 3, 4, 5$, connected in parallel of the type ($k|c$) is schematized Voigt- Kelvin (Fig. 6.1). The composite rheological model E-V represented by the stiffness *k* and the viscous damping coefficient c_j connected in series of (*K-C*) type is Maxwell schematized (Fig. 6.2).

6.2.2 Amplitude-Frequency Characteristic

For the Voigt-Kelvin rheological model the amplitude of the forced vibrations as a function of the angular frequency ω as a continuous variable and the viscous damping coefficient c with discrete variation is of the form:

$$A^V = \frac{m_0 r\omega^2}{\sqrt{(k - m\omega^2)^2 + c^2\omega^2}} \tag{6.1}$$

The amplitude of the vibrations in the case of the Maxwell model as a function of the angular frequency ω and the discrete variation of the damping c is given by the relation

$$A^M = m_0 r\omega^2 \sqrt{\frac{k^2 + c^2\omega^2}{m^2k^2\omega^4 + c^2\omega^2(k - m\omega^2)^2}} \tag{6.2}$$

In resonance regime for $k - m\omega^2 = 0$, that is for $\omega = \omega_n = \sqrt{\frac{k}{m}}$ results the following expressions of amplitudes at resonance A^V_{rez} and A^M_{rez} as:

$$A^V_{rez} = \frac{m_0 r}{m}\frac{\sqrt{km}}{c} \tag{6.3}$$

$$A^M_{rez} = \frac{m_0 r}{m}\sqrt{1 + \frac{c^2}{km}} \tag{6.4}$$

Noting $A_0 = \frac{m_0 r}{m}$ and $c = 2\zeta\sqrt{km}$, where ζ is the depreciation rate, relations (6.3) and (6.4) can still be written in the form:

$$A^V_{rez} = A_0\frac{1}{2\zeta} \tag{6.5}$$

$$A^M_{rez} = A_0\sqrt{1 + 4\zeta^2} \tag{6.6}$$

At the limit of the depreciation rate $\zeta_0 = \frac{\sqrt{2}}{2}$ from relations (6.5) and (6.6) results the value corresponding to the resonance amplitude $\overline{A} = A^V_{rez}$ and $\overline{\overline{A}} = A^M_{rez}$ as follows:

- for the Voigt-Kelvin model we have:

$$\overline{A} = A_0\frac{1}{2\zeta_0}$$

and for $\zeta_0 = \frac{\sqrt{2}}{2}$ results

$$\overline{A} = \frac{\sqrt{2}}{2}A_0 \tag{6.7}$$

- for the Maxwell model results

$$\overline{\overline{A}} = A_0\sqrt{1 + 4\zeta_0^2}$$

Hence, for $\zeta_0 = \frac{\sqrt{2}}{2}$, we have

$$\overline{\overline{A}} = \sqrt{3}A_0 \tag{6.8}$$

In conclusion, for values of $\zeta < \zeta_0$ the amplitudes A_{rez}^V according to Voigt-Kelvin tend to high values specific to the resonance state, and for the Maxwell model the amplitudes tend to a minimum value without amplifying the A_0 amplitudes, which degenerates the resonance state.

For values of $\zeta > \zeta_0$ the resonance state is attenuated by the Voigt-Kelvin model and amplified by the Maxwell model.

6.2.3 *Variation Curves of Resonance Amplitudes as a Function of Viscous Damping*

For the case study of a vibratory equipment, the following physical–mechanical quantities are given, namely: mass $m = 4000\,\text{kg}$, the static moment of the eccentric masses $m_0 r = 2\,\text{kg}\,\text{m}$, stiffness $k = (1, 2, 3, 4, 5) \times 10^8\,\frac{\text{N}}{\text{m}}$ and continuous variation of the coefficient of damping $c = (0, 1...50) \times 10^5\,\frac{\text{Ns}}{\text{m}}$. In this case, were increased the variation curves $A_{rez}^V(k, c)$ and $A_{rez}^M(k, c)$, in accordance with relations (6.3) and (6.4) respectively. The graphical representation is given in Figs. 6.3 and 6.4.

6.2.4 *Amplitude-Frequency Characteristic for Discrete Damping Variation*

At the discrete change of the damping $c = (1, 3, 5, 7, 9, 11, 13, 15, 17, 21) \times 10^5\,\frac{\text{Ns}}{\text{m}}$ and at the monotonous-increasing variation of the excitation pulsation (angular frequency) $\omega = (0...500)\,\frac{\text{rad}}{\text{s}}$ the families of curves for the amplitudes $A^V(\omega, c)$ and $A^M(\omega, c)$ as in Figs. 6.5 and 6.6.

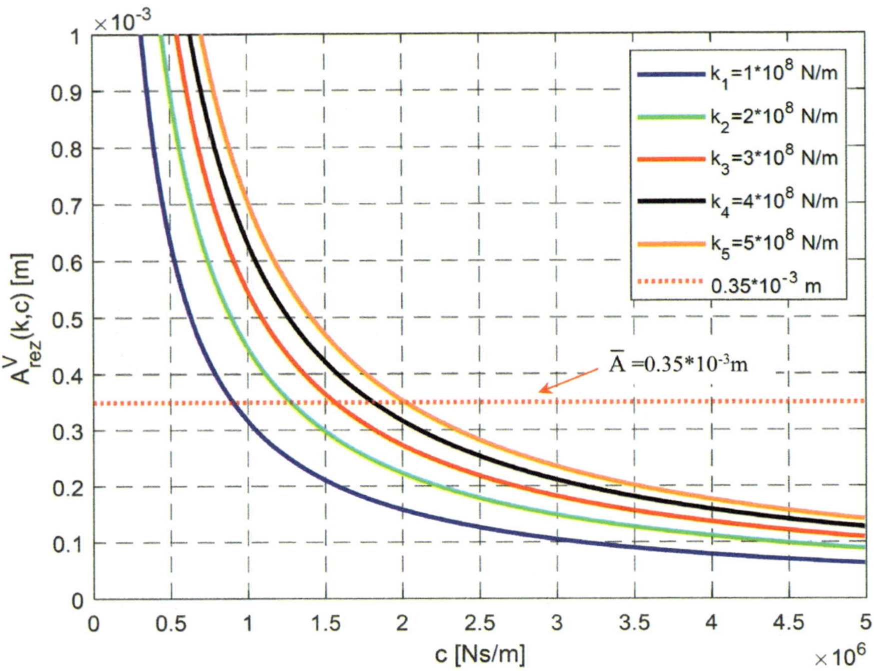

Fig. 6.3 Resonance amplitude curve A^V_{rez} according to continuously varying damping c

6.3 Conclusions

Based on the experimental and numerical results for the Voigt-Kelvin and Maxwell rheological materials, processed in the harmonic vibration regime, an essentially different behavior at resonance was found. Thus, by raising the amplitude and frequency characteristic curves and taking into account at least two distinct and measurable values of the viscous damping, the Voigt-Kelvin or Maxwell rheological behavior can be ascertained.

Taking into account the behavior in the resonance state as well as for the post-resonance technological state, the following conclusions can be formulated:

- For the Voigt-Kelvin rheological model, after each stage of processing with the modification of stiffness, it is found that at resonance with the increase of damping over $\zeta_0 = \frac{\sqrt{2}}{2}$ the amplitude decreases to such small values that the amplitude can no longer represent a peak. In this case, the resonance is attenuated until the phenomenon is completely reduced.
- For the Maxwell model after each processing step, the stiffness increases and for the resonance states the increase in damping leads to a significant increase in the resonance amplitude with maximum points specific to the phenomenon. Thus, for

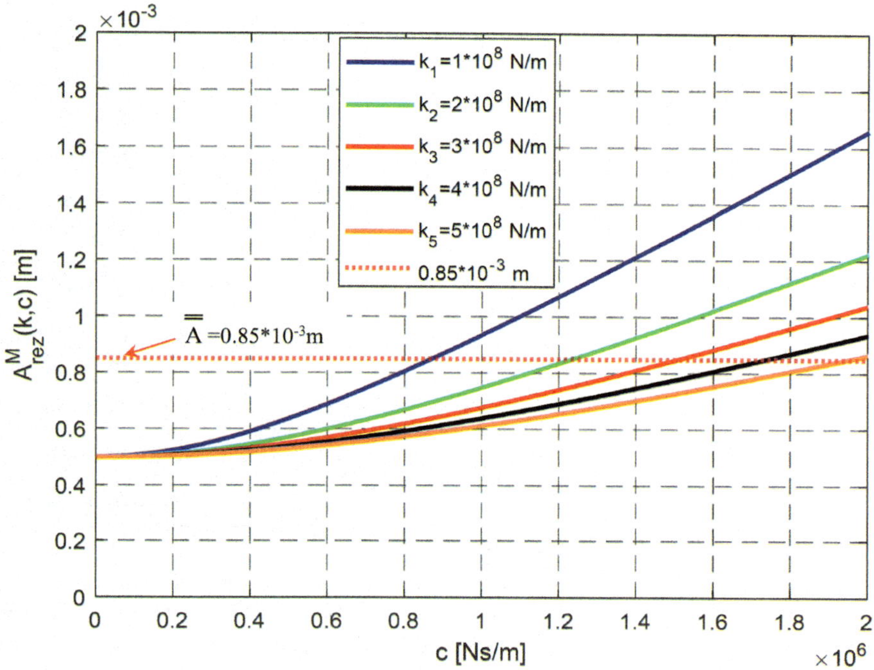

Fig. 6.4 Resonance amplitude curve A_{rez}^{M} according to continuous variation in damping c

values of $\zeta > \frac{\sqrt{2}}{2}$ the resonance is highlighted and leads to peak values of the amplitude.

- The amplitude—frequency characteristic for the Voigt-Kelvin model highlights the fact that for $\zeta_0 = \frac{\sqrt{2}}{2}$, where $A_{rez}^{V} = \overline{A} \leq \left(\frac{\sqrt{2}}{2}\right) A_0$ that is, it is less than 0.35×10^{-3} mm, the curves for $c_1 > 7 \times 10^5 \, \frac{\text{Ns}}{\text{m}}$ are flattened without the maximum point (peak). This shows that the resonance has degenerated.
- In the case of the Maxwell model, the amplitude-frequency characteristic highlights the non-existence of resonance at values of $c_1 < 5 \times 10^5 \, \frac{\text{Ns}}{\text{m}}$, while for high damping values over $6 \times 10^5 \, \frac{\text{Ns}}{\text{m}}$ the state of resonance with remarkable peaks is manifested compared to the above, it can be appreciated that for viscoelastic materials processed with harmonic vibrations, the Voigt-Kelvin or Maxwell rheological behavior can be established by the dynamic analysis in resonance using the amplitude-frequency characteristic. In the laboratory, on categories of samples taken, vibration systems can be used with the continuous change of frequency and the elevation of the amplitude-frequency curve using an appropriate instrumental equipment.

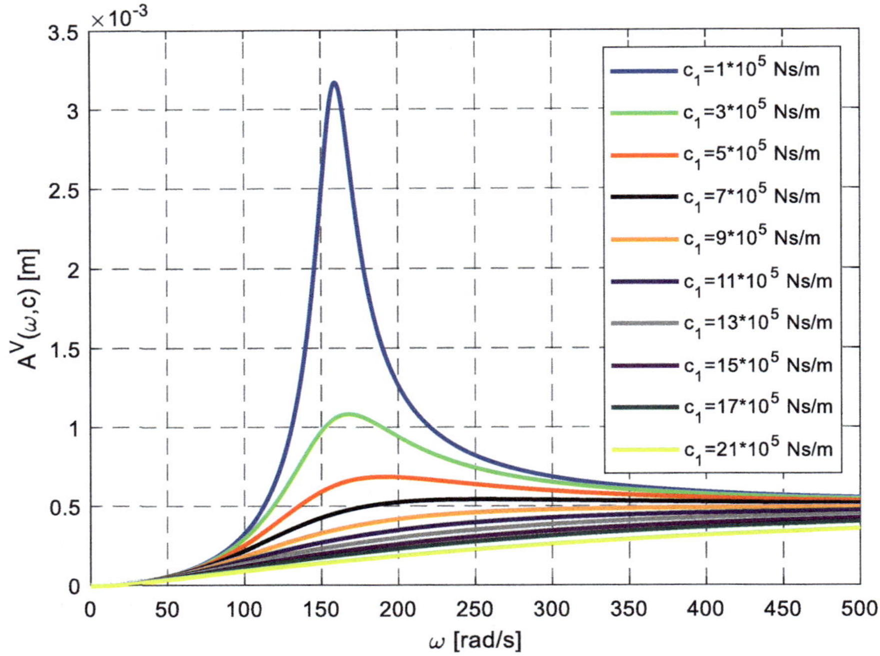

Fig. 6.5 Family of $A^V - \omega$ curves for the discrete variation of c

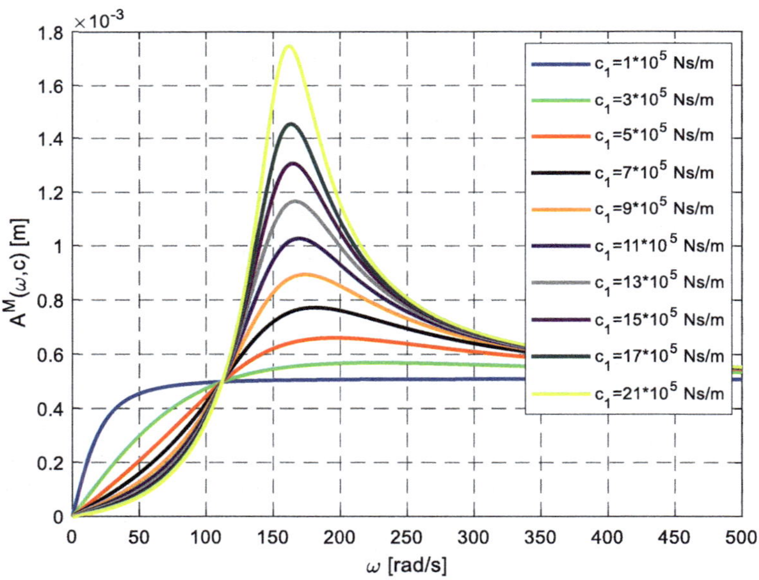

Fig. 6.6 Family of $A^M - \omega$ curves for the discrete variation of c

References

1. X. Hu, Q. Chen, D. Weng, R. Zhang, X. Ren, Estimation of additional equivalent amortization ratio of the damped structure based on energy dissipation, Hindawi. Adv. Civ. Eng. **2019**, 8052413 (2019)
2. L.E. Suarez, C.A. Gaviria, Equivalent frequencies and amortization ratios of buildings with viscous fluid dampers: a closed form formulation, 10 NCEE, 21–25 July 2014, Anchorage, Alaska (2014)
3. J. Butterworth, H.J. Lee, B. Davidson, Experimental determination of modal amortization from full scale testing. 13 WCEE 1–6 Aug 2004, Vancouver, B.C. Canada (2004)
4. Q. Jiang, Z. Zhou, W. Huang, Investigation on the modal strain energy for dynamic analysis of steel-concrete vertically. J. Asian Arch. Build. Eng. **14**(3), 671–678 (2015). https://doi.org/10.3130/jaabe.14.671
5. A.K. Chopra, Dynamics of structures: theory and applications to earthquake engineering, 4th ed. (Upper Saddler River, NJ, 2012)
6. H.V. Tsai, J.M. Kelly, Seismic response of early damped base isolation systems. Earthq. Eng. Struct. Dyn. **22**(7), 633–645 (1993)
7. J.M. Kelly, H.C. Tsai, Seismic response of light internal equipment in base-isolated structures. Earthq. Eng. Struct. Dyn **13**(6), 711–712 (1985)
8. P. Bratu, N. Dragan, L'analyse des mouvements désaccouplés appliquée au modèle de solide rigide aux liaisons élastiques. Annale, Univ. "Dunarea de Jos" XIV, Galati (2011)
9. N. Dragan, Some considerations about the dynamics of the reinforced concrete modeled as a rigid solid with viscous-elastic bearings. Ann. Oradea Univ. **IX**(XIX), 3.42–3.48 (2010). ISSN 1583-0691
10. C.F. Dobrescu, Highlighting the change of the dynamic response to discrete variation of soil stiffness in the process of dynamic compaction with roller compactors based on linear rheological modeling, in *Applied Mechanics and Materials, Vol. 801: Acoustics & Vibration of Mechanical Structures II*, pp. 242–248 (2015)
11. C.F. Dobrescu, E. Braguta, Optimization of vibro-compaction technological process considering rheological properties, in *Acoustics and Vibration of Mechanical Structures—AVMS-2017* (Springer Proceedings in Physics, 2018), pp. 287–293
12. J.M. Kelly, The role of amortization in seismic isolation. Earthq. Eng. Dyn. **28**(1), 3–20 (1999)
13. H. Anajafi, R.A. Medina, E. Santini-Bell, Effects of the improper modeling of viscous amortization on the first-mode and higher-mode dominated responses of base-isolated buildings. Earthq. Eng. Struct. Dyn. **2020**(49), 51–73 (2020)
14. O. Vasile, Experimental evaluation of the damping variation of an elastomeric device harmonically excited, in *12th International Symposium Acoustics and Vibration of Mechanical Structures (AVMS 2013), Applied Mechanics and Materials*, 430, pp. 329–334 (2013)
15. O. Vasile, Experimental evaluation of the hysteretic damping of elastomeric systems at low-cyclic harmonic kinematic displacements, in *12th International Symposium Acoustics and Vibration of Mechanical Structures (AVMS 2013), Applied Mechanics and Materials*, 430, pp. 323–328 (2013)
16. C.F. Dobrescu, Analysis of the dissipative energy performance in the process of compaction for stabilized earth through technological vibrations. ACTA Electrotehnica **60**(1–2), 114–123 (2019)
17. P. Bratu, C. Dobrescu, Evaluation of the dissipated energy in vicinity of the resonance, depending on the nature of dynamic excitation. Romanian J. Acoust. Vibr. **16**(1), 66–71 (2019)
18. O. Tonciu, C.F. Dobrescu, Optimal dynamic regimes of vibration compaction equipment. ACTA Tehnica Napocensis, Ser. Appl. Math. Mech. Eng. **63**(1), 101–106 (2020)
19. C.F. Dobrescu, Analysis of dynamic earth stiffness depending on structural parameters in the process of vibration compaction. Romanian J. Acoust. Vibr. **16**(2), 174–177 (2020)

20. C.F. Dobrescu, The dynamic response of the vibrating compactor roller, depending on the viscoelastic properties of the soil. Appl. Syst. Innov. **3**(2), 25 (2020)
21. C.F. Dobrescu, The assessment of the dynamic compaction effect with vibratory rollers based on the soil rheological behaviour. ACTA Tehnica Napocensis Ser. Appl. Math. Mech. Eng. **63**(2), 191–196 (2020)

Chapter 7
Considerations Regarding the Dynamic Stability in Case of a Damage for Tower Cranes

Maria Aurora Potirniche and Gigel-Florin Capatana

Abstract Stability is an important indicator when sizing cranes. According to the way forces are considered, there is static stability and dynamic stability. Ensuring stability is done by determining the so-called stability coefficient, in two extreme cases of possible departures from stability, respectively with load and without load. Static stability is a necessary condition for the stability of the crane, but not sufficient. Due to the dynamic demands, the dynamic stability must also be taken into consideration. But dynamic solicitations are, unlike static ones, much more difficult to evaluate. In general, the problems raised by dynamic stability are difficult problems, and their treatment is often done under simplifying and covering assumptions. Using modern means of calculation and algorithms developed especially for laborious and in large numbers calculations, the problem of dynamic stability can be approached from positions very close to reality. But, before making this investment of intellectual effort, it must be tested if the problem of dynamic stability is a real, important problem or, being a minor problem, it can be included in the conditions that ensure static stability. The present work presents a study on the dynamic stability in case of a damage for tower cranes. It is analysed in detail the case of dynamic stability when the load suddenly falls into the hook, when the load falls from the hook and, finally, an energetic interpretation of the stability for tower cranes is formulated.

M. A. Potirniche (✉) · G.-F. Capatana
Faculty of Engineering and Agronomy in Brăila, Research Center for Mechanics of Machines and Technological Equipments, "Dunărea de Jos" University of Galați, Galați, Romania
e-mail: aurora.potirniche@ugal.ro

G.-F. Capatana
e-mail: gcapatana@ugal.ro

N. Herisanu and V. Marinca (eds.), *Acoustics and Vibration of Mechanical Structures—AVMS-2023*, Springer Proceedings in Physics 302,
https://doi.org/10.1007/978-3-031-48087-4_7

7.1 Introduction and Theoretical Approaches

The dynamic actions that usually occur during the operation of lifting machines in general have the effect of not only stressing the mechanisms' organs and the elements of their load-bearing structures, but also on their overturning stability. In the current designing, the verification of the cranes' overturning stability is done in accordance with the calculation rules—in the case of our country SR ISO 4304 and 4305. The rules prescribe that the ratio between the moment of stability of the crane, i.e. the moment of its own weight in relation to the overturning edge and the overturning moment determined by the load and possibly by the other external loads (such as inertial forces and wind action) should be superunitary [1, 4].

All these forces are considered to be applied statically, but the load is amplified by a superunitary factor, called the load coefficient. The value of this coefficient depends on the type of crane and the conditions related to the crane in which the checks are made.

In practice, there are numerous cases of the action of forces on construction machines when the forces significantly change their size, meaning, direction or position during application, so that under the action of these forces the elements of the construction machine are in movement. In this case, the accelerations impressed on different points of the construction equipment are not negligible, the process of soliciting the machine is thus also characterized by the appearance of inertial forces. Such loads applied to construction machines are called dynamic loads [2].

As stated previously, dynamic loads have an effect on the overturning stability of construction machines.

Sensitive to overturning are the jib cranes leaned freely on the ground or on the running track, a category in which the tower cranes that can be moved on self-propelled carriages, which are the object of study of this work, also enter [3].

7.2 Case Study

In this subchapter, aspects will be presented regarding the loss of dynamic stability in the event of an accident when the load suddenly falls into the hook, the load falling from the hook, and it will be concluded with an energetic interpretation of the stability. The sudden fall of the load into the crane hook is a dynamic shock solicitation, characterized by a force that is applied to the crane in a very short time interval. At the time of the shock, phenomena appear that can be divided into two large categories:

- local phenomena—with implications in the size of the deformed area in the contact area as well as on the stresses that appear in this area;
- global phenomena—which refer to aspects related to the transmission and propagation of the shock wave in the mass of the struck body [4].

A complete solution to the shock involves treating both local and global aspects. The complete resolution of the shock, however, induces major difficulties in general caused by the ignorance of the aspects related to the shock wave. The speed with which this wave propagates is generally known, this being equal to the speed of sound propagation in the medium of which the struck body is made up. What is not known is the function according to which the shock wave can be described, this function having relatively simple shapes only for bodies with simple geometry.

It should be noted that the same phenomena that occur in the struck body also occur in the body which strikes.

In general, the shock problem is solved based on the law of conservation of energy, according to which the kinetic energy of the body that hits immediately before the shock is equal to the deformation energy accumulated by the body hit immediately after the shock.

It is assumed that due to the shock the stresses produced in the struck body, respectively the crane hook, do not exceed the elastic limit of its material and, in this case, the deformation energy accumulated by it is equal to the potential deformation energy [4, 5].

The deformation energy, which is equal to the kinetic energy, has two components:

- a deformation energy accumulated by the struck body U_1;
- a deformation energy accumulated by the body that strikes U_2.

From the above it follows that:

$$E_c = \frac{1}{2} m \cdot v^2 = U \tag{7.1}$$

where U = deformation energy accumulated by the system of the two bodies,

$$U = U_1 + U_2 \tag{7.2}$$

where U_1, U_2 are the deformation energies accumulated by the two bodies.

The shock calculation can also be done on the basis of the impact multiplier, which is also based on the law of conservation of energy.

The calculation method based on the impact multiplier assumes that the dyna-mic quantities (deformation, force, stress) can be calculated based on the static ones by multiplying the latter with a coefficient called the impact multiplier.

In cases where the mass of the body subjected to the shock is small in relation to the mass of the load that produces the shock, this is neglected, but in this way the kinetic energy consumed for transmitting the speed and setting in movement the mass elements of the struck body is neglected. This is equivalent to considering that during the impact the velocity of the falling body does not change. In reality, at the moment of impact, the falling body loses speed, and the part of the hit body in contact with the falling body picks up speed, the speed of the fallen body changing until it and the hit body at the point of contact have the same speed [4, 6].

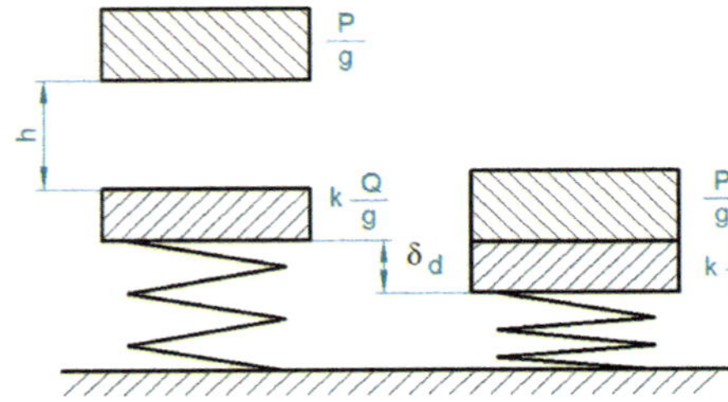

Fig. 7.1 Model for determining the dynamic coefficient taking into account the mass of the body subjected to shock. Reprinted from [4] with permissions. © Engineering and Agronomy Faculty in Brăila, all rights reserved

Therefore, in the model that will be considered in Fig. 7.1 it will try to determine the dynamic coefficient if the mass of the body subjected to the shock is taken into account. In Fig. 7.1, the notations are the following: P = striking force; P/g = mass of the striking body; h = fall height; Q = weight of struck body (real); Q/g = mass of the struck body (real); δ_d = dynamic displacement (from shock).

Let v be the velocity of the force P that falls immediately before contact. Also, let $(k_1 \cdot Q)/g$ be the mass of the equivalent body having the same amount of movement as the real body subjected to the shock and $(k_2 \cdot Q)/g$ the mass of the equivalent body having the same kinetic energy as the real body subjected to the shock; k_1 and k_2 are the reduction coefficients of the mass for the amount of movement, respectively for the kinetic energy [1, 4].

In the process of shock solicitation, three characteristic moments are distinguished:

- the moment that immediately precedes the contact of the two bodies and in which the speed of the force P is $v = \sqrt{2gh}$ and that of the equivalent body is zero;
- the moment after making the contact, when the force P together with the equivalent body have the same speed v_1; the speed v_1 is deduced by applying the momentum conservation theorem, written as follows:

$$\frac{P \cdot v}{g} = \frac{P + k_1 \cdot Q}{g} \cdot v_1 \Rightarrow v_1 = \frac{P}{P + k_1 \cdot Q} \cdot v \tag{7.3}$$

- the moment when the two bodies set in motion reach their maximum displacements, and the speed of the force P and of the struck body is zero.

The energy conservation theorem can be used, expressed in the form:

$$E_2 - E_1 = L_{ext} - L_{int} \tag{7.4}$$

where E_2 = final kinetic energy; E_1 = initial kinetic energy; L_{ext} = the mechanical work of the external forces; L_{int} = the mechanical work of the internal forces.

It will be considered that $E_2 = 0$, because it represents the kinetic energy at the moment of maximum displacement, so when the speed is zero, and E_1 represents the kinetic energy after the collision at the beginning of the movement, when the two bodies have the common speed v_1.

The terms into the energy conservation theorem are evaluated:

$$E_1 = \frac{P + k_2 \cdot Q}{2 \cdot g} \cdot v_1^2 = \frac{P^2 \cdot (P + k_2 \cdot Q)}{2 \cdot g \cdot (P + k_1 \cdot Q)^2} \cdot v^2$$

$$L_{ext} = (P + k_2 \cdot Q) \cdot \delta_d, L_{int} = -\frac{1}{2} \cdot P_d \cdot \delta_d - k_2 \cdot Q \cdot \delta_d. \tag{7.5}$$

The energy conservation formula will be written in the form [4]:

$$\frac{P^2 \cdot (P + k_2 \cdot Q)}{2 \cdot g \cdot (P + k_1 \cdot Q)^2} \cdot v^2 + P \cdot \delta_d = \frac{1}{2} \cdot P_d \cdot \delta_d \tag{7.6}$$

From this formula, introducing the dynamic coefficient Φ, we obtain:

$$\frac{P^2 \cdot (P + k_2 \cdot Q)}{2 \cdot g \cdot (P + k_1 \cdot Q)^2} \cdot 2 \cdot g \cdot h + \Phi \cdot P \cdot \delta_{st} = \frac{1}{2} \cdot \Phi^2 \cdot P \cdot \delta_{st} \tag{7.7}$$

Some notations are introduced:

$$W_0 = P \cdot h; W_{st} = \frac{1}{2} \cdot P \cdot \delta_{st} \tag{7.8}$$

where W_0 = kinetic energy of force P at the beginning of contact; W_{st} = potential energy of deformation of the body obtained by the static application of P. It is obtained:

$$\Phi^2 - 2 \cdot \Phi - \frac{P \cdot (P + k_2 \cdot Q)}{(P + k_1 \cdot Q)^2} \cdot \frac{W_0}{W_{st}} = 0 \tag{7.9}$$

Solving the equation and eliminating the solution without physical meaning, we will obtain:

$$\Phi = 1 + \sqrt{1 + \frac{2 \cdot h}{\delta_{st}} \cdot \frac{1 + k_2 \cdot \frac{Q}{P}}{\left(1 + k_1 \cdot \frac{Q}{P}\right)^2}} = 1 + \sqrt{1 + \frac{v^2}{g \cdot \delta_{st}} \cdot \frac{1 + k_2 \cdot \frac{Q}{P}}{\left(1 + k_1 \cdot \frac{Q}{P}\right)^2}} \tag{7.10}$$

An important observation should be made in the particular case where h = 0, representing a sudden application of force P on the body without height. It is obtained, by customizing the last formula obtained Φ = 2, therefore, through the sudden application of forces, tensions and deformations are produced in structures that are at least twice greater than if these forces were applied statically [4].

Below are some aspects regarding the loss of dynamic stability in the event of an accident when the load falls from the hook. The sudden release of the crane from the load located in its holding device can be:

- an accidental situation encountered with hook cranes, due to a faulty hanging or the breaking of the auxiliary organs for suspending the load;
- a normal operating situation for cranes equipped with a lifting device that releases the load suddenly, such as cranes with electromagnets, cranes with bucket grapple and cranes with self-unloading buckets.

In all these cases, a bottom-up impulse appears on the arm at the point of suspension of the load, which affects the stability of the crane as a whole or only of the arm, if this is attached non-stiff to the crane (with flexible tie rods and cable hoist).

Through the theoretical approach to the problem in the case of the rigid attachment of the arm to the crane, it is found that the dynamic moment produced by the sudden detachment of the load has an important value, so that checking the stability of the cranes when they overturn in this situation is always necessary. The stability condition is formulated as follows: the crane is considered stable if the supports located on the opposite side of the overturning edge do not detach from the support surface.

However, it should be mentioned that this stability condition is extremely restrictive because by detaching from the track or the supporting ground the supports not located on the overturning edge, the crane is still far from the overturning situation, which occurs only if the amplitude of the movement is high enough that the center of gravity of the crane exceeds the vertical that passes through the tipping axis. Thus, by the accepted definition of stability, the detachment of the supports is allowed, provided that it is of short duration and small, so that the return to the track or the ground does not occur with a shock that endangers the resistance of the crane through its support organs.

For the energetic interpretation of stability, both the mechanical overturning work and the deformation potential energy must be evaluated.

In a previous work [5] the concrete case of a tower crane was considered for which the mechanical overturning work was calculated. The considered crane is shown in Fig. 7.2.

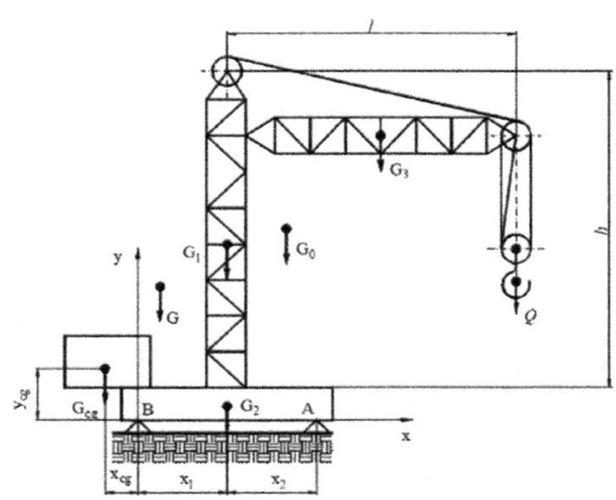

Fig. 7.2 Tower crane model used to exemplify the status of dynamic stability/ instability in transient regimes

Fig. 7.3 The graph of the variation of G_{cg} weight according to the x_{cg} coordinate

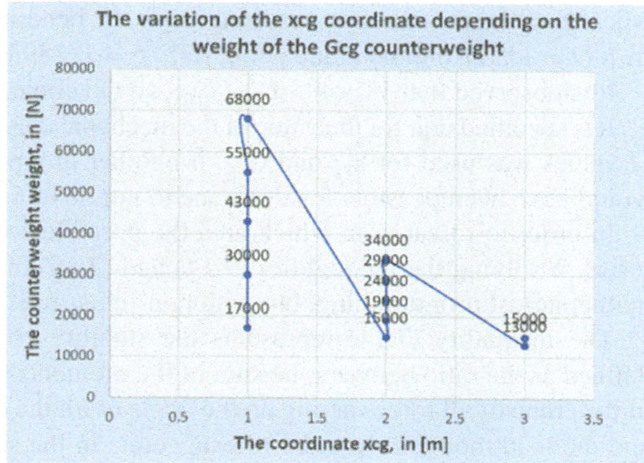

In Fig. 7.2, the notations are the following: G_0 = crane weight, without load and without counterweight, [N]; G_{cg} = counterweight weight, [N]; x_{cg} = coordinate on the abscissa axis of the force G_{cg}.

To determine the mechanical overturning work, three groups of values for x_{cg} and G_{cg} were considered:

$$
\begin{aligned}
x_{cg} &= 1 \Rightarrow 16545 \leq G_{cg} \leq 68501 \\
x_{cg} &= 2 \Rightarrow 14477 \leq G_{cg} \leq 34251 \\
x_{cg} &= 3 \Rightarrow 12868 \leq G_{cg} \leq 22834
\end{aligned}
\tag{7.11}
$$

Based on the ranges of values proposed above for G_{cg} and x_{cg}, a graph of the variation of the G_{cg} weight according to the x_{cg} coordinate was made. The graph is presented in Fig. 7.3.

Following the calculation, a set of values was chosen:

$$
\begin{cases}
x_{cg} = 2\,\text{m} \\
G_{cg} = 30{,}000\,\text{N}
\end{cases}
\tag{7.12}
$$

A value was obtained for the mechanical overturning work $L_{mecrast} = 916.468$ J. This mechanical work is carried out by the elastic metal structure of the crane at the sudden loss of the load from the hook, when the potential deformation energy W_p stored by the metal structure is released [4, 5]. The effect of this mechanical work is the displacement of the crane or some parts of its structure or, in extreme cases, the mechanical work can cause the crane to overturn.

The potential deformation energy accumulated in the structure is:

$$
W_{pdef.} = \frac{Q \cdot f}{2}
\tag{7.13}
$$

where f = the deformation resulting from the bending solicitation of the structure, [m]; Q = load from the crane hook, [N]; $f = 0.1893m$, $W_{pdef.} = 1893\ J$.

It is observed that $W_{pdef.} > L_{mec.rast.}$, so the crane can overturn.

It is specified that for the value of the mechanical overturning work a certain group of values was used for x_{cg} and G_{cg}. If another group of values had been chosen, it would have been possible for the crane to not be in a situation of overturning.

In order to research in which area the possible solutions can be found, one can resort to solving the inequalities (7.14) and (7.15) through the specific methods of mathematical programming, but with non-linear functions.

The inequality (7.14) represents the stability coefficient under load which is defined as the ratio between the sum of the moments relative to the overturning axis of the crane of all forces acting on the crane (with the exception of the load moment) and the load moment. In order to exaggerate in the sense of ensuring stability, it is chosen a higher value of the coefficient, namely 1.2 instead of 1.15, as proposed by the specialized literature [4, 6].

The inequality (7.15) represents the stability in void coefficient which is defined as the ratio between the sum of the moments relative to the overturning axis of the crane of the forces given by the weights of the elements that tend to maintain the crane in a state of equilibrium and the sum of the moments (relative to the same axis) of the forces given by the weights tending to overturn the crane [4, 7].

$$\frac{G_{cg} \cdot \left(x_{cg} + x_1 + x_2\right) + G_0 \cdot \left(x_1 + x_2 - x_{G_0}\right)}{Q \cdot (l - x_2)} \geq 1.2 \tag{7.14}$$

$$\frac{G_0 \cdot x_{G_0}}{G_{cg} \cdot x_{cg}} \geq 1.2 \tag{7.15}$$

In the coordinates of the two independent variables, x_{cg} and G_{cg}, the inequalities above, to which other restrictions can be added (of gauge, weight and pressure on supports, etc.) delimit a solutions space, represented in Fig. 7.4.

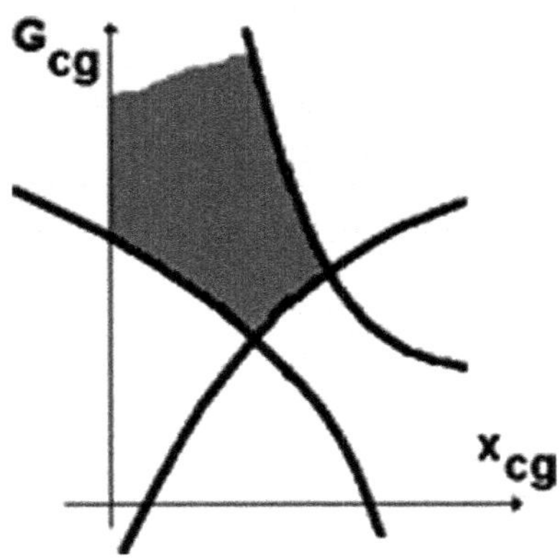

Fig. 7.4 The solutions space represented in the coordinates of the independent variables x_{cg} and G_{cg}. Reprinted from [4] with permissions. © Engineering and Agronomy Faculty in Brăila, all rights reserved

Usually, the choice of the solution is not subject to any criteria. In order to have a certain criterion, additional conditions are imposed in the form of functions called objective functions.

The objective function is a function that tends towards an extreme (minimum or maximum). This objective function, linear or non-linear, has a variable position in the graph and its minimum value, for example, means the position closest to the origin keeping at least one common point with the solutions space. This way of solving the stability problem is called nonlinear mathematical programming because at least one function is nonlinear.

It was shown that in the present work, the overturning situation of the crane resulted from the calculation.

To be sure that the crane does not turn over, the angle α must be as large as possible. The $\alpha \rightarrow$ max condition can be considered an objective function in the model of static stability [4].

With this objective function, after ensuring the static stability, the potential deformation energy stored by the structure is calculated and the comparison is again made with the mechanical overturning work.

If this energy is still greater than the mechanical overturning work, there being no certainty of a conclusion regarding the overturning (case of doubt), a dynamic calculation of the return of the structure to its movement produced by the discharge of potential energy is required.

7.3 Conclusions

Research into the dynamic stability of cranes shows that there are possible cases in which the crane overturns due to the too sudden release of the potential energy stored in its structure. In the paper, this was shown on a concrete example.

Ensuring dynamic stability can be done directly, researching cases of dynamic requests and their effect on stability. Energy methods are the most appropriate, indicating almost directly the danger situations, and for their application, a finite element analysis, carried out in a specialized software, is necessary.

Another way to ensure dynamic stability is to ensure static stability through methods specific to mathematical programming with objective functions. Unfortunately, non-linear functions appear which, in the space with more than two dimensions, are almost impossible to analyze. Although it offers an alternative to dynamic stability, achieving static stability by optimizing some objective functions is impossible to achieve on the set of independent variables greater than two.

Thus, the use of energy methods, as shown above, completed with the calculation of the release speed of the potential energy and with the dynamics of the overturning movement, remains the safe method.

As for the mathematical programming in the two-dimensional space, it can of-fer at most an optimized dimensioning of the crane, according to certain criteria, such as for example $\alpha \rightarrow$ max.

And if it is desired that the static stability also ensures the dynamic stability, those values of the static stability coefficients must be determined which are also covering the dynamic stability.

References

1. G.D. Muscă, *Mașini de ridicat și transportat - Note de curs* (Partea I. Galaţi University Press, Galați, 2020)
2. G.D. Muscă (Anghelache), Static and dynamic stability for floating cranes. IOP Conf. Ser.: Mater. Sci. Eng. **400**, 082015 (2018)
3. D.G. Muscă (Anghelache), S.M. Năstac, Parametric modelling and finite elements study for the boom of a truck crane. IOP Conf. Ser.: Mater. Sci. Eng. **1182**, 012050 (2021)
4. A. Potîrniche, Stabilitatea dinamică a macaralelor turn. Dissertation work supported at "Dunărea de Jos" University of Galati, Engineering and Agronomy Faculty in Brăila for the Master's degree in specialization "Dynamic Analysis of Machines and Technological Equipments" (2004)
5. A. Potîrniche, G. Căpățână, Aspects regarding static stability and dynamic stability of the tower cranes. Int. J. Mod. Manuf. Technol. **15**(2), (to appear) (2023)
6. J. Weiguang, D. Lieyun, Z. Cheng, Digital twin: stability analysis for tower crane hoisting safety with a scale model. Autom. Constr. **138**, 104257 (2022)
7. P. Swuste, A 'normal accident' with a tower crane? An accident analysis conducted by the Dutch Safety Board. Saf. Sci. **57**, 276–282 (2013)

Chapter 8
Application of Spectral Element Method and Machine Learning Technique in the Composite Concrete Beam Damage Prediction

Maciej Dutkiewicz, Marcela R. Machado, and Jefferson da Silva Coelho

Abstract The paper presents the simulation results of the monitoring and prediction of damage of a cement composite beam using the analysis of changes in resonant frequencies with Spectral Element Method and the Machine Learning technique. For a beam with random generated crack, the effect of cracking on the change of resonance frequencies was shown by analyzing resonance curves comparing to the health beam. For the four basic resonance frequencies, frequencies were randomly generated, constituting the frequency database for the Machine Learning—Support Vector Machine classification. Damage Index was calculated. The presented methodology is a procedure involving the simulation and prediction of beam damage.

8.1 Introduction

Contemporary building structures are exposed to loads that violate their integrity. Structures made of concrete, e.g. beams, slabs, columns, which are cracked in the initial phase are the example of elements that requires the structural health monitoring (SHM). There is a lot of research on damage detection through analysis of frequency, performed in the aerospace, machinery, bridge, and other engineering structures. Adewuyi et al. [1] presents the vibration-based damage identification methods using displacement and distributed strain measurements. The authors state that long-gage distributed strain measurements are efficient for the reliable civil structural health monitoring. In [2] the modal shape identification of large structure by operational modal analysis technique is presented. The article [3] provides a summary

---

M. Dutkiewicz (✉) · M. R. Machado
Faculty of Civil, Architecture and Environmental Engineering and Architecture, Bydgoszcz University of Science and Technology, Al. Prof. S. Kaliskiego 7, 85-796 Bydgoszcz, Poland
e-mail: macdut@pbs.edu.pl

M. R. Machado · J. da Silva Coelho
Department of Mechanical Engineering, University of Brasilia, Brasil 70910-900, Brazil

N. Herisanu and V. Marinca (eds.), *Acoustics and Vibration of Mechanical Structures—AVMS-2023*, Springer Proceedings in Physics 302,
https://doi.org/10.1007/978-3-031-48087-4_8

of modal testing and structural model validation of wind turbine blades. The theory and practice of the modal testing, theoretical basis, measurement techniques, models are studied in the [4]. Based on the vibration features, the damage identification methods are classified into four major categories [5]. Pandey and Biswas [6] present the flexibility difference method for locating damage in structures. They emphasize that damage diagnosis can be divided into three sub-problems: damage detection, i.e. determining the presence of damage, damage location, i.e. determining the location of damage, damage quantification, i.e. determining the amount of damage. Interesting methods of damage detection are presented in [7–9]. The damage of the structure is progressing with the decrease in the fundamental frequency, and near the ultimate load the frequency may decrease by about 70% [10]. There has been a lot of works on the analysis of the frequency response functions (FRFs) as a source of information on the state of the structure based on SEM. The methodology is described in [11–17]. Machine Learning (ML) is the tool which support the SHM in the scope of qualification of elements as damage or non-damage. In particular ML is useful in the prediction of the structural state of analyzed systems. In the last period the ML is observed as the useful tools which is used in many industrial sectors as the algorithms of ML can be applied in wide scope of engineering [18].

8.2 Spectral Element Model and Support Vector Machine Technique

The dynamic analysis of the one-sided fixed composite-cement beam was carried out by SEM. The influence of the height of the crack on the resonance frequency was investigated. For each frequency the damage index (DI) was calculated with the reference to mean value of the natural frequency. The Support Vector Machine (SVM) algorithm was used to train the system to qualify the beam as damage or health. 80% of data was used for training and 20% for testing.

8.2.1 Spectral Element Model

Considering a simplified model, the governing differential equation for the undamped Euler–Bernoulli for free vibration can be written as [17]:

$$EI\frac{\partial^4 v(x,t)}{\partial x^4} + \rho A\frac{\partial^2 v(x,t)}{\partial t^2} = 0 \tag{8.1}$$

where ρA is mass per unit length, EI the uniform bending rigidity and $v(x,t)$ is the beam displacement as a function of the position x and time t. By considering a constant coefficient, a displacement solution can be assumed in the form:

$$v(x,t) = v_0 e^{-i(kx-\omega t)} \tag{8.2}$$

where v_0 is a amplitude, ω is the frequency and k is the wave number.

The dynamic stiffness matrix for the spectral beam element under axial tension can be determined as:

$$\boldsymbol{S}(\omega) = \boldsymbol{K}(\omega) - \omega^2 \boldsymbol{M}(\omega) \tag{8.3}$$

By solving the integral, the dynamic stiffness matrix is:

$$\boldsymbol{S}(\omega) = \frac{\text{EI}}{\Delta}\begin{bmatrix} s_{11} & s_{12} & s_{13} & s_{14} \\ & s_{22} & s_{23} & s_{24} \\ & & s_{33} & s_{34} \\ sym & & & s_{44} \end{bmatrix} \tag{8.4}$$

where $\Delta = \cos(\text{k L})\cosh(\text{k L}) - 1$ and the components of element matrix (Eq. 8.4) are given as:

$$\begin{aligned}
s_{11} &= -k^3(\cos(kL)\sinh(kL) + \sin(kL)\cosh(kL)), \\
s_{12} &= -k^2\sin(kL)\sinh(kL), \\
s_{13} &= k^3(\sin(kL) + \sinh(kL)), \\
s_{14} &= k^2(\cos(kL) - \cosh(kL)), \\
s_{22} &= k(\cos(kL)\sinh(kL) - \sin(kL)\cosh(kL)), \\
s_{23} &= k^2(\cosh(kL) - \cos(kL)), \\
s_{24} &= k(\sin(kL) - \sinh(kL), \\
s_{33} &= -k^3(\cos(kL)\sinh(kL) + \sin(kL)\cosh(kL)), \\
s_{34} &= k^2\sin(kL)\sinh(kL), \\
s_{44} &= k(\cos(kL)\sinh(kL) - \sin(kL)\cosh(kL)).
\end{aligned} \tag{8.5}$$

The crack flexibility coefficient θ is calculated using Castigliano's theorem, where the flexibility in the crack position for a one-dimensional beam spectral element can be obtained by $c = \partial^2 U/\partial P^2$[19], where U denotes the elastic deformation energy due to cracking, P is the nodal force on the element. Considering an open crack in the beam element, the elastic deformation energy can be expressed as [19]

$$U = \frac{1-\nu^2}{E}\int_{S_c} K_I^2 dS_c, \tag{8.6}$$

which ν is the Poisson's ratio, S_c is the cracked area, and K_I is a stress intensity factor corresponding to mode I of the crack, which can be represented by:

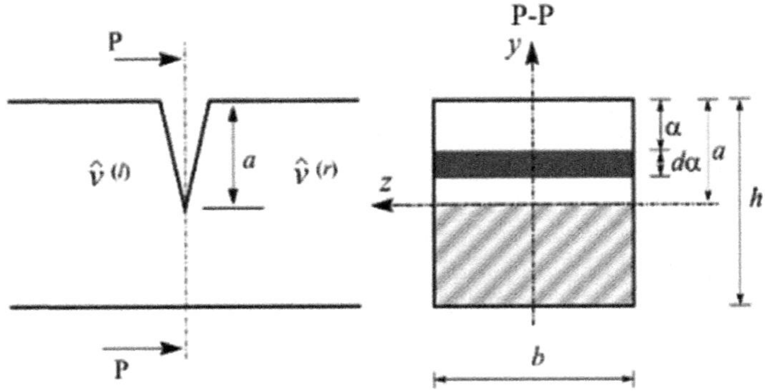

Fig. 8.1 Cross-section of the cracked beam in the crack position

$$K_I = \frac{6M}{bh^2}\sqrt{\pi\alpha f\left(\frac{\alpha}{h}\right)}, \tag{8.7}$$

where b is the base and h the height of the beam, α is the variation in the crack depth, as shown Fig. 8.1, M is the bending moment in crack position, and f is a correction function of the mode I stress intensification factor, which can be written as:

$$f\left(\frac{\alpha}{h}\right) = \sqrt{\frac{2h}{\pi\alpha}\tan\left(\frac{\pi\alpha}{2h}\right)}\frac{0.923 + 0.199\left(1 - \sin\left(\frac{\pi\alpha}{2h}\right)\right)^4}{\cos\left(\frac{\pi\alpha}{2h}\right)}. \tag{8.8}$$

The coefficient c that calculates the crack flexibility is expressed as follows

$$c = \frac{72\pi}{bh^2}\int_0^{\alpha} \alpha f^2(\alpha)d\alpha \tag{8.9}$$

where $\alpha = a/h$. Finally, the dimensional local flexibility is given by $\theta = EIc/L$.

8.2.2 Machine Learning and Support Vector Machine Technique

Machine Learning (ML) is part of Artificial Intelligence (AI). ML is the learning through experience and based on the measured data. Part of ML is Deep Learning (DL), i.e. algorithms that allow to filter the data. In ML we can distinguish: supervised, unsupervised and reinforcement learning. Supervised Learning includes classification and regression. The classification includes: Support Vector Machines (SVM), Naïve Bayes Classifier (NBC), Decision Trees (DT), Random Forest (RF) and K – Nearest Neighbors (K-NN). In the SVM algorithm, the line that separates the data must be in the optimal place. The distances between the line and the nearest set points

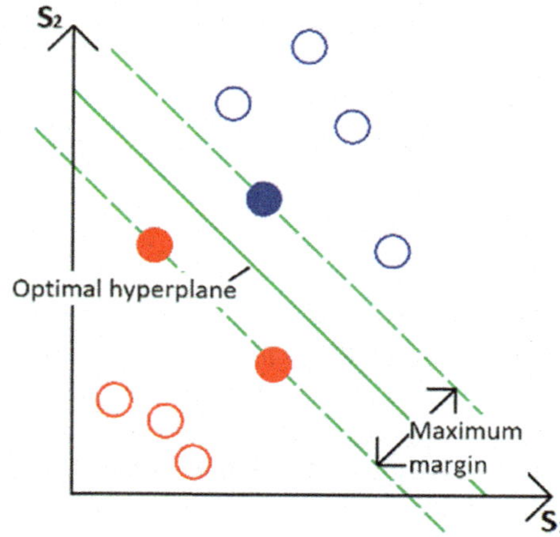

Fig. 8.2 SVM for two dimensional training set

should be as small as possible. These distances are the Support Vectors. SVM idea is presented in Fig. 8.2 where s_1 and s_2 mean the analyzed plane.

The advantages of SVM are finding the optimal distances (margins) between groups of points, it is a computationally efficient method—the complexity increases only linearly with the number of dimensions, it solves both linear and non-linear problems.

8.3 Numerical Analysis and Results

The dynamic analysis of the one-sided fixed composite-cement beam was carried out by SEM. In the simulations, the reinforced concrete beam with a basic parameters of length $l = 2$ m, area of $A = 0.02$ m2 is considered, width of 0.10 m, height of 0.20 m (Fig. 8.3.). The material properties are modulus of elasticity $E = 30.0$ GPa, the material density of $\rho = 2500$ kg/m^3.

The influence of the height of the crack on the resonance frequency was investigated. The height of the crack was generated as a random number from the normal distribution with mean parameter equals to 0.3 h, where h is the height of the beam, and standard deviation parameter equals to 0.03 h. 10 resonance curves for the cracked

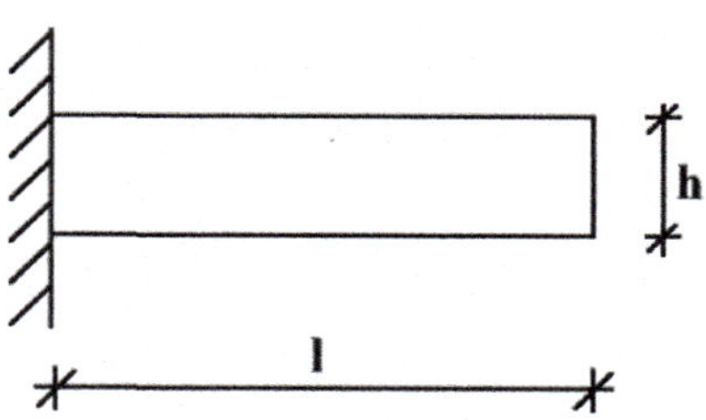

Fig. 8.3 Scheme of the analysed beam

Table 8.1 Crack height in the beam

Cracked beam no	1	2	3	4	5	6	7	8	9	10
Crack height [m]	0.0632	0.0710	0.0464	0.0652	0.0619	0.0522	0.0574	0.0621	0.0815	0.0766

beam and a resonance curve for the health beam were investigated. Average values of resonance frequencies were calculated. For the average resonant frequencies, a distribution of 100 frequencies were determined for each fundamental frequency as a random number from the normal distribution with randomness of 5% and 10%. For each frequency the damage index (DI) was calculated with the reference to mean value of the natural frequency. The beam was qualified based on DI as damage beam if DI $\leq$ 0.99, otherwise the beam was qualified as health. Then, the Support Vector Machine (SVM) algorithm was used to train the system to qualify the beam as damage or health. 80% of data was used for training and 20% for testing.

The generated crack heights are shown in Table 8.1.

Figure 8.4 presents the inertance response obtained for cement composite beam, and the height of the crack in simulation analysis are shown in Table 8.1. FRF for cracked element are compared with health beam.

Samples generated with 5%, 10% randomness of frequencies [27, 164, 489, 925] Hz for ML training are presented in Fig. 8.5.

For damage and health state the accuracy of training algorithm, 5% randomness (Fig. 8.6), for:

- the 27 Hz natural frequency is 95.0%,
- the [164, 489, 925] Hz natural frequencies are 100.0%.

For damage and health state the accuracy of training algorithm, 10% randomness for: the [27, 164, 489, 925] Hz natural frequencies are 100.0%.

8.4 Conclusions

The subject of the analysis was the use of SEM and ML in damage prediction. The dynamic response of a cantilevered concrete composite beam was tested. The FRF analysis showed how cracks affect the shift of resonance frequency. The obtained results confirm the effectiveness of SEM and FRF in damage detection.

The ML technique was used to classify the beam element. Structural elements were described by a number of parameters. Based on the data obtained from the simulation, the elements were classified depending on the resonance frequency depending on the size of the crack. The learning process was carried out using the SVM algorithm for randomly generated data with a distribution of 5 and 10%. The obtained classification results confirm the high efficiency of the learning process.

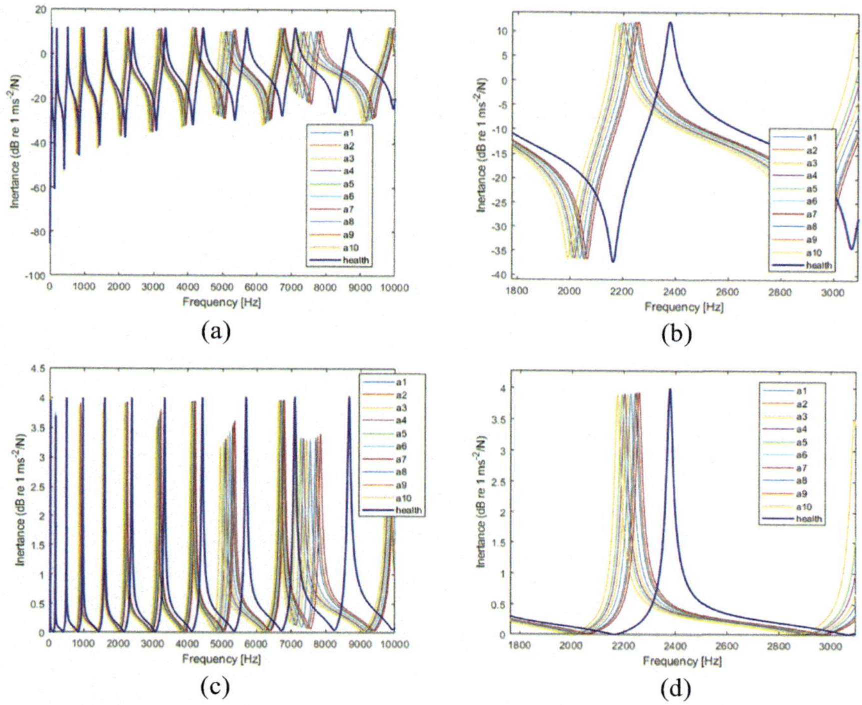

Fig. 8.4 Inertance response for composite concrete beam, **a** full range of analysed frequency, log scale, **b** zoom for frequency range <1800, 3000> Hz, log scale, **c** full range of analysed frequency, nonlog scale, **d** zoom for frequency range <1800, 3000> Hz, nonlog scale

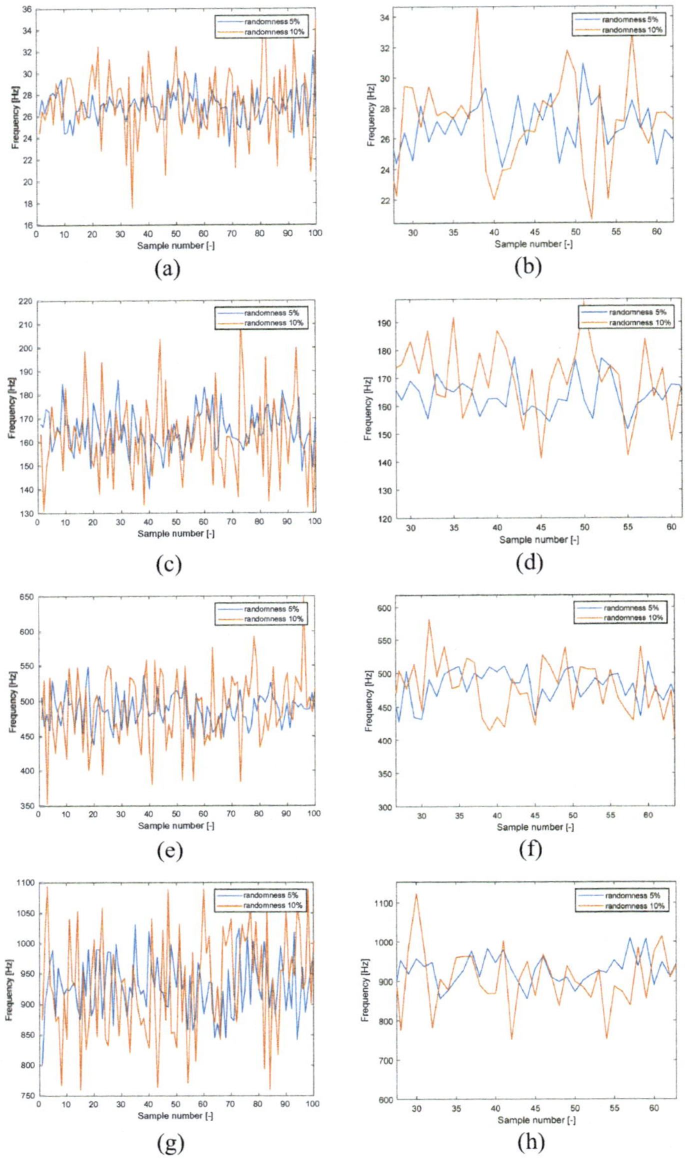

Fig. 8.5 Samples generated with 5%, 10% randomness of frequencies [27, 164, 489, 925] Hz for ML training; **a**, **c**, **e**, **g** full range samples; **b**, **d**, **f**, **h** zoom of analysed samples

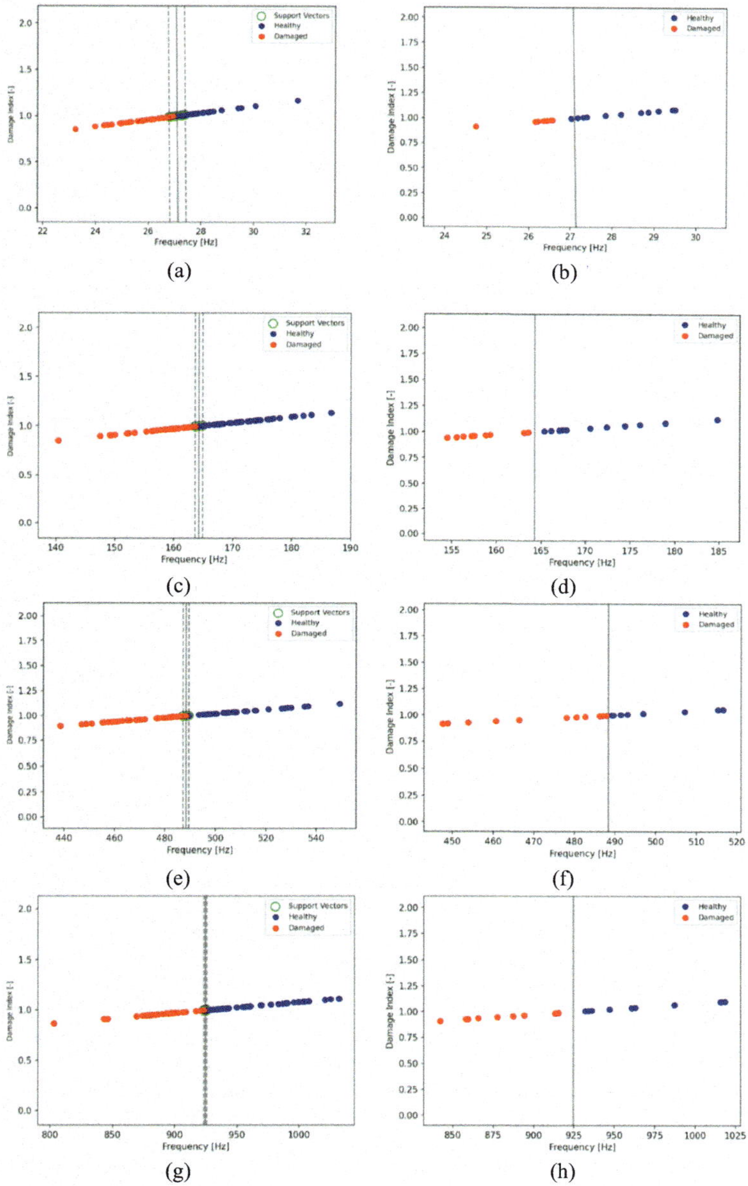

Fig. 8.6 ML for 5% randomness samples; **a**, **c**, **e**, **g** training; **b**, **d**, **f**, **h** testing

References

1. A. Adewuyi Philips, Z.H. Serker, N.H.M. Kammrujaman, Assessment of vibration-based damage identification methods using displacement and distributed strain measurements. Struct. Health Monit. **8**, 443–461 (2009)
2. R.A. De Vivo, C. Brutti, J.L. Leofanti, Modal shape identification of large structure exposed to wind excitation by operational modal analysis technique. Mech. Syst. Signal Process. **19**, 195–206 (2013)
3. T.D. Griffith, T.G. Carne, Experimental modal analysis of research-sized wind turbine blades. Sound Vibr. **44**, 8–12 (2010)
4. D.J. Ewins, *Modal Testing Theory and Practice* (John Wiley and Sons, New York, 1984)
5. F. Wei, P.Z. Qiao, Vibration-based damage identification methods: a review and comparative study. Struct. Health Monit. **10**, 83–111 (2011)
6. A.K. Pandey, M. Biswas, Experimental verification of flexibility difference method for locating damage in structures. J. Sound Vib. **184**, 311–328 (1995)
7. A.K. Pandey, M. Biswas, M.M. Samman, Damage detection from changes in curvature mode shapes. J. Sound Vib. **145**(2), 321–332 (1991)
8. E. Parloo, P. Verboven, P. Guillaume, M. Van Overmeire, Autonomous structural health monitoring-part ii: vibration based in-operation damage assessment. Mech. Syst. Signal Process. **16**(4), 659–675 (2002)
9. L.M. Khoo, P.R. Mantena, P. Jadhav, Structural damage assessment using vibration modal analysis. Struct. Health Monit. Int. J. **3**(2), 177–194 (2004)
10. M.S. Mirza, O. Ferdjani, A. Hadj-Arab, K. Joucdar, A. Khaled, A.G. Razaqpur, An experimental study of static and dynamic responses of prestressed concrete box girder bridges. Can. J. Civ. Eng. **17**(3), 481–493 (1990)
11. M. Dutkiewicz, M. Machado, Spectral element method in the analysis of vibrations of overhead transmission line in damping environment. Struct. Eng. Mech. Int. J. **71**(3), 291–303 (2020)
12. M. Machado, M. Dutkiewicz, C.F.T. Matt, D.A. Castello, Spectral model and experimental validation of hysteretic and aerodynamic damping in dynamic analysis of overhead transmission conductor. Mech. Syst. Signal Process. **136**, 106483 (2020)
13. M. Dutkiewicz, M.R. Machado, Spectral approach in vibrations of overhead transmission lines. IOP Conf. Ser. Mater. Sci. Eng. **471**(5) (2019). https://doi.org/10.1088/1757-899X/471/5/052029
14. M. Dutkiewicz, M. Machado, Measurements in situ and spectral analysis of wind flow effects on overhead transmission lines. Sound Vibr. **53**(4), 161–175 (2020)
15. M. Dutkiewicz, M. Machado, Dynamic response of overhead transmission line in turbulent wind flow with application of the spectral element method. IOP Conf. Ser. Mater. Sci. Eng. **471**(5), 052031 (2019). https://doi.org/10.1088/1757-899X/471/5/052031
16. M. Dutkiewicz, M. Machado, Vibration characterisation of a damaged cable by the spectral element method, acoustics and vibration of mechanical structures—AVMS-2021, in *Proceedings of the 16th AVMS*, Timişoara, Romania, 28–29 May 2021
17. M. Dutkiewicz, M. Machado, Spectral approach of reinforced concrete column dynamics, acoustics and vibration of mechanical structures—AVMS-2021, in *Proceedings of the 16th AVMS*, Timişoara, Romania, 28–29 May 2021
18. J. Coelho da Silva, A. Rodrigues de Sousa, M. Machado, M. Dutkiewicz, Study of machine learning techniques for damage identification in a beam, recent trends in wave mechanics and vibrations. Proc. WMVC (2022)
19. I. Matsko, I. Javorskyj, R. Yuzefovych, Z. Zakrzewski, Forced oscillations of cracked beam under the stochastic cyclic loading. Mech. Syst. Signal Process. **104**, 242–263 (2018)

Chapter 9
Analysis of Structures with Three Identical Parts

Calin Itu, Polidor Bratu, Sorin Vlase, Nicusor Dragan, Maria Aurora Potirniche, Adrian Mihai Goanta, Eliza Chircan, and Paul Nicolae Borza

Abstract The work aims to determine how the existence of repetitive substructures existing in the component of a structure can influence the vibration behavior of the system. The particular case of a structure that has four identical bars as a component is analyzed in the paper. The use of systems with repetitive elements allows the ease of design and practical implementation and the reduction of the cost price of the project. Some vibration properties will be revealed that allow the calculation for such systems to be easier. The modeling of the system was done using The finite element method (FEM) as used for the vibration analysis.

C. Itu · S. Vlase · E. Chircan (✉) · P. N. Borza
Department of Mechanical Engineering, Transilvania University of Brașov, Brașov, Romania
e-mail: chircan.eliza@unitbv.ro

C. Itu
e-mail: calinitu@unitbv.ro

S. Vlase
e-mail: svlase@unitbv.ro

P. N. Borza
e-mail: borzapn@unitbv.ro

P. Bratu · S. Vlase
Romanian Academy of Technical Sciences, Bucharest, Calea Victoriei, Romania
e-mail: icecon@icecon.ro

P. Bratu
Institutul de Mecanica Solidelor, Bucuresti, Romania

N. Dragan · M. A. Potirniche · A. M. Goanta
University Dunarea de Jos, Galati, Romania
e-mail: nicusor.dragan@ugal.ro

M. A. Potirniche
e-mail: Potarniche.Aurora@ugal.ro

N. Herisanu and V. Marinca (eds.), *Acoustics and Vibration of Mechanical Structures—AVMS-2023*, Springer Proceedings in Physics 302,
https://doi.org/10.1007/978-3-031-48087-4_9

9.1 Introduction

In all engineering works and achievements, symmetries are a common thing, often found and used. For example, they can be found in structures with large roofs, in stadiums and sports complexes, industrial warehouses and in general in a very diverse range of industrial or public buildings.

The use of symmetries is determined by the advantages they offer in reducing the design time, the reduction of the calculations necessary to be done in the modeling process, the number of different elements used in the construction of these buildings decreases, so in the end there are important benefits in terms of the time of achievement and cost price.

Advantages also appear in the modeling stage, both in static and dynamic calculation. It is found that if less information is used due to symmetries and the number of necessary calculations, therefore the working time, decreases [1, 2]. The advantages that symmetries can bring in the vibration calculation of a structure have been observed and used by designers [3]. Different simple particular cases of using symmetries in the calculation of structures have been addressed by different authors in [4–9].

Some research has been done on the way in which these symmetries can simplify the writing of the equations. Research performed in this field focused more on how can be used these symmetries to write the equations of motion [10–12]. Some users of the common FEM software know some advantages of existence of symmetries and use them in their work. The existence of symmetries is not a prerogative of civil engineering or architecture, but they can be found in all industrial branches such as automotive engineering, aerospace, etc. where the parts of a structure, product, machine, and devices are repetitive or identical.

In aerospace industry the existence of repetitive elements is used in the aerospace industry where the researchers or with substructure (parts of the main structure). Once a substructure is built, creating the model for the analyzed problem is greatly simplified [13]. In the analysis of the vibrations, the evolution equations are differential equations of the second order. Using the symmetries offers us the possibility to write more easily the equations of motion that leads to a simpler method to solve the problem [14]. In civil engineering, the symmetry of structures is analyzed in [15] where the symmetries of the configurations of stratified spatial networks are analyzed. The group theory is used by researchers in field of physics and chemistry, but it is also useful in engineering mechanics. The symmetry properties of a mechanical system can be described by symmetry groups. An analysis carried out to identify the symmetry groups in industrial applications is carried out in [16, 17]. Research has been developed in this direction which can be found in [18–20].

9.2 Modeling the Structure

From the multitude of aspects of a structure with repetitive elements, the paper focused to determine the response to vibrations. In civil engineering, symmetrical structures are frequently used due to a multitude of factors such as: ease of design, ease of manufacture, lower cost of design, and lower cost of manufacture of these structures, made up of repetitive elements. Given such a wide area of use, the question arises whether the structural symmetries that appear in these systems could not be used to ease the calculations related to their design and dimensioning. The main field of applicability is in the field of civil engineering, but in all engineering domain can be found such problems.

The paper aims to study the consequences that the existence of repetitive elements, with the same properties, within some structures can have. In engineering, in civil engineering but also in numerous other fields, such as for example the car or machine building, automotive or aerospace industry, there are elements with identical, repetitive parts or having symmetries of different types. Buildings, works of art, halls, etc. have symmetries or identical parts. Symmetries of various kinds that provide specific properties to structures have been observed for a long time [21]. The studied properties are found both in discrete systems and in continuous systems. Several important properties of the vibrations of these systems with symmetric elements have been highlighted. The use of highlighted properties allows for easier calculation in the design phase of the structure. The advantages of the symmetries existence were little analyzed, however some works covering the subject exist [22–26].

The study will be done for a structure made of bars with known geometry and properties. There is a central bar, fixed by three identical bars, placed at an angle of 120°. All three bars are made of the same material and have identical dimensions. The central bar is fixed to the lower part and can only have a torsional movement. It is considered that the three upper bars can deform only in an horizontal plane. The stiffness in the vertical plane is high enough to be able to neglect the deformation in this plane. The truss DO is a vertical bar. It has length *L1*, section S1, moment of inertia Iz1 and Young's modulus E1. The other three beams OA, OB and OC are rigidly fixed at point O to the central beam. The other ends are clamped (in A, B, C). The bars OA, OB, OC are therefore arranged identically symmetrically around the point O. They have the Young's modulus E2, the length *L2*, the moment of inertia Iz2 and the area of the section *S2*. This structure presents a certain simplicity, to be able to present these properties more suggestively.

For the trusses system in Fig. 9.1 the equations of undamped free vibrations are:

$$\begin{bmatrix} M_{11} & 0 & 0 & M_{12} \\ 0 & M_{11} & 0 & M_{12} \\ 0 & 0 & M_{11} & M_{12} \\ M_{12}^T & M_{12}^T & M_{12}^T & M_{22} \end{bmatrix} \begin{Bmatrix} \ddot{X}_{OA} \\ \ddot{X}_{OB} \\ \ddot{X}_{OC} \\ \ddot{X}_{OD} \end{Bmatrix} + \begin{bmatrix} K_{11} & 0 & 0 & K_{12} \\ 0 & K_{11} & 0 & K_{12} \\ 0 & 0 & K_{11} & K_{12} \\ K_{12}^T & K_{12}^T & K_{12}^T & K_{22} \end{bmatrix} \begin{Bmatrix} X_{OA} \\ X_{OB} \\ X_{OC} \\ X_{OD} \end{Bmatrix} = \{0\} \quad (9.1)$$

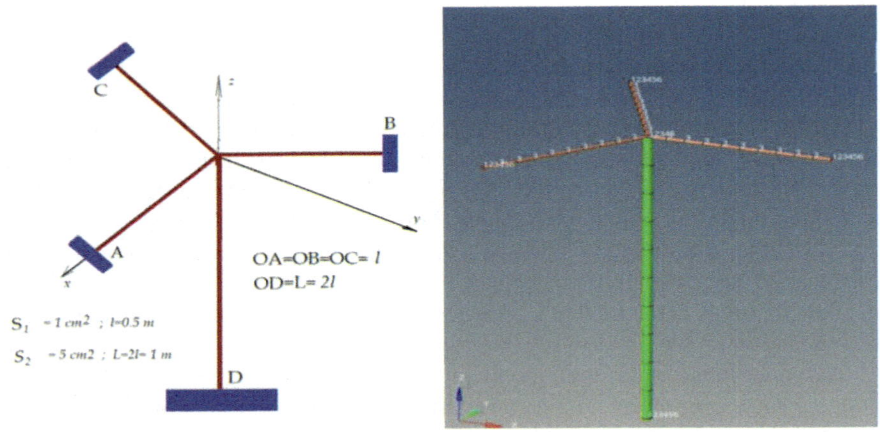

Fig. 9.1 The structure with three identical parts

To determine the eigenvalues leads us to the characteristic equation:

$$\begin{vmatrix} K_{11}-\omega^2 M_{11} & 0 & 0 & K_{12}-\omega^2 M_{12} \\ 0 & K_{11}-\omega^2 M_{11} & 0 & K_{12}-\omega^2 M_{12} \\ 0 & 0 & K_{11}-\omega^2 M_{11} & K_{12}-\omega^2 M_{12} \\ K_{12}^T-\omega^2 M_{12}^T & K_{12}^T-\omega^2 M_{12}^T & K_{12}^T-\omega^2 M_{12}^T & K_{22}-\omega^2 M_{22} \end{vmatrix} = 0 \tag{9.2}$$

and determining the eigenmodes leads to solving the linear homogeneous system:

$$\begin{bmatrix} K_{11}-\omega^2 M_{11} & 0 & 0 & K_{12}-\omega^2 M_{12} \\ 0 & K_{11}-\omega^2 M_{11} & 0 & K_{12}-\omega^2 M_{12} \\ 0 & 0 & K_{11}-\omega^2 M_{11} & K_{12}-\omega^2 M_{12} \\ K_{12}^T-\omega^2 M_{12}^T & K_{12}^T-\omega^2 M_{12}^T & K_{12}^T-\omega^2 M_{12}^T & K_{22}-\omega^2 M_{22} \end{bmatrix} \begin{Bmatrix} \Phi_{OA} \\ \Phi_{OB} \\ \Phi_{OC} \\ \Phi_{OD} \end{Bmatrix} = \{0\} \tag{9.3}$$

Consider one of the identical truss OA, OB, OC, clamped at both ends. The undamped free vibration equations for this truss are:

$$[M_{11}]\{\ddot{X}_{OA}^*\} + [K_{11}]\{X_{OA}^*\} = \{0\} \tag{9.4}$$

with the characteristic equation:

$$|M_{11} - \omega^2 K_{11}| = 0 \tag{9.5}$$

and the eigenmodes offered by the linear system:

$$[M_{11} - \omega^2 K_{11}]\{\Phi^*_{OA}\} = 0 \tag{9.6}$$

9.3 Properties of the Structure

Theorem 1 *The eigenvalues for the trusses OA, OB, OC, clamped at the both ends, are also eigenvalues for the whole system. The theorem has been proved in a more general case, in a previous paper* [5]. *Using this result it follow that the eigenvalues considering only one truss, clamped at both ends, are too, eigenvalues of the entire system.*

Theorem 2 *For the eigenvalues obtained in Theorem 1 the eigenvectors are of the form*

$$\{B\} = \begin{Bmatrix} B^{\mathrm{OA}} \\ B^{\mathrm{OB}} \\ B^{\mathrm{OC}} \\ 0 \end{Bmatrix} \tag{9.7}$$

with

$$B^{\mathrm{OA}} + B^{\mathrm{OB}} + B^{\mathrm{OC}} = 0. \tag{9.8}$$

Proof For the eigenvalues obtained from (9.3) the following system must be solved:

$$[A]\{B\} = 0 \tag{9.9}$$

where

$$[A] = \begin{bmatrix} A_{11} & 0 & 0 & A_{12} \\ 0 & A_{11} & 0 & A_{12} \\ 0 & 0 & A_{11} & A_{12} \\ A_{12}^T & A_{12}^T & A_{12}^T & A_{22} \end{bmatrix} \tag{9.10}$$

and $[A_{11}]$ satisfies the condition (9.5).

Condition (9.9) implies that a vector B^{OA}, B^{OB}, B^{OC} can be found such that:

$$A_{11}B^{\mathrm{OA}} = 0;\ A_{11}B^{\mathrm{OB}} = 0,\ A_{11}B^{\mathrm{OC}} = 0 \tag{9.11}$$

Equation (9.9) becomes:

$$A_{12}B^{\mathrm{OD}} = 0; \tag{9.12}$$

$$A_{12}B^{\mathrm{OD}} = 0; \tag{9.13}$$

$$A_{12}B^{\mathrm{OD}} = 0. \tag{9.14}$$

$$A_{12}^{T}\left(B^{\mathrm{OA}} + B^{\mathrm{OB}} + B^{\mathrm{OC}}\right) + A_{22}B^{\mathrm{OD}} = 0 \tag{9.15}$$

or

$$A_{12}B^{\mathrm{OD}} = 0; \tag{9.16}$$

$$A_{12}^{T}\left(B^{\mathrm{OA}} + B^{\mathrm{OB}} + B^{\mathrm{OC}}\right) + A_{22}B^{\mathrm{OD}} = 0. \tag{9.17}$$

From (9.16), because in general det (A12) $\neq 0$, it follows immediately:

$$B^{\mathrm{OD}} = 0 \tag{9.18}$$

and introducing in (9.17) it results $B^{\mathrm{OA}} + B^{\mathrm{OB}} + B^{\mathrm{OC}} = 0$.

Theorem 3 *For the other eigenvalues, the eigenvectors are of the form*:

$$\Phi = \begin{Bmatrix} \Phi_1 \\ \Phi_1 \\ \Phi_1 \\ \Phi_4 \end{Bmatrix} \tag{9.19}$$

where

$$\Phi_5 = -3A_{22}^{-1}A_{21}^{T}\Phi_1. \tag{9.20}$$

Proof Equation (9.3) becomes, after performing some simple calculations:

$$A_{11}B^{\mathrm{OA}} + A_{12}B^{\mathrm{OD}} = 0; \tag{9.21}$$

$$A_{11}B^{\mathrm{OB}} + A_{12}B^{\mathrm{OD}} = 0; \tag{9.22}$$

$$A_{11}B^{\mathrm{OC}} + A_{12}B^{\mathrm{OD}} = 0. \tag{9.23}$$

$$A_{21}^{T}\left(B^{\mathrm{OA}} + B^{\mathrm{OB}} + B^{\mathrm{OC}}\right) + A_{22}B^{\mathrm{OD}} = 0, \tag{9.24}$$

Subtracting (9.22) from (9.21), (9.23) from (9.21) we obtain:

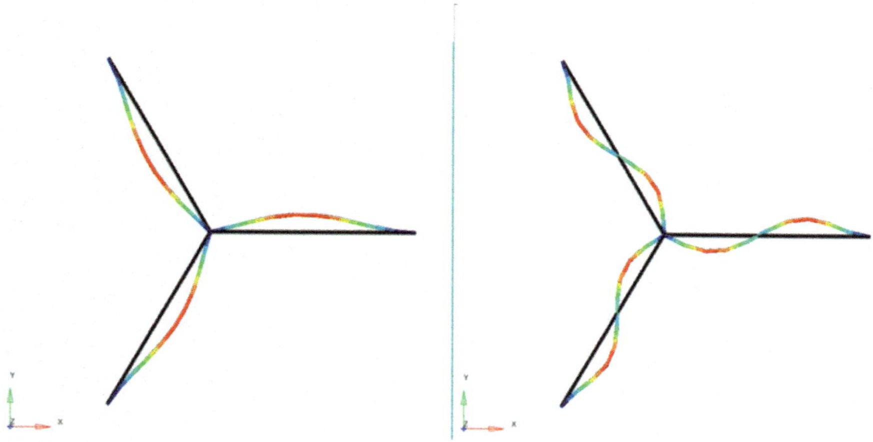

Fig. 9.2 Eigenmodes 1 and 2 of vibration

$$A_{11}\left(B^{\mathrm{OA}} - B^{\mathrm{OB}}\right) = 0; \tag{9.25}$$

$$A_{11}\left(B^{\mathrm{OA}} - B^{\mathrm{OC}}\right) = 0; \tag{9.26}$$

from where:

$$B^{\mathrm{OA}} = B^{\mathrm{OB}} = B^{\mathrm{OC}} = \Phi_1.$$

Equation (9.24) becomes:

$$3A_{21}^T\Phi_1 + A_{22}B^{\mathrm{OD}} = 0,$$

or B^{OD} can be expressed as:

$$B^{\mathrm{OD}} = -3A_{22}^{-1}A_{21}^T\Phi_1.$$

Some eigenmodes of vibration for the structure are shown in Figs. 9.2, 9.3 and 9.4, where the highest value for the intensity is highlighted in red.

9.4 Conclusions

The use of symmetries is a current practice in civil, mechanical, aerospace, automotive, etc. engineering. The existence of repetitive elements makes the information necessary for the mechanical description of the system to be reduced and also allows

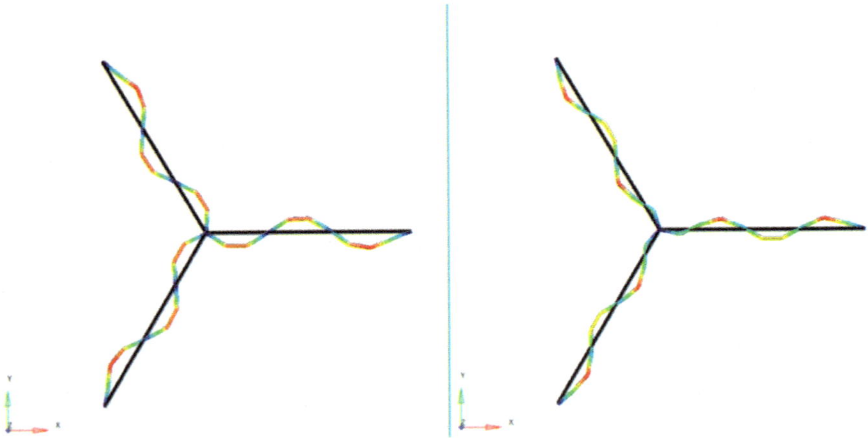

Fig. 9.3 Eigenmodes 3 and 4 of vibration

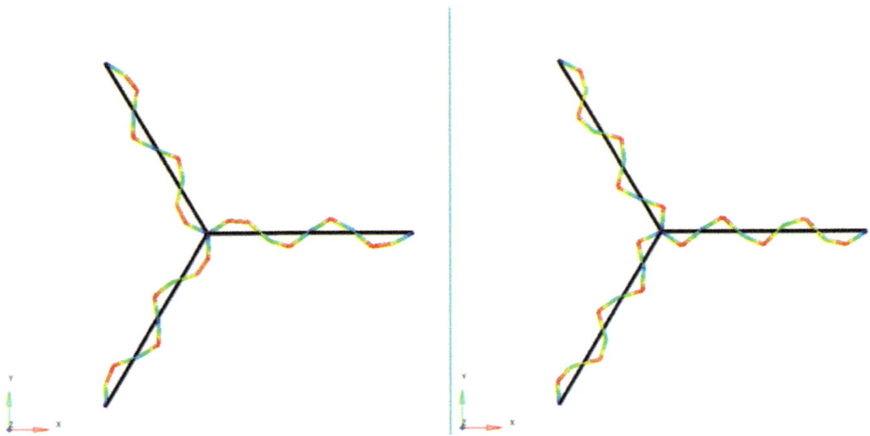

Fig. 9.4 Eigenmodes 5 and 6 of vibration

reducing the design, modeling and manufacturing time of a mechanical structure. As a result, the physical realization costs become lower. Repetitive elements or structures with symmetrical parts can be found practically everywhere. In the case of studying the vibrations of such a structure, the existence of repetitive elements or certain types of symmetries can offer some significant advantages. In the work, using the FEM, some properties were presented that can reduce the calculation time for such a structure and also help in a deeper understanding of the phenomena that take place. The work complements previous studies on these properties.

In the work FEM was used to determine the natural frequencies and modes of vibration for a structure. The work represents a starting point for further research on mechanical structures with symmetries used in civil and industrial engineering. Other types of symmetries as well as other constructive solutions used in construction can be taken into account. In the case of forced vibrations of such systems, little studied in this paper, it is found that certain terms of the motion equations cancel. It would be interesting to study how this could lead to reduced stresses in the structure and, as an important application, whether these constructive solutions will naturally reduce mechanical stresses in events such as earthquakes.

References

1. S. Vlase, M. Marin, A. Oechsner, Considerations of the transverse vibration of a mechanical system with two identical bars. Proc. Inst. Mech. Eng. Part L J. Mater. Des. Appl. (2017)
2. S. Vlase, M. Marin, M.L. Scutaru, R. Munteanu, Coupled transverse and torsional vibrations in a mechanical system with two identical beams. AIP Adv. **7**, AR 065301 (2017)
3. L. Meirovitch, *Analytical Methods in Vibrations* (McMillan, 1967)
4. J. Lin, S. Jin, C. Zheng, Z.M. Li, Y.H. Liu, Compliant assembly variation analysis of aeronautical panels using unified substructures with consideration of identical parts. Comput.-Aided Des. **57**, 29–40 (2014)
5. M. Mihalcica, M. Paun, S Vlase, The use of structural symmetries of a U12 engine in the vibration analysis of a transmission. Symmetry-Basel **11**(10), 1296 (2019)
6. C.Z. Shi, R.G. Parker, Modal structure of centrifugal pendulum vibration absorber systems with multiple cyclically symmetric groups of absorbers. J. Sound Vibr. (2013)
7. D.N. Paliwal, R.K. Pandey, Free vibrations of circular cylindrical shell on Winkler and Pasternak foundations. Int. J. Press. Vessels Pip. **69**(1), 79–89 (1996)
8. Z. Celep, On the axially symmetric vibration of thick circular plates. Ingenieur-Archiv. **47**(6), 411–420 (1978)
9. A. Zingoni, A group-theoretic finite-difference formulation for plate eigenvalue problems. Comput. Struct. **94–95**, 34–44 (2012)
10. D.D. Holm, C. Stoica, D.C.P. Ellis, *Geometric Mechanics and Symmetry* (Oxford University Press, 2009)
11. J.E. Marsden, T.S. Ratiu, *Introduction to Mechanics and Symmetry: A Basic Exposition of Classical Mechanical Systems* (Springer, 2003), 586 p.
12. S.F. Singer, *Symmetry in Mechanics* (Springer, 2004)
13. J. Lin, S. Jin, C. Zheng, Z.M. Li, Y.H. Liu, Compliant assembly variation analysis of aeronautical panels using unified substructures with consideration of identical parts. Comput. Aided Des. **57**, 29–40 (2014)
14. V.I. Elkin, General-solutions of partial-differential equation systems with identical principal parts. Differential Eqs. **21**(8), 952–959 (1985)
15. A. Zingoni, On the symmetries and vibration modes of layered space grids. Eng. Struct. **7**(4), 629–638 (2005)
16. A. Zingoni, On the best choice of symmetry group for group-theoretic computational schemes in solid and structural mechanics. Comput. Struct. **223**, 106101 (2019)
17. A. Zingoni, Group-theoretic exploitations of symmetry in computational solid and structural mechanics. Int. J. Numer. Meth. Eng. **79**, 253–289 (2009)
18. E.M. Abd-Elaziz, M. Marin, M. Othman, On the effect of Thomson and initial stress in a thermo-porous elastic solid under GN electromagnetic theory, Symmetry **11**(3), 413 (2019)
19. M.M. Bhatti, et al., Swimming of motile gyrotactic microorganisms and nanoparticles in blood flow through anisotropically tapered arteries. Front. Phys. **2020**, 1–12 (2020)
20. A.A. Khan, et al., Effects of chemical reaction on third-grade MHD fluid flow under the influence of heat and mass transfer with variable reactive index. **50**(11), 1061–1080 (2019)

21. S. Vlase, M. Marin, M.L. Scutaru, R. Munteanu, Coupled transverse and torsional vibrations in a mechanical system with two identical beams. AIP Adv. **7**(6) (2017)
22. S. Vlase, C. Itu, O. Vasile, S. Nastac, M.D. Stanciu, M.L. Scutaru, Vibration analysis of a mechanical system composed of two identical parts. Romanian J. Acoust. Vibr. **15**(1), 58–63 (2018)
23. C. Itu, S. Vlase, M. Marin, A. Toderita, Use of the symmetries in the study of vibration response of a hollow cylinder. Symmetry **13**(11) (2021)
24. P. Bratu, et al., Modal analysis of the inertial platform of the laser ELI-NP facility in Magurele-Bucharest. Romanian J. Acoust. Vibr. **19**(2) (2022)
25. F. Boussem, A. Belounar, L. Belounar, L. Fortas, Strain-based mindlin finite element for vibration analysis of rectangular plates coupled with a fluid. Romanian J. Acoust. Vibr. **19**(2), 93–100 (2022)
26. M.L. Besliu-Gherghescu, N.D. Stanescu, On the reduction of a system of planar cylindrically jointed bars with small deformations. Romanian J. Acoust. Vibr. **19**(2), 101–104

Chapter 10
Measurement and Prediction of Highway Noise—Case Study from Serbia

Momir Praščević, Darko Mihajlov, and Aleksandar Gajicki

Abstract The paper deals with the assessment of the environmental quality of one urban area near the section of the A4 highway that passes through the city of Niš based on the values of noise indicators determined by computation and measurement. The measurement and computation values of noise indicators were compared for two measurement location near the A4 highway in the residential zone of the city of Niš. The conformity assessment showed that computation results obtained using the computation method CNOSSOS-EU and LimA software are in good agreement with the measurement results for both measurement locations.

10.1 Introduction

Environmental noise, as one of the basic indicators of the environment quality, is nowadays serious problem for the social communities of major cities around the world. In order to assess the environment quality, it is necessary to know the values of noise indicators as a physical scale for description of environmental noise, which has a relationship with a harmful effect of environmental noise.

The estimations of population exposure to environmental noise inside urban areas given in EEA report [1] show that 81.7 million are exposed to L_{den} levels of 55 dB and higher due to road traffic, 10.7 million due to railway traffic, 3.1 million due to air traffic and 0.8 million due to industry. Therefore, 57.5 million are exposed to L_{night} levels of 50 dB and higher due to road traffic, 8.1 million due to rail traffic, 0.9 million due to air traffic and 0.4 million due to industry.

There is also a considerable number of people exposed to noise outside urban areas. The estimations of population exposure to environmental noise inside urban areas given in EEA report [1] show that 31.1 million are exposed to L_{den} levels of

M. Praščević (✉) · A. Gajicki
Faculty of Occupational Safety, University of Niš, Carnojevica 10a, 18000 Niš, Serbia
e-mail: momir.prascevic@znrfak.ni.ac.rs

D. Mihajlov · A. Gajicki
Institute of Transportation CIP, ad, Nemanjina 6/IV, 11000 Belgrade, Serbia

N. Herisanu and V. Marinca (eds.), *Acoustics and Vibration of Mechanical Structures—AVMS-2023*, Springer Proceedings in Physics 302,
https://doi.org/10.1007/978-3-031-48087-4_10

55 dB and higher due to road traffic, 10.9 million due to railway traffic and 1.1 million due to air traffic. Therefore, 21.1 million are exposed to L_{night} levels of 50 dB and higher due to road traffic, 9 million due to rail traffic and 0.4 million due to air traffic.

These results show that at least 20% of Europeans are exposed to high L_{den} levels and more than 15% of Europeans are exposed to high L_{night} levels considering road traffic noise only, from which adverse health effects can occur [1]. The presented results indicate that road traffic is the biggest source of environmental noise inside and outside urban areas. Main roads of more than 3,000,000 vehicle passages a year inside urban areas have a particularly significant impact on population exposure to environmental noise.

The results of strategic noise mapping of the city of Niš, as third largest city in Serbia, are presented in AECOM report [2] available at Serbian Environmental Protection Agency web pages. In this report is given the number of people living in dwellings at the most exposed building façade that are exposed to noise levels in band values recommended by Directive 2002/49/EC of the European Parliament and of the Council [3]. The estimations of population exposure to environmental noise given in AECOM report show that 198,200 are exposed to L_{den} levels of 55 dB and higher due to road traffic, 21,200 due to railway traffic, 100 due to air traffic and 300 due to industry. Therefore, 182,000 are exposed to L_{night} levels of 50 dB and higher due to road traffic, 18,000 due to rail traffic, 200 due to industry and nobody due to air traffic [2]. These results show that at least 87.9% of population of the city of Niš are exposed to high L_{den} levels and more than 80.7% of population of the city of Niš are exposed to high L_{night} levels considering road traffic noise only, from which adverse health effects can occur.

The paper deals with the assessment of the environmental quality of one urban area near the section of the A4 highway (major road) that passes through the city of Niš based on the values of noise indicators determined by computation and measurement, two methods for noise indicators determination discussed within the Directive 2002/49/EC of the European Parliament and of the Council [3]. The measurement and computation values of noise indicators were compared for two measurement location near the A4 highway in the residential zone of the city of Niš.

10.2 Description Area to Be Noise Mapped

According to the objective of the research, a suitable residential area for noise indicators computation was selected. The selected residential area shown in Fig. 10.1 covers the area of dimensions 1.21 km $\times$ 0.77 km and total area about 0.93 km^2 with the A4 highway section and the location of measurement and computation points (MT 01 and MT 02).

The A4 highway section is main road with more than 3 000 000 vehicle passages a year and dominant noise source for the selected residential area. The A4 highway is a state road of the first A order in the eastern part of Serbia that connects Niš and the

Fig. 10.1 Selected residential area near A4 highway section with noise terminal locations

Table 10.1 Annual average daily traffic (AADT) for the A4 highway section [4]

Year	PC	BUS	LT	MT	HT	TT/AT	Total
2013	3456	130	74	69	49	838	4616
2015	5600	170	110	140	75	930	7025
2016	6235	169	121	147	75	995	7742
2021	7879	302	313	400	187	1609	10,690

Legend PC—passenger car, BUS—autobus, LT—light truck, MT—medium track, HT—heavy truck, TT/AT—truck with trailer and auto truck

Nišava region with Bulgaria and represents the eastern branch of the Pan-European Corridor 10. Annual average daily traffic (AADT) is shown in Table 10.1.

Two locations are selected for noise measurement and computation—MT 01 for long-term measurements and MT 02 for short-term measurements (see Fig. 10.1).

10.3 Methods for Noise Indicators Determination

10.3.1 Measurement Methods

Environmental noise measurement is specified by SRPS ISO 19961-1 [5] and SRPS ISO 1996-2 [6]. SRPS ISO 1996-2 specify two measurement strategies [6]:

- long-term unattended measurements with automated sound monitoring system sufficiently long to encompass all emission situations and meteorological conditions which are needed to obtain a representative average; long-term time intervals that represent some significant fraction of a year should be used (e.g. 3 months, 6 months, and 1 year); and
- short-term attended measurements with portable sound level meters during time intervals with well-defined emission and meteorological conditions (minimum 5 min. if Eq. 11 from [6] comply with or the propagation is favorable).

Long-Term Unattended Measurement

Long-term unattended measurement of environmental noise was carried out on measurement point MT 01 in period from August 2021 to July 2022 (365 days). Brüel and Kjær's Environmental Noise Management System [7, 8] was used for long-term unattended measurement. This system consists of Environmental Noise Management System Software, type 7843, two Noise Monitoring Terminals (NMT), type 3639-B-203 based on Sound Level Meter, type 2550 and one Weather Station, type WXT520. One Noise Monitoring Terminal was located on MT 01, on house terrace (see Fig. 10.2). The height of microphone was 4 m above surrounding terrain.

Sampling of sound pressure levels was performed with a sampling frequency of 0.5 s. Based on these values, the values of equivalent continuous sound pressure levels were calculated for a duration of 15 min and one hour. On the basis of these

Fig. 10.2 Location of noise monitoring terminal

values, the daily, monthly and annual values of the noise indicators were determined, as well as the noise indicators during weekdays, during the weekend and during the week.

Short-Term Attended Measurement

Short-term unattended measurements of environmental noise were carried out on measurement point MT 02 during 2013, 2015 and 2016. The measurements were carried out using the Sound Level Meter, type Brüel and Kjær 2250. After the calibration, the Sound Level Meter was placed at the measurement location on a tripod, at the height of 1.5 m above the surrounding terrain.

One 15-minutes measurement of the equivalent continuous sound pressure levels was carried out quarterly during one year in the following time periods: 09:00–12:00, 13:00–16:00, 16:00–22:00, 22:00–01:00 and 02:00–05:00. The time periods were chosen in such a way that they encompass the whole cycle of noise level changes during whole day.

The obtained measurement results in four quarters were statistically processed and the mean values of the equivalent continuous sound pressure levels were calculated for each time period separately. Then, the values of noise indicator were determined according to SRPS ISO 1996-2 [6].

10.3.2 Computation Method

For comparing the results of long-term unattended measurements and short-term attended measurements, the values of noise indicator were computed based on the data used to create strategic noise maps for the agglomeration of the city of Niš [6]. The computation of noise indicators and their graphical representation in the form of noise maps was done using the software package Predictor-LimA (V2023) using the CNOSSOS-EU method [3]. The computation was performed on a grid of points with dimensions of 10 $\times$ 10 m and with a height of 4 m, using only the first reflection. The dynamic error during the calculation was 0 dB. In order to include all relevant sources of noise, the computation included a buffer zone with a width of 100 m around area to be noise mapped (see Fig. 10.1).

The acoustic model included 190 noise sources (entire and/or parts of traffic roads), 2257 objects (buildings), 673 topographical elements (isohypses and/or terrain lines) and 45 areas with different terrain characteristics from the aspect of sound absorption (CORINE Land Cover). About 4,500 inhabitants live in 1600 housing units in the observed area. For the purposes of acoustic computation, it was adopted that the absorption of building facades is 1 dB.

AADT data for the section of A4 highway from 2021 (see Table 10.1) were used for the hourly flow assessment. The flow per hour for all vehicle category, in accordance with the requirements of the CNOSSOS-EU method, was determined as follows:

- 8% of the total PGDS for day period (06:00–18.00),
- 6% of the total PGDS for evening period (18.00–22.00), and
- 2% of the total PGDS for night period (22:00–06:00).

The flow of traffic on the parallel roads to the A4 highway section and the rest of the road network included in the computation was not known. For the parallel roads, it was adopted that the traffic flow is 50% of the A4 highway section traffic flow. Data from 2019 used to create strategic noise maps for agglomeration of the city of Niš were entered in the road mode. Traffic speeds are adopted on the basis of designed traffic signals, i.e. speed limits depending on the category of roads and their location.

For comparing the measured and computed values of the noise indicators at the measurement points MT 01 and MT 02, the noise indicator values were computed at the same locations and heights. All elements of the acoustic model largely corresponded to the conditions during the measurement, except for the traffic flow, composition and speed.

10.4 Results

10.4.1 Long-Term Unattended Measurement Results

Table 10.2 shows the annual values of the noise indicator (second raw, in bold), as well as the mean values of the noise indicators during month, week, weekdays and weekend and corresponding the standard deviations. The monthly values of noise indicators are shown at Fig. 10.3.

Table 10.2 The values of noise indicators and standard deviations in dB

Period		L_{day}	$L_{evening}$	L_{night}	L_{den}
Year	**Total**	**69.5**	**67.7**	**64.0**	**72.0**
Month	Main	69.5	67.7	64.0	72.0
	Std. dev.	0.31	0.49	0.63	0.42
Week	Main	69.5	67.7	63.9	72.0
	Std. dev.	0.86	0.60	0.79	0.60
Weekdays	Main	69.7	67.8	63.9	72.0
	Std. dev.	0.51	0.61	0.81	0.63
Weekend	Main	68.7	67.8	64.0	71.9
	Std. dev.	0.66	0.62	0.80	0.65

The values for the year are in bold as a true values of annual indicators

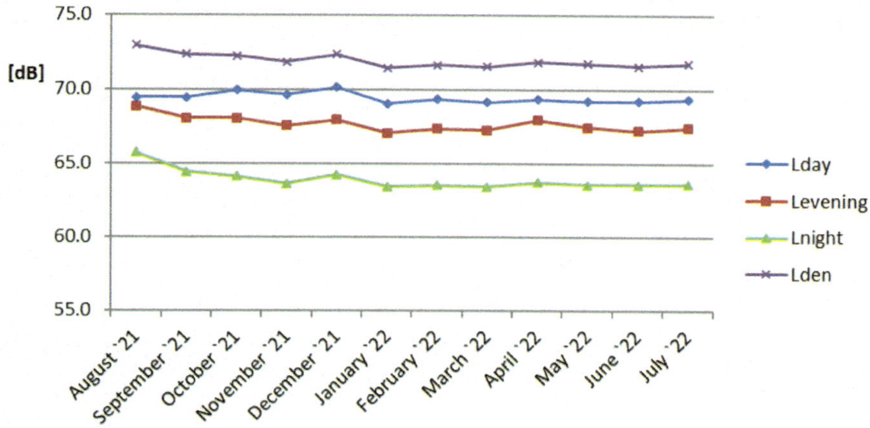

Fig. 10.3 The monthly values of noise indicators

10.4.2 Short-Term Attended Measurement Results

The mean values of the equivalent continuous sound pressure levels calculated separately for each time period based on measurement results in all four quarters are shown in Table 10.3.

Table 10.4 shows the estimated values of noise indicator determined based on the mean values of the equivalent continuous sound pressure levels according to the equations of SRPS ISO 1996-2 [6].

Table 10.3 The mean values of the equivalent continuous sound pressure levels per time period

Year	9:00–12:00	13:00–16:00	18:00–22:00	22:00–01:00	02:00–05:00
2013	66.7	66.5	66.3	60.4	58.3
2015	67.0	68.5	67.0	63.3	59.1
2016	68.0	69.0	67.0	63.0	61.0

Table 10.4 The estimated values of noise indicators in dB

Year	L_{day}	$L_{evening}$	L_{night}	L_{den}
2013	67	66	59	68.6
2015	68	67	62	70.4
2016	69	67	62	70.4

Table 10.5 The computed values of noise indicators based on data from strategic noise map of the city of Niš and the values of noise indicators obtained based on the measurement results

	Computed values				Difference of computed and estimated values			
Meas. point	L_{day}	$L_{evening}$	L_{night}	L_{den}	ΔL_{day}	$\Delta L_{evening}$	ΔL_{night}	ΔL_{den}
MT 01	70.5	69.4	64.6	73.0	1.0	1.7	0.6	1.0
MT 02	70.1	68.9	64.3	72.3	1.1	1.9	2.3	1.9
MT 02*					– 0.3	0.5	0.9	0.5

Table 10.6 The computed values of noise indicators based on data from noise map of selected residential area near A4 highway section and the values of noise indicators obtained based on the measurement results

	Computed values				Difference of computed and estimated values			
Meas. point	L_{day}	$L_{evening}$	L_{night}	L_{den}	ΔL_{day}	$\Delta L_{evening}$	ΔL_{night}	ΔL_{den}
MT 01	70.1	68.8	64.1	72.5	0.6	1.1	0.1	0.5
MT 02	69.6	68.1	63.6	71.9	0.6	1.1	1.6	1.5
MT 02*					– 0.8	– 0.3	0.2	0.1

10.4.3 Computation Results

The computed values of noise indicators are compared with the values of noise indicators obtained based on the measurement results. The comparison is done for the both measurements points MT 01 and MT 02 for the different cases (see Tables 10.5 and 10.6). The computed values of noise indicators for MT 02 are compared with the estimated values of noise indicators for 2016 as well as with the corrected values according to AADT rise from 2016 to 2021 marked with * in tables. The correction was 1.4 dB.

10.5 Conclusion

Prediction of environmental noise exposure will be one of the important tool for assessing and quantification of the noise burden of the population in the future. It is used not only for assessing the noise health risk, but it also provides information required for designing all types of measures for decreasing the harmful effect of noise.

The conformity assessment showed that computation results obtained using the computation method CNOSSOS-EU and LimA software are in good agreement with the evaluated results for both measurement locations, especially in case of long-term unattended measurements. The differences between the computation results and the estimated values of the noise indicators may come from the difference in the traffic flow, the difference in the composition of the traffic and the difference in the speed of the traffic that were used in the computation and that existed during the measurement procedures. The good agreement between computed and estimated values by short-term attended measurements that this strategy can be successfully used next to highways or other roads with a relatively large traffic flow and with small changes in traffic characteristics during different periods of the day. In doing so, special attention should be paid to the selection of relevant time intervals for short-term attended measurements.

Acknowledgements This research is conducted under the auspices of Ministry of Science and Technological Development of the Republic of Serbia (Contract no. 451-03-47/2023-01/ 200148).

References

1. European Environmental Agency webpage. https://www.eea.europa.eu/publications/environmental-noise-in-europe. Last accessed 7 Apr 2023
2. AECOM, Strategic noise map of Niš agglomeration (in Serbian), Serbian Environmental Agency webpage. http://www.sepa.gov.rs/index.php?menu=307&id=230&akcija=showAll. Last accessed 19 Apr 2023
3. Directive 2002/49/EC of the European Parliament and the Council relating to the assessment and management of environmental noise. Official J. Euro. Commun. L **189**, 45 (2002)
4. https://www.putevi-srbije.rs/index.php/sr/brojanje-saobraćaja
5. SRPS ISO 1996-1:2019, *Acoustics. Description, Measurement and Assessment of Environmental Noise. Part 1: Basic Quantities and Assessment Procedures* (2019). (in Serbian)
6. SRPS ISO 1996–2:2019, *Acoustics. Description, Measurement and Assessment of Environmental Noise. Part 2: Determination of Sound Pressure Levels* (2019). (in Serbian)
7. D. Mihajlov, M. Prascevic, Permanent and semi-permanent road traffic noise monitoring in the city of Niš (Serbia). J. Low Frequency Noise Vibr. Active Control **34**(3), 251–268 (2016). https://doi.org/10.1260/0263-0923.34.3.251
8. M. Prascevic, D. Mihajlov, Noise indicators determination based on long-term measurements. Facta universitatis Ser. Work. Living Environ. Protect. **11**(1), 1–11 (2014)

Chapter 11
Comparative Analysis of the Application of Different Metaheuristic Methods for Determining the Coefficients of Empirical Models for Determining Sound Absorption

Tanja Miodragović, Branko Radičević, Goran Miodragović, and Mišo Bjelić

Abstract Metaheuristics are widely used in various fields due to their ability to solve complicated optimization problems. In acoustics, it is possible to apply metaheuristics to determine the optimal values of coefficients in empirical models of acoustic impedance. In this paper, different metaheuristic approaches (i.e. gray wolf optimization algorithm, marine predator algorithm, beluga whale optimization, sand cat swarm optimization) are presented for solving the above problem. They are based on different biological phenomena and with these methods no experience is required in determining the initial values of the variables as it is possible to set a wide range for the initial values of the variables. The applied methods provide a satisfactory prediction of the sound absorption coefficient of open-cell polyurethane foam, compared to the experimental results obtained in an impedance tube. The paper provides a comparative analysis of recent metaheuristics: gray wolf optimization algorithm, beluga whale optimization algorithm, marine predator algorithm and sand cat algorithm.

11.1 Introduction

In the modern world, noise is increasingly affecting our everyday lives, and sound quality is becoming more important in architecture, construction, industry, and transportation. Sound absorption of materials, which absorbs sound energy instead of reflecting or transmitting it, is a crucial parameter in acoustics [1]. Empirical models

T. Miodragović (✉) · B. Radičević · M. Bjelić
Faculty of Mechanical and Civil Engineering in Kraljevo, University of Kragujevac, Kraljevo, Serbia
e-mail: miodragovic.t@mfkv.kg.ac.rs

G. Miodragović
Academy of Applied Study Šumadija, Department Trstenik, Trstenik, Serbia

N. Herisanu and V. Marinca (eds.), *Acoustics and Vibration of Mechanical Structures—AVMS-2023*, Springer Proceedings in Physics 302,
https://doi.org/10.1007/978-3-031-48087-4_11

with coefficients related to material characteristics are used for the design and optimization of materials [2]. However, determining the exact values of these coefficients can be a challenging task due to the complexity of materials and the multitude of factors that can affect their acoustic performance. In acoustics, metaheuristics are particularly attractive as they are capable of solving complex optimization problems without the need for information about the problem or deterministic models. These techniques simulate biological processes to find the best solution in a large search space [3]. In acoustics, metaheuristics have shown their efficiency in determining optimal coefficient values in empirical models for predicting sound absorption [4, 5]. This study compares different metaheuristic methods for determining the coefficients of empirical models used to predict the sound absorption of open-cell polyurethane foams. Experimental results obtained from impedance tube measurements for the polyurethane foam material S255 [6] are used as reference values to evaluate the algorithm's performance. The goal is to analyze the performance of these metaheuristic algorithms in various frequency ranges and determine which one yields the best results.

11.2 Theoretical Basis

11.2.1 Empirical Models

Empirical models for predicting sound absorption are key tools for designing and optimizing materials with desired acoustic properties. They rely on experimentally determined coefficients that describe material characteristics such as density, thickness, and porosity to predict material behavior at different sound frequencies [2]. In 1970, Delany and Bazley [7] introduced a model that allows for the calculation of characteristic impedance (Zc) and sound propagation constant (γ) in absorbing materials. In this study, the transformed Delany-Bazley model was used according to standard requirements [8] to demonstrate the justification for applying a biologically inspired algorithm to determine the characteristic acoustic impedance.

The dependencies in this model are given by the following Eqs. (11.1)–(11.4) [6].

$$Z_{CR} = \rho_0 c_0 \left[1 + C_1 \left(\frac{\sigma}{\rho_0 f} \right)^{C_2} \right] \tag{11.1}$$

$$Z_{CI} = -\rho_0 c_0 \left[C_3 \left(\frac{\sigma}{\rho_0 f} \right)^{C_4} \right] \tag{11.2}$$

$$\alpha = \left(\frac{2\pi f}{c_0} \right) \left[C_5 \left(\frac{\sigma}{\rho_0 f} \right)^{C_6} \right] \tag{11.3}$$

$$\beta = \left(\frac{2\pi f}{c_0}\right)\left[1 + C_7\left(\frac{\sigma}{\rho_0 f}\right)^{C_8}\right] \tag{11.4}$$

where

Z_{CR} и Z_{CI}	real and imaginary parts of the characteristic acoustic impedance, Z_C,
α and β	real and imaginary parts of the propagation constant, Γ,
σ	air flow resistance,
f	frequency,
ρ_0	air density and
c_0	sound velocity.

The sound absorption coefficient at normal incidence, α_n, for a solid-backed layer of material with thickness d, can be obtained using well-known expressions (11.5) and (11.6), along with knowledge of the characteristic acoustic impedance and propagation constant Γ.

$$Z_S = Z_C coth \Gamma d \tag{11.5}$$

$$\alpha_n = 1 - \left|\frac{Z_S - \rho_0 c_0}{Z_S + \rho_0 c_0}\right|^2 \tag{11.6}$$

The values of the sound absorption coefficient are provided in [6], and they were obtained by measuring in an impedance tube using the transfer function method between two microphones, as described in the standard SRPS EN ISO 10534-2. This method allows for obtaining the absorption coefficient values at normal incidence within the frequency range defined by the physical dimensions of the tube and the distance between the microphones. [6]

A new empirical model [6] for determining the sound absorption of PU foam was developed using the least squares method by utilizing the measured sound absorption coefficient values at normal incidence in the impedance tube. In the model [6], new values for the coefficients of the characteristic acoustic impedance and propagation constant were determined (Table 11.1).

These coefficients can be determined using metaheuristic algorithms and experimental data on the sound absorption of materials. The coefficients may vary for different types of materials, as their absorption behavior depends on the structure and composition.

Table 11.1 New values of the coefficients of the characteristic acoustic impedance and propagation constant [6]

C_1	C_2	C_3	C_4	C_5	C_6	C_7	C_8
0.342	0.0853	0.0899	0.482	0.197	0.463	0.333	0.249

11.2.2 Metaheuristics

Metaheuristics are universal optimization algorithms used to find global optima in complex problems. They are efficient when classical optimization methods do not yield satisfactory results and focus on the global optimum rather than local solutions [9]. Metaheuristics can be defined as universal heuristics that guide problem-specific heuristics towards regions of the search space where good solutions are likely to be found [3]. There are various types of metaheuristics, based on population, memory, inspiration from natural or social processes, employing approximate search methods. Biologically inspired algorithms are optimization methods that use principles from biological systems as the basis for solving complex problems. They mimic natural mechanisms such as evolution, insect swarms, or neural networks to find optimal solutions. These algorithms have proven to be effective in various fields, including engineering, computer science, robotics, and bioinformatics [10].

11.3 Material and Methods

11.3.1 Material

Polyurethane foams are commonly used as sound-absorbing materials due to their ability to convert mechanical sound energy into heat through friction. Open-cell foams allow sound to freely pass through them, while closed-cell foams are dense and rigid, providing better sound insulation. Physical properties such as thickness, density, and elasticity also play an important role in sound absorption capability. The design of polyurethane foams is tailored to specific application requirements and desired levels of sound absorption [11].

11.3.2 Methods

Various metaheuristic algorithms will be used for optimizing the objective function. The algorithms used in this research include the grey wolf algorithm, the beluga whale algorithm, the marine predator algorithm, and the sand cat algorithm. The coefficients determined using these algorithms will be compared with the NMI model [6].

Grey Wolf Algorithm. The Grey Wolf Optimizer (GWO) is an optimization algorithm inspired by the hunting behavior of grey wolves in packs. Each solution to the problem represents a wolf, and the algorithm uses tracking, chasing, and attacking to simulate their behavior. The alpha, beta, and delta wolves represent the best solutions in the population, while the omega wolves represent the remaining solutions. These phases allow the wolves to track the best solution, improve convergence, and maintain search diversity [12]. The GWO algorithm is easy to implement and demonstrates high efficiency and convergence in solving various types of optimization problems.

Beluga Whale Optimization. The Beluga Whale Optimization (BWO) is a novel metaheuristic algorithm inspired by the behavior of beluga whales. It consists of exploration, exploitation, and whale fall phases. The exploration phase allows the population to move towards new search areas, while the exploitation phase focuses on promising solutions. The whale fall phase introduces randomness to avoid getting trapped in local optima. The self-adaptive balance factor maintains a balance between exploration and exploitation, and the Levy flight in the exploitation phase enhances global convergence and the ability to avoid local optima [13]. The BWO algorithm shows promising results in solving optimization problems and has potential for applications in various field.

Marine Predator Algorithm. In the paper "Marine Predators Algorithm: A nature-inspired metaheuristic," a new metaheuristic algorithm inspired by the feeding behavior of marine predators is introduced. The algorithm consists of three optimization phases that take into account the velocity ratio between predators and prey. Each phase is applied based on this ratio to achieve a balance between exploration and exploitation, allowing the algorithm to adapt to different characteristics of optimization problems. MPA is effective in solving various types of optimization problems, including engineering problems, indicating its potential for practical applications [14].

Sand Cat Swarm Optimization. Sand Cat Swarm Optimization (SCSO) is a novel metaheuristic approach to global optimization inspired by the characteristics of the sand cat. The algorithm utilizes low-frequency detection and digging ability, which are key features of the cat. Low-frequency detection is used to attract particles toward optimal solutions, while the digging ability is simulated to discover better solutions. SCSO enables a balance between exploration and exploitation, leading to improved performance compared to other metaheuristic algorithms, especially in situations with fewer parameters and operations [15].

Further details about the characteristic equations of the mentioned optimization algorithms and the pseudo-code of the algorithms can be found in the literature [12–15].

11.4 Results and Discussion

In Table 11.2, comparative results of the obtained coefficients for determining the sound absorption coefficient are presented. The proposed algorithms were applied without any modifications. The boundary values of the search space were defined as an infinite search space ($lb_i = -\infty$, $ub_i = +\infty$; $i = 1, \ldots, 8$). The fundamental parameters of the algorithms were the same, with 10,000 iterations and 50 search agents.

Table 11.2 Results obtained by the proposed algorithms

Coefficients	Sand cat SCSO [15]	Beluga whale [13]	Marine predator [14]	Gray wolf [12]
C1	0.72643	0.29593	0.54563	0.51626
C2	0.00000	0.29323	0.14903	0.076275
C3	0.00000	0.07011	0.00000	0.087325
C4	0.08986	0.24197	4.86188	0.00015551
C5	0.39886	0.26121	0.37773	0.3306
C6	0.00000	0.28044	0.02624	0.17521
C7	0.14259	0.34410	0.11698	0.31105
C8	0.90821	0.42631	0.90398	0.49521
Objective function	**0.28954**	**0.3111928505**	**0.2855971025**	**0.2896**
Worst solution	**9.0855**	**0.5581956437**	**5.1665905042**	**0.8818**
Average solution	**0.31165**	**0.3459531712**	**0.3107880293**	**0.29911**
Standard deviation	**0.30735**	**0.0536763920**	**0.2077109150**	**0.029188**
Algorithm execution time [s]	**204.88080**	**251.2792203**	**327.1297053**	**222.08088**

Bold represents the optimal solutions

From Fig. 11.1, it can be observed that the GW, SC, and MP algorithms converge very quickly, within a few passes, while this is not the case for the BW algorithm. Convergence is a good indicator but can also pose a risk as the algorithm may get trapped in a local minimum. The best result was obtained with the MP algorithm, which also had the longest duration, t = 327.1297053 s.

By incorporating the obtained coefficients, Ci, i = 1, …, 8, into Eqs. (11.1–11.6) for determining the sound absorption coefficient, α_n, diagrams depicting the dependence of the sound absorption coefficient, α_n, on material thickness and frequency for the given foam are obtained (Fig. 11.2).

Tables 11.3, 11.4 and 11.5 provide the values of the sound absorption coefficient, α_n, for the obtained coefficient values, Ci, i = 1, …, 8, as well as the absolute differences from the values obtained by NMI (least squares method) and the experiment. Table 11.4 shows that the Marine Predator algorithm yields the best results, especially at lower frequencies up to 400 Hz, as illustrated in Fig. 11.3. Figure 11.3 compares the values of the sound absorption coefficient obtained through the experiment and the proposed algorithms (a), and specifically the Marine Predator algorithm (b), for a foam thickness of 5 cm.

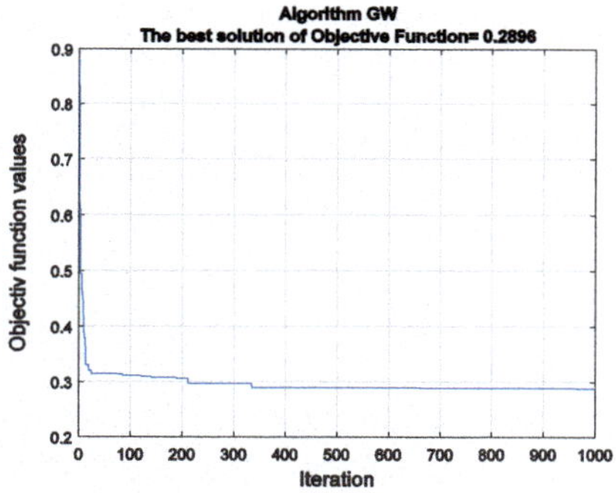

Gray Wolf algorithm, GW

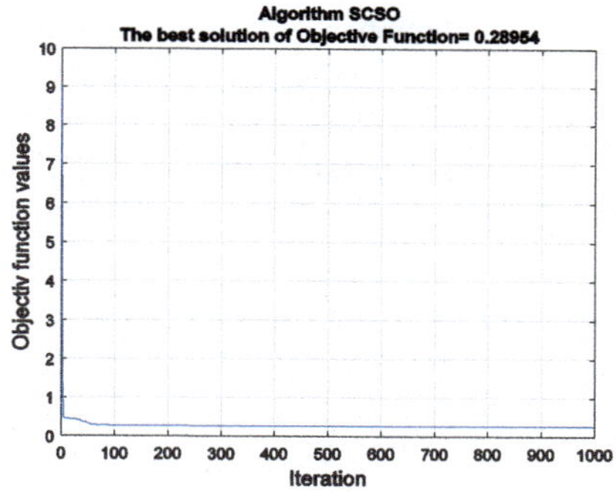

Sand Cat algorithm, SC

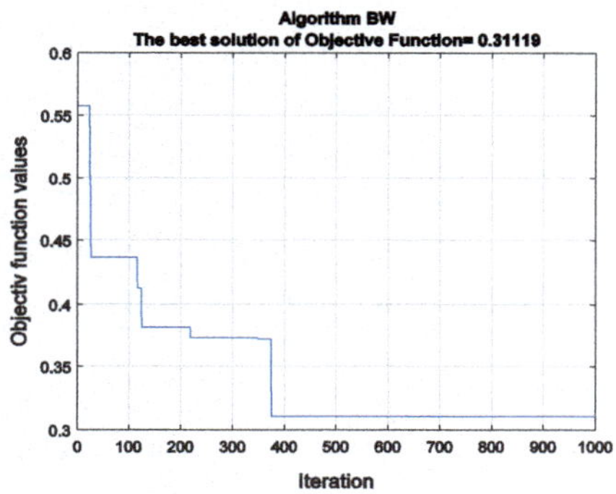

Beluga Whale algorithm, BW

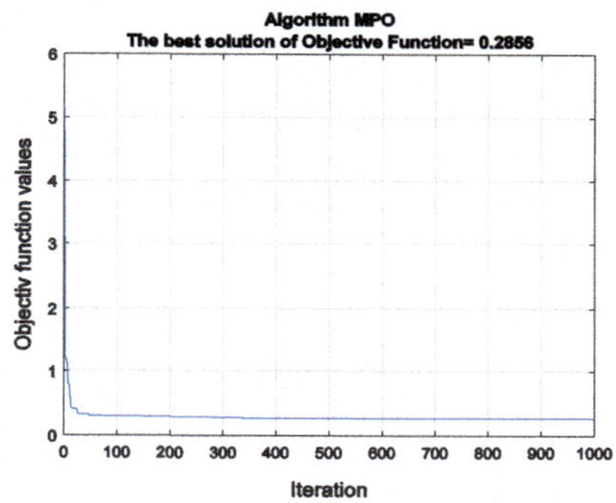

Marine predator algorithm, MP

Fig. 11.1 Iterative process for determining the coefficients Ci, i = 1, …, 8

Fig. 11.2 Sound absorption coefficient of foam S 2535 for sample thicknesses from 10 to 100 mm

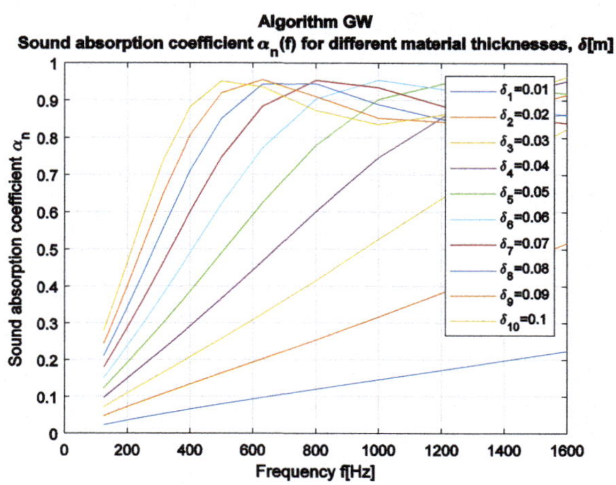

Table 11.3 Absorption coefficient for material thickness 0.05 [m]

		Thickness δ = 0.05 [m] Algorithm				
f (Hz)	Impedance tube [6]	NMI model [6]	Sand cat SCSO [15]	Beluga whale [13]	Marine predator [14]	Gray wolf [12]
125	0.12554	0.108820249	0.113465775	0.116135049	0.104618710	0.125257006
160	0.13374	0.133499080	0.146056775	0.146329024	0.135735276	0.157309428
200	0.15279	0.160976982	0.183780903	0.181002681	0.171857210	0.194022555
250	0.18509	0.195250406	0.231690134	0.225074550	0.217782855	0.240574698
315	0.22428	0.240895702	0.295222096	0.284085572	0.278647762	0.302802242
400	0.26599	0.303790700	0.380156430	0.364342484	0.359841397	0.387471503
500	0.35765	0.383049161	0.481573399	0.461846358	0.456569249	0.490769824
630	0.47585	0.492528983	0.611588070	0.587451320	0.580726803	0.624959343
800	0.62189	0.635597285	0.765573058	0.732185525	0.729948254	0.780762055
1000	0.76989	0.777276564	0.899808802	0.848749439	0.866869616	0.904034229
1250	0.87787	0.878340310	0.976893806	0.906124428	0.959831964	0.955447518
1600	0.90703	0.887349120	0.962215446	0.889546505	0.972015833	0.918738322
Objective function			**0.28954**	**0.3111928505**	**0.2855971025**	**0.2896**

Bold represents the optimal solutions

Table 11.4 Absolute difference for material thickness 0.05 [m] in relation to the results obtained on the impedance tube

		Thickness δ = 0.05 [m] Absolute difference between the obtained results and the experiment Δ			
f (Hz)	Impedance tube [6]	Sand cat SCSO [15]	Beluga whale [13]	Marine predator [14]	Gray wolf [12]
125	0.12554	0.012074225	0.009404951	*0.02092129*	0.000282994
160	0.13374	0.012316775	0.012589024	*0.001995276*	0.023569428
200	0.15279	0.030990903	0.028212681	*0.01906721*	0.041232555
250	0.18509	0.046600134	0.03998455	*0.032692855*	0.055484698
315	0.22428	0.070942096	0.059805572	*0.054367762*	0.078522242
400	0.26599	0.11416643	0.098352484	*0.093851397*	0.121481503
500	0.35765	0.123923399	0.104196358	0.098919249	0.133119824
630	0.47585	0.13573807	0.11160132	0.104876803	0.149109343
800	0.62189	0.143683058	0.110295525	0.108058254	0.158872055
1000	0.76989	0.129918802	0.078859439	0.096979616	0.134144229
1250	0.87787	0.099023806	0.028254428	0.081961964	0.077577518
1600	0.90703	0.055185446	0.017483495	0.064985833	0.011708322

Italics represent the optimal solutions

Table 11.5 Absolute difference for material thickness 0.05 [m] compared to the results obtained by the NMI model

		Thickness δ = 0.05 [m] Absolute difference between the obtained results and the experiment Δ			
f (Hz)	NMI model [6]	Sand cat SCSO [15]	Beluga whale [13]	Marine predator [14]	Gray wolf [12]
125	0.108820249	0.004645526	0.0073148	*0.004201539*	0.016436757
160	0.133499080	0.012557695	0.012829944	*0.002236196*	0.023810348
200	0.160976982	0.022803921	0.020025699	*0.010880228*	0.033045573
250	0.195250406	0.036439728	0.029824144	*0.022532449*	0.045324292
315	0.240895702	0.054326394	0.04318987	*0.03775206*	0.06190654
400	0.303790700	0.07636573	0.060551784	*0.056050697*	0.083680803
500	0.383049161	0.098524238	0.078797197	0.073520088	0.107720663
630	0.492528983	0.119059087	0.094922337	0.08819782	0.13243036
800	0.635597285	0.129975773	0.09658824	0.094350969	0.14516477
1000	0.777276564	0.122532238	0.071472875	0.089593052	0.126757665
1250	0.878340310	0.098553496	0.027784118	0.081491654	0.077107208
1600	0.887349120	0.074866326	0.002197385	0.084666713	0.031389202

Italics represent the optimal solutions

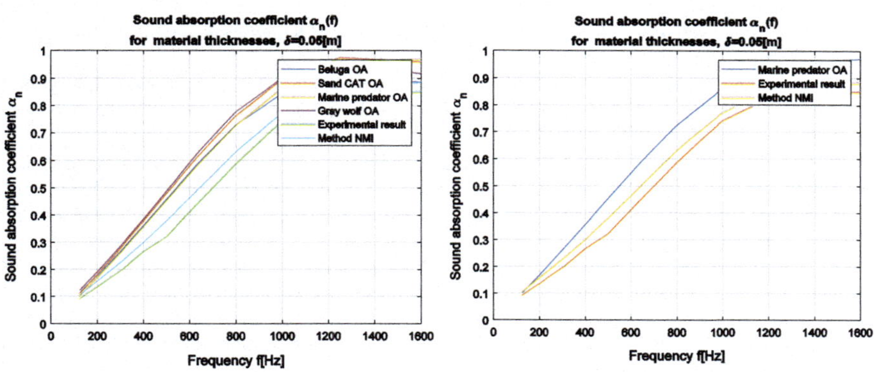

(a) (b)

Fig. 11.3 Comparative analysis

11.5 Conclusion

This paper presents the application of metaheuristic algorithms for determining the coefficients of sound absorption. The algorithms were applied as proposed by the authors, without any modifications.

All four algorithms gave approximately the same results with very excellent convergence and the duration of the search process. The algorithms quickly entered the space of potential optimal solutions. The best result was achieved by the Marine Predator algorithm (with the least deviation), particularly at lower frequencies up to 400Hz, as shown in Fig. 11.3 and Table 11.3, 11.4 and 11.5.

In order to achieve even better results, it is necessary to perform modifications to the metaheuristic algorithms. Accordingly, in further research, the possibility of algorithm hybridization should be considered, as well as improvements in the areas of exploration and exploitation.

Acknowledgements This work is co-financed by the Ministry of Education, Science and Technological Development of the Republic of Serbia on the base of the contract whose record number is 451-03-47/2023-01/200108. The authors thank the Ministry of Education, Science and Technological Development of the Republic of Serbia for supporting this research.

References

1. L.E. Kinsler, A.R. Frey, A.B. Coppens, J.V. Sanders, *Fundamentals of Acoustics*, 4th edn. (Wiley, New York, 2000)
2. J.F. Allard, N. Atalla, *Propagation of Sound in Porous Media: Modelling Sound Absorbing Materials*, 2nd edn. (Wiley, Chichester, 2009)
3. G. Miodragović, *Advanced Bio-Inspired Algorithms Development for Solving Optimization Problems in Applied Mechanics*. Ph.D. thesis (University of Kragujevac, Faculty of Mechanical and Civil Engineering in Kraljevo, 2015)
4. T. Miodragović, B. Radičević, G. Miodragović, M. Ivanović, Hybrid optimization algorithm for determining sound absorption, in *27th Proceedings of the International Conference Noise and Vibration* (Niš, 2022), pp. 143–149
5. M. Kolarević, B. Radičević, M. Bjelić, T. Miodragović, G. Miodragović, Application of biologically inspired algorithms for determining the coefficients of empirical models for determining sound absorption, in *10th Proceedings of the International Conference "Heavy Machinery-HM 2021* (Vrnjačka Banja, 2021), pp. F.13–F.19
6. B. Radičević, *Development of a Decision-Making Model for the Selection of the Optimal Mixture of Sound Absorbing Materials*. Ph.D. thesis (University of Kragujevac, Faculty of Mechanical and Civil Engineering in Kraljevo, 2016)
7. M.E. Delany, E.N. Bazley, Acoustical properties of fibrous absorbent materials. Appl. Acoust. **3**(2), 105–116 (1970)
8. EN 12354-6:2003, *Building Acoustics—Estimation of Acoustic Performance of Buildings from the Performance of Elements—Part 6: Sound Absorption in Enclosed Space* (European Committee for Standardization, Brussels, Belgium, 2003)
9. E.G. Talbi, *Metaheuristics: From Design to Implementation*, 1st edn. (Wiley, New York, 2009)
10. I.J. Fister, X.S. Yang, I. Fister, J. Brest, D. Fister, A brief review of nature-inspired algorithms for optimization. Elektrotehniški vestnik **80**(3), 116–122 (2013)
11. J.G. Gwon, S.K. Kim, J.H. Kim, Sound absorption behavior of flexible polyurethane foams with distinct cellular structures. Mater. Des. **89**, 448–454 (2015)
12. S. Mirjalili, S.M. Mirjalili, A. Lewis, Grey wolf optimizer. Adv. Eng. Softw. **69**, 46–61 (2014)
13. C. Zhong, G. Li, Z. Meng, Beluga whale optimization: a novel nature-inspired metaheuristic algorithm. Knowl.-Based Syst. **251**, 109215 (2022). https://doi.org/10.1016/j.knosys.2022.109215

14. A. Faramarzi, M. Heidarinejad, S. Mirjalili, A.H. Gandomi, Marine predators algorithm: a nature-inspired metaheuristic. Expert Syst. Appl. **152**, 113377 (2020). https://doi.org/10.1016/j.eswa.2020.113377
15. A. Seyyedabbasi, F. Kiani, Sand cat swarm optimization: a nature-inspired algorithm to solve global optimization problems. Eng. Comput.Comput. (2022). https://doi.org/10.1007/s00366-022-01604-x

Chapter 12
Road Traffic, Source of Urban Noise

Sorin Simion, Angela Găman, Izabella Kovacs, Marius Kovacs, and Lorand Toth

Abstract Noise pollution can disrupt human communication, activity and rest, causing various psychological, physiological and even pathological reactions in certain situations, when exposure is prolonged and noise level exceeds allowed limits. Sources of urban noise are varied and randomly distributed in space and time. Industry generates occupational exposure of workers, the latter being permanently exposed to noise coming from machinery and industrial operations, also generating exposure to noise pollution in the vicinity of industrial areas. Noise caused by road traffic comes from vehicle engines and from the contact between roadway and tires, also the intensity of road traffic and the type of vehicles, whether they are light or heavy vehicles, also play an important role. Even if it is not permanent, noise coming from fire, rescue and police intervention vehicles adds to the other sources and amplifies noise pollution especially on heavily trafficked arteries. Although the human body has a high tolerance for noise, it still experiences various effects of physiological or psychological nature that affect its health and activities. The current paper analyses the level of noise generated by road traffic, over a monitoring period of 4 months, for the main intersections in the Eastern area of Jiu Valley. Prevention of exposure to noise must be based on both determination of noise level and its evolution over time, in order to determine the best measures to reduce discomfort caused by road traffic. In this sense, the paper aims to analyze the level of noise pollution created by road traffic and the discomfort created by to residents in the vicinity of traffic arteries.

S. Simion (✉) · A. Găman · I. Kovacs · M. Kovacs · L. Toth
National Institute for Research and Development in Mine Safety and Protection to Explosion—INSEMEX Petroşani, 32-34 G-Ral Vasile Milea Street, 332047 Petroşani, Romania
e-mail: sorin.simion@insemex.ro

N. Herisanu and V. Marinca (eds.), *Acoustics and Vibration of Mechanical Structures—AVMS-2023*, Springer Proceedings in Physics 302,
https://doi.org/10.1007/978-3-031-48087-4_12

12.1 Introduction

The activities of transporting goods and road transport are the main sources generating noise in urban agglomerations. Placement of transport routes near residential areas, respectively the increase in traffic, generates an increasingly extensive exposure of population to ambient noise [1].

Determining ambient noise is a complex task because there are a large number of variables to consider when planning and making measurements. Each measurement is subject to source operating conditions, namely meteorological conditions (which cannot be controlled by the operator) which generate considerable measurement uncertainty. Uncertainty can be determined after measurements, based on an analysis of acoustic measurements, collected data related to source operating conditions and meteorological parameters important for sound propagation.

Along with urban development, noise has become a component of daily life, affecting more and more people but also the fauna in the area where it occurs. Noise pollution has multiple causes:

- traffic sounds such as: cars, buses, ambulances, pedestrians;
- construction activities;
- airports and air traffic;
- railway transport;
- music at a very high level, concerts;
- sounds generated by industrial activities (fans, compressors or generators, etc.) [2].

Even though there are many sources of noise, generated by everyday activities, road traffic remains the main source of noise with a major impact on the activities carried out in urban agglomerations.

Noise pollution generated by road traffic is accentuated by the state of roadways, technical condition of vehicles, respectively their speed.

Increases in the number of vehicles led to intensification of noise level generated in the ambient environment that affects neighbourhoods.

To quantify effects and reduce the level of noise pollution, the following steps must be taken:

- Identifying major sources of noise and the contribution of each to noise level;
- Measurement the noise produced by road traffic in points of interest through sound measurements carried out according to applicable standards;
- Evaluation of noise level generated by road traffic by comparison with the maximum allowed values according to applicable national legislation and standards in force;
- Suggestions on works and measures to reduce noise level;
- Evaluating the efficiency of various technical options suggested or forecasted in order to reduce noise level [3, 4].

In order to identify places susceptible of exposure to noise levels above admissible limits at the limit of functional areas, limits set by the SR 10009:2017 standard were used.

Noise pollution has a lot of negative effects and affects millions of people every day. The most common problem it causes is hearing impairment and in extreme cases it can cause deafness, but it can also generate other neuropsychic and physical effects, which can lead to:

- hypertension;
- headaches;
- gastritis;
- myocardial infarction;
- stress;
- sleeping disorders;
- memory disorders;
- behaviour disorders;
- depression, fatigue, anxiety;
- decrease in concentration skills;
- aggressive behaviour and irritability.

12.2 Working Tools

To measurement noise levels, Danish-made Bruel and Kjaer type 2250 integrating sound level meters equipped with filters with a bandwidth of one octave and/or 1/3 octave, Bruel and Kjaer type 4231 calibrators and the BZ 5503 software for viewing measurements were used. To ensure the quality of results, sound level meters and calibrators used were metrologically verified in accordance with requirements of legislation in force [4].

The noise analyser type 2250 (Fig. 12.1) is used for performing individual measurements or series of measurements for laboratory and in-situ analysis.

The acoustic calibrator (Fig. 12.2) is used for verifying sound level meters used for measurements, to obtain results with a high degree of accuracy (verification is performed both before and after a series of measurements).

To perform the acoustic calibration, the acoustic calibrator type 4231 was used. It provides a stable sound pressure at 1 kHz and is the least influenced by environmental factors.

If the calibration differs by more than $\pm$ 1.5 dB from the original one, then the sound level meter used is checked to see if it corresponds from a metrological point of view.

When reporting results of measurements, an estimate of measurements' global uncertainty was included, taking into account the influence of factors such as:

- measuring equipment;
- microphone location;

Fig. 12.1 Analyser type 2250A

Fig. 12.2 Acoustic calibrator type 4231

- number of measurements;
- meteorological conditions (temperature, humidity, wind speed, its direction, etc.)
- variation in time and space of the noise source [5].

12.3 Case Study

As part of research activities carried out by NIRD INSEMEX Petrosani, a series of noise measurements produced by road traffic in the eastern area of the Jiu Valley were carried out, measurements being carried out in accordance with the national legislation in force. The monitoring points were located in the main intersections (Fig. 12.3), especially in areas crossed by national roads (DN 66) as well as county roads (DJ 709 K) and boulevards (December 1, 1918) (Figs. 12.4 and 12.5).

Table 12.1 shows noise level values for each monitoring point.

In order to quantify noise level, the obtained values were compared with values specified by SR 10009:2017 [6] (Fig. 12.6).

Exposure to noise is a risk factor for health. Sleep disturbances generated by exposure to noise can contribute to: cardiovascular diseases, neuroses, fear, aggression. Noise can create difficulties in the learning process, especially in schools, where it is necessary to ensure a very low noise level.

Although being constantly surrounded by sounds, in most cases, the activity can be carried out ignoring the "noise", but with the increase in the noise level, it becomes a polluting factor of the living and working environment.

Fig. 12.3 Location of monitoring points

Fig. 12.4 Location of monitoring points (points 7 and 8)

Fig. 12.5 Location of monitoring points (points 7 and 10)

The influence of noise on the human body depends on several characteristic factors such as:

- noise type: intensity, frequency, exposure time, continuous or intermittent character;
- individual characteristics: age, activity, state of fatigue, habit, mood, sensitivity, culture, education;
- environmental factors: size of space, architectural structure, etc.

Table 12.1 Values obtained during the monitoring period

No.	Location	February	March	April	May
1	Gyratory "general trans"	70.2	66.6	71.3	71.4
2	Livezeni street	64.2	71.2	70.1	69.3
3	Gyratory avion	69.6	68.9	69.8	69.3
4	University	52.3	51.8	53.6	52.1
5	Gyratory "Piața Victoriei"	59.9	70.1	68.9	65.3
6	Petrila gyratory	56.3	67.5	66.3	68.9
7	Petrila Townhall street	69.6	70.9	72.3	71.3
8	Petrila Police	65.0	67.5	65.1	68.2
9	Lonea Mine crossroad	64.1	65.3	63.5	65.1
10	Lonea Mine	43.3	49.2	53.6	44.5
11	Lonea Firefighters	65.2	64.1	63.5	64.5
12	Jieț Lonea crossroad	57.4	53.8	53.1	54.1

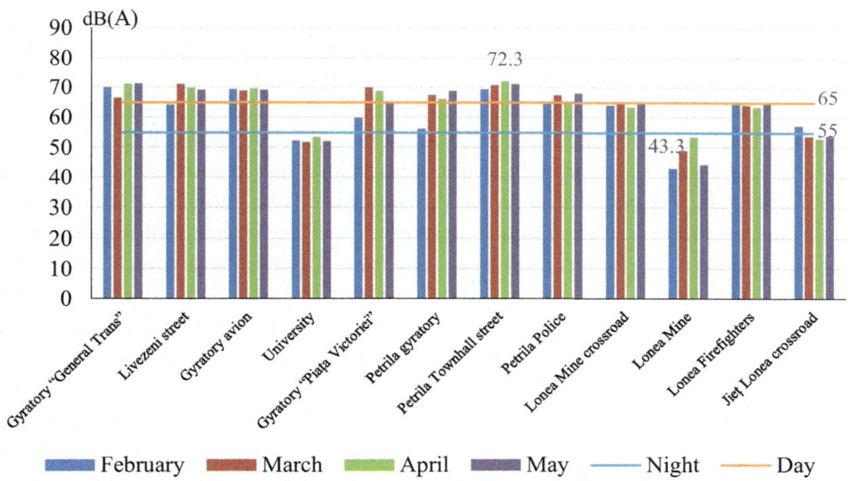

Fig. 12.6 Graphic representation of results obtained and limits established by legislation in force for night and day period, respectively

Within the European Union, almost 40% of the population is exposed to road traffic noise with levels exceeding 55 dB, as A-weighted sound pressure level, during a day, and 20% of the population is exposed to levels exceeding 65 dB. If noise generated by all sources of transport is taken into account, it turns out that almost half of the citizens of the European Union live in areas where acoustic comfort is not ensured [7].

For the night-time period, it is estimated that more than 30% of the population is exposed to levels exceeding 55 dB, that disturb sleep [7].

The quality of environmental factors and especially urban noise influences the health of population, therefore monitoring noise levels outside buildings and assessing the impact on health is an essential component of preventive activities.

In accordance with provisions of Order MS 119/2014, chapter I, art. 16,

(a) during the day, the A-weighted equivalent continuous sound pressure level, measured outside the home, according to SR ISO 1996/2-08 standard, at 1.5 m above the ground, should not exceed 55 dB and the Cz 50 curve of noise.
(b) during the night, between 11:00 p.m. and 7:00 a.m., the A-weighted equivalent continuous sound pressure level, measured outside the home according to SR ISO 1996/2-18 standard, at 1.5 m above the ground, is not to exceed 45 dB and the Cz 40 noise curve, respectively.

12.4 Conclusions

Following the analysis of noise level of generated by road traffic, the following conclusions materialized:

- Analysis of noise measurements carried out in areas with heavy traffic, showed that the level of noise is between 43.3 $\div$ 72.3 dB and in most of the monitoring points exceeds the limit of 65 dB for the monitoring period of 4 months.
- Considering the fact that reducing the noise generated by road traffic is a system problem, the goal of noise reduction is to obtain acoustic comfort at an acceptable price; the primary variables of the system are the acoustic characteristics and the cost price. In order to have a system to reduce exposure of population to noise generated by car traffic, it is necessary to adopt some technical and organizational measures such as: asphalting the streets, introducing one-way traffic on adjacent streets, promoting electric vehicles, planting ornamental shrubs near traffic arteries, installing soundproof barriers near noisy traffic arteries, etc.
- Reduction of the most active and widespread cause of urban noise pollution, that created by road traffic, is mainly achieved by building bypasses for cities transited by national or European roads, by redirecting urban traffic on one-way routes, by definitively or periodically banning traffic in the central perimeters of large cities.

In conclusion, road traffic is an increasingly pressing problem in most localities, not only in large urban agglomerations (LAW no. 121 of July 3, 2019 on the assessment and management of ambient noise) with over 100,000 inhabitants [8].

The measures to reduce the level of exposure to noise must be applied uniformly for all localities affected by road traffic, not only for large urban agglomerations that are subject to monitoring in the form of noise maps made in accordance with Law 121/2019.

Acknowledgements The current paper was carried out through the Nucleu Program within the National Research Development and Innovation Plan 2022–2027, carried out with the support of MCID, project no. PN 23320101.

References

1. A.F. Simion, C. Drebenstedt, M. Lazăr, Environmental impact on soil and water because of mining activities in the Eastern Part of Jiu Valley. Mining Revue **27**(3), 1–16 (2021)
2. https://stratos.ro/en/poluarea-fonica-ce-este-si-care-sunt-cauzele/
3. S. Simion, A.N. Găman, A.F. Simion, M.E. Kovacs, L. Toth, Exposure of workers to occupational Noxae. MATEC Web Conf. **343**, 10008 (2021)
4. S. Sorin, V. Ciprian, K. Marius, et al., Exposure of workers to noise in mining industry, in 12th *International Symposium Acoustics and Vibration of* Mechanical Structures (AVMS 2013), ed. by N. Herisanu, V. Marinca, Timisoara, Romania, 23–24 May 2013
5. SR 6161-1:2022 Acoustics in construction. Part 1: determination of the noise level in civil constructions and urban settlements. Methods of determination
6. SR 10009:2017/C91:2020 Acoustics. Admissible limits of the noise level in the ambient environment
7. https://www.eea.europa.eu/ims/exposure-of-europe2019s-population-to
8. LAW no. 121 of July 3, 2019 regarding the assessment and management of ambient noise, issued by the Parliament of Romania Published in the Official Gazette no. 604 of July 23, 2019

Chapter 13
Acoustic Analysis of Violins in the Concert Hall—The Stainer Violins Case Study

Mariana Domnica Stanciu, Manuela Micu, Alina Maria Nauncef, Ioan Calin Rosca, Vasile Ghiorghe Gliga, and Silviu-Marian Nastac

Abstract The purpose of this article is to study comparatively the acoustic signals of an original Jacobus Stainer violin, from 1716, versus a Stainer Copy, not dated, within a concert hall. The experimental study conducted within the con-cert hall of the Brasov Philharmonic, comprised in the recording of musical signals with five microphones placed in various positions and processing signals. The musical signals were performed by the same violinist and consisted on four parts (open strings excitation, pizzicato, an excerpt from Max Bruch—Concerto no. 1 in G minor op. 26, PI (first cadence of the solo violin and an excerpt from Jules Massenet—Meditation for violin and orchestra from the Opera Thaïs). The obtained acoustic signals have been processed using a program developed in MatLab. The results in the form of acoustic spectrograms, frequency spectrum, magnitude and power spectrum density for each receiver were analyzed comparatively, highlighting the fact that de-pending on the position of the receivers in the concert hall, the sounds emit-ted by the violins are amplified or dampened. The spectra of amplified frequencies in different positions within the concert hall was established based on the statistical analysis and compared

M. D. Stanciu (✉) · M. Micu · I. C. Rosca · V. G. Gliga · S.-M. Nastac
Departament of Mechanical Engineering, Transilvania University of Brașov, B-Dul Eroilor 29, 500036 Brasov, Romania
e-mail: mariana.stanciu@unitbv.ro

M. Micu
Siemens Industry Software Romania, Brasov, Nine, Bulevardul Gării 13A, 500227 Brașov, Romania

A. M. Nauncef
Faculty of Music, Transilvania University of Brașov, 500036 Brașov, Romania

V. G. Gliga
S.C. Gliga Instrumente Muzicale S.A. Reghin, Mures, Romania

S.-M. Nastac
Faculty of Engineering and Agronomy in Braila, "Dunarea de Jos" University of Galati, Galati, Romania

N. Herisanu and V. Marinca (eds.), *Acoustics and Vibration of Mechanical Structures—AVMS-2023*, Springer Proceedings in Physics 302,
https://doi.org/10.1007/978-3-031-48087-4_13

in order to observe the power of the resonance body for the original violin and for the copy ones.

13.1 Introduction

Violin acoustics is part of the field of musical acoustics, the latter being a branch of the science of sounds. In general, the acoustics of the violin are associated with the musical sound produced by it, the carrier of an aesthetic-psychological information, capable of generating artistic impressions. Over time, numerous theoretical and experimental studies have been carried out on musical instruments, in order to optimize their design and musical performance [1–3]. Regarding the acoustics and vibrations of the violin body, there are some experiments aimed at finding the transfer function, which revealed that a number of specific signature modes can be attached and proposed as evaluation modes [4–6]. These modes, which constitute the instrument's acoustic signature, are an important criterion for quality control, or for dating and determining its origin.

The paper presents a comparative analysis of the acoustic analysis of two violins—a historical violin, dated Jacobus Stainer 1716 and a Stainer copy violin, whose signals were recorded in the concert hall of the Philharmonic in Brasov, the musical passage being performed by the same violinist.

13.2 Experimental Set-Up

13.2.1 Materials

In this study, two violins were analyzed, one belonging to the school of Jacobus Stainer, (1619–1683) a well-known luthier of the Austrian-German schools, being born in Absam, respectively a Stainer copy violin. In Fig. 13.1 shows the two studied violins: the front view of the violins, with all the accessories, an image obtained by X-ray radiography and a section through the body of the violin obtained by a computer tomograph. Table 13.1 shows the anatomical characteristics of the wood used in the construction of the two violins. The structure of the spruce wood of the Stainer Copy violin is different from that of the Stainer 1716 violin, and from a constructive point of view, there are also differences [7, 8].

Although Stainer's early designs appear to have been inspired by Amati's designs, he later created his own style, producing exceptional instruments that rivaled or even surpassed the works of his 17th-century Cremonese contemporaries. As constructive features, it is worth noting: the bowing of the front plates is greater than that of

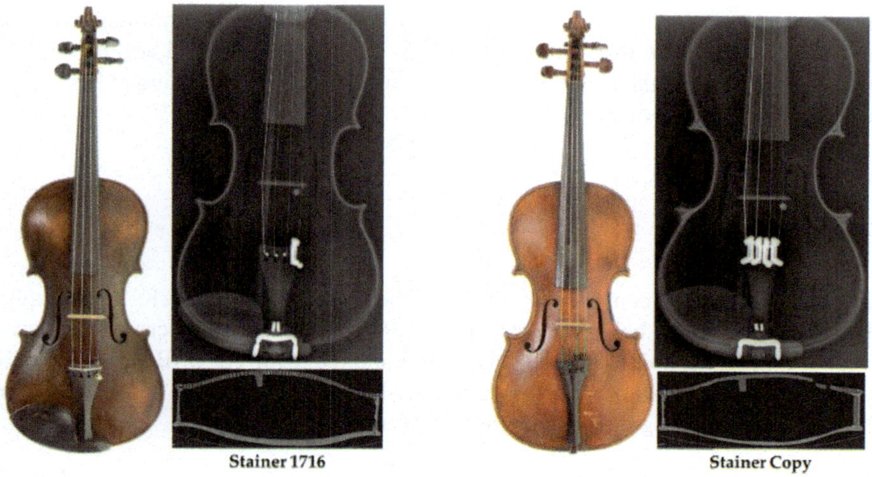

(a) (b)

Fig. 13.1 The external and internal characteristics of the studied violins: **a** Stainer 1716 violin; **b** Stainer copy violin

Table 13.1 Anatomical characteristics of the spruce and maple wood of the studied violins

Anatomical characteristics (average and STDV)	Plate	Stainer 1716	Copia Stainer
The width of the annual rings (mm)	Spruce wood (top plate)	2.247 (0.567)	0.985 (0.527)
Width of early wood (mm)		1.676 (0.518)	0.689 (0.450)
Width of late wood (mm)		0.496 (0.178)	0.300 (0.118)
Proportion of early wood (%)		76.184 (9.152)	66.127 (11.286)
Proportion of late wood (%)		23.816 (9.152)	33.873 (11.286)
Gleichläufigkeit correlation coefficient between the right and left half ring series (%)		62.7	60.6
The width of the annual rings (mm)	Sycamore maple (back plate)	1.908 (0.531)	1.277 (0.297)
The wavelength of the curly fiber (mm)		4.021 (1.577)	4.984 (1.589)

the back plates, the increase being maintained until half the length; the finish used is yellow, with a pale rose tint [5–7], and the soundbar is finished at both ends by chamfers which extend over the soundboard to which it is glued, being placed below the G string of the instrument.

13.2.2 Methods

Dynamic Analysis consisted of exciting the violins with a B&K 8204 impact hammer, the signals being recorded with a PCB 130E20 ICP microphone placed near the acoustic holes. The signals were captured with NI USB-9233 dynamic data acquisition board produced by National Instruments (Austin, USA), and later processed with a program developed in MatLab.

Recording acoustic signals in the concert hall consisted of the interpretation by a professional violinist of four musical fragments: the excitation of the free strings with the bow, the excitation of the free strings in pizzicato style, the interpretation of a fragment from Max Bruch—Concerto no. 1 in G minor op. 26, PI (first Cadence for solo violin) and an excerpt from Jules Massenet—Méditation for violin and orchestra from Opera Thaïs. The recording was made with five microphones located in the concert hall according to Fig. 13.2.

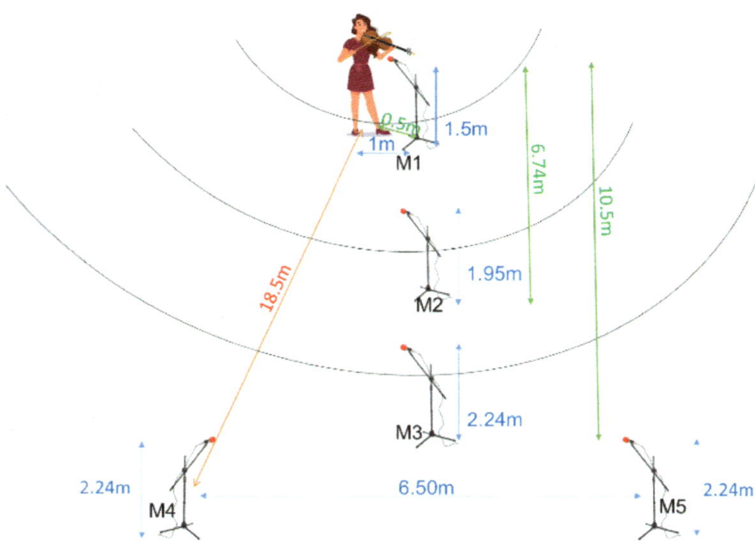

Fig. 13.2 The positions for the violinist and the microphones within the Concert Hall

Signal processing. The processing of the acoustic signals consisted of sampling the signals on the domains of interest, applying the FFT function, graphical representation in the frequency domain, extracting frequencies with significant amplitudes; the statistical analysis of the occurrence of frequencies in the determined frequency spectrum and finally the comparison of the frequency spectra obtained for the investigated violins [3, 8].

13.3 Results and Discussions

13.3.1 Dynamic Analysis

As a result of the impact hammer tests, both the graphs of signal damping over time and their frequency analyzes resulted, as shown in Fig. 13.3. For each violin, the eigenvalues specific to violins were extracted (the first five values, considered in the specialized literature signature mode or "specific timbre" and the dominant frequency, which were subsequently analyzed comparatively (Table 13.2). The quality factor was calculated based on the data extracted from the frequency analysis, applying the calculation relations presented by [3, 7].

The first mode is the A0 mode, known as the air vibration mode of the violin body, the Stainer copy violin having a frequency approximately 10.65 Hz higher than the historical violin, but with a magnitude twice lower than the Stainer 1716 violin. In the case of mode 2, around the frequency of 400 Hz, the rhomboidal vibration mode (known in the literature as the CBR mode) is formed. Thus, the closest CBR mode frequency values are recorded for the Stainer 1716 violin and the Stainer copy (frequency 389 Hz in both cases). For the next resonance mode (B1−), the two violins show values at a difference of 8 Hz, and for the B1+ mode, the differences between the two violins are notable (approx. 60 Hz). The quality factor Q defines the musical instrument's ability to damp vibrations, or how quickly the resonance will fade after the excitation of the system has ceased. In this study, the lowest value of the quality factor is recorded for the Stainer copy violin (7.64) and the Stainer 1716 violin. In the early studies of [7], a value of 20 for Q for the main air resonance, A0, in the case old Italian violins was identified. For mode B1−, both studied violins register a quality factor of 30 (Stainer 1716) and 32 (Stainer Copy), respectively.

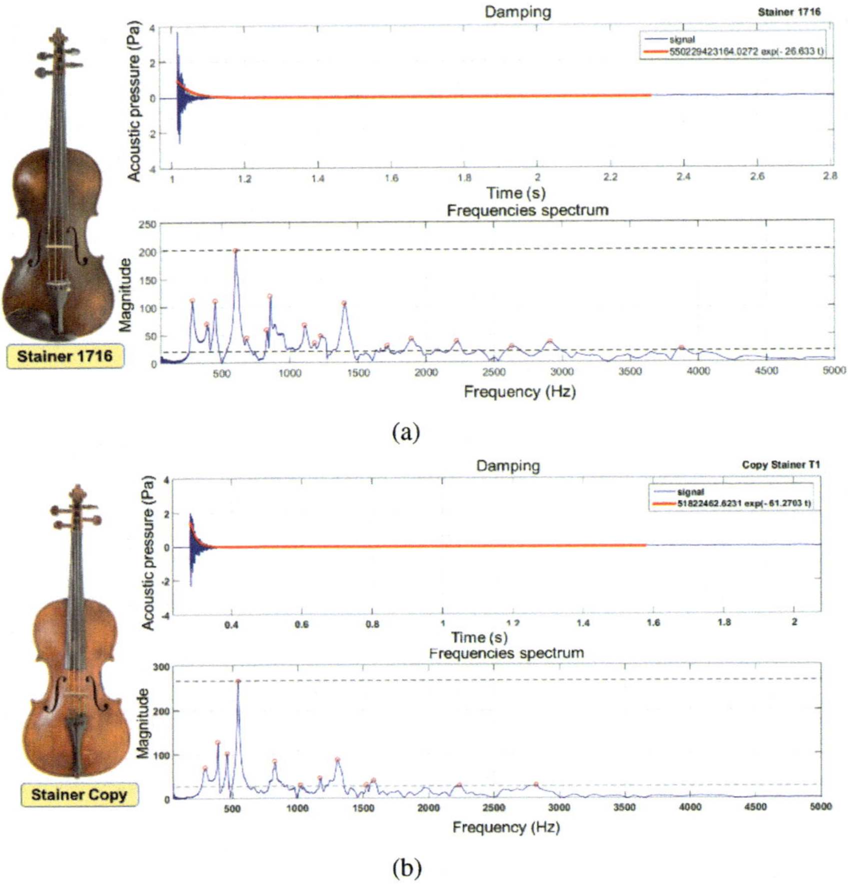

Fig. 13.3 The frequency response of violins: **a** Stainer 1716 violin; **b** Stainer copy violin

Table 13.2 Resonance frequencies for the first five vibration modes of the studied violins

	Signature modes (Hz)				
Violins	f1 (A0)	f2 (CBR)	f3 (B1−)	f4 (B1+)	f5
Stainer 1716					
Frequency (Hz)	283.55	389.70	449.50	604.95	857.00
Quality factor *Q*	11.94	14.32	30.08	22.28	77.99
Copy Stainer					
Frequency (Hz)	294.20	389.75	458.65	544.30	823.35
Quality factor *Q*	7.64	43.77	32.63	33.42	30.08

13.3.2 Acoustic Analysis of Violins in the Concert Hall

The recorded and processed musical signals of open strings are shown in Fig. 13.4. In these figures, the dynamics of musical sounds can be observed, characterized by transient phenomena ADSR (the period of attack (denoted A) 1–2, decay 2–3 (denoted D), extinction or sustain (3–4) (denoted S) and release 4–5 (denoted R), of the sound emission), the continuous variation of the sound emission [3, 5]. The transient processes of musical sounds consist of two stages (initial and final)—the first stage is that of the "sound attack", stage in which it goes from the state of rest to the state of vibration, the intensity of the sound gradually increasing up to the desired value, and in the second stage, of extinction, the attenuation of the sound occurs, the intensity gradually decreasing to zero. Between the two stages, sound stability occurs, this being a stationary process, when harmonics and other phenomena related to acoustic emission and interpretative techniques intervene [3, 5, 7].

The open strings of violin are: G (Sol 196 Hz), D (Re 293.7 Hz), A (La 440 Hz), E (Mi 659.3 Hz). In Fig. 13.5, the spectrograms of musical signals are presented. The octave analysis consisted of applying a set of filters, each filter having a center frequency and a bandwidth. Thus, analysis was used in a single octave band (noted 1/1) and in thirds of an octave (1/3), applying the short-time Fourier transform (STFT) [7, 8]. The central (nominal) frequencies of the frequently used octaves were: 22.1; 44.2; 88.4; 176.8; 353.6; 707.1; 1414.2; 2828.4; 5656.9; 11,313.7; 22,627.4 Hz, center frequencies covering one octave (Fig. 13.6).

13.4 Conclusions

According to research, the factors that influence the timbre of the investigated violins were: the number of harmonics—the higher it is, the richer the timbre, as can be seen in Fig. 13.6; the intensity of harmonics—a warm sound is played by harmonics 2 to 6 that form the perfect major chord; the position of the foundation on the height scale; the intensity of the fundamental—it was found that intense sounds produce auditory sensations that give rise to subjective harmonics, which can alter the objective harmonics; the noise factor—unavoidable especially during transient processes or due to the friction of the bow on the strings. It can be concluded that each instrument has its acoustic and constructive personality, but there are also common elements from an acoustic point of view, which can group these violins according to their acoustic characteristics, in the same category—Stainer violins.

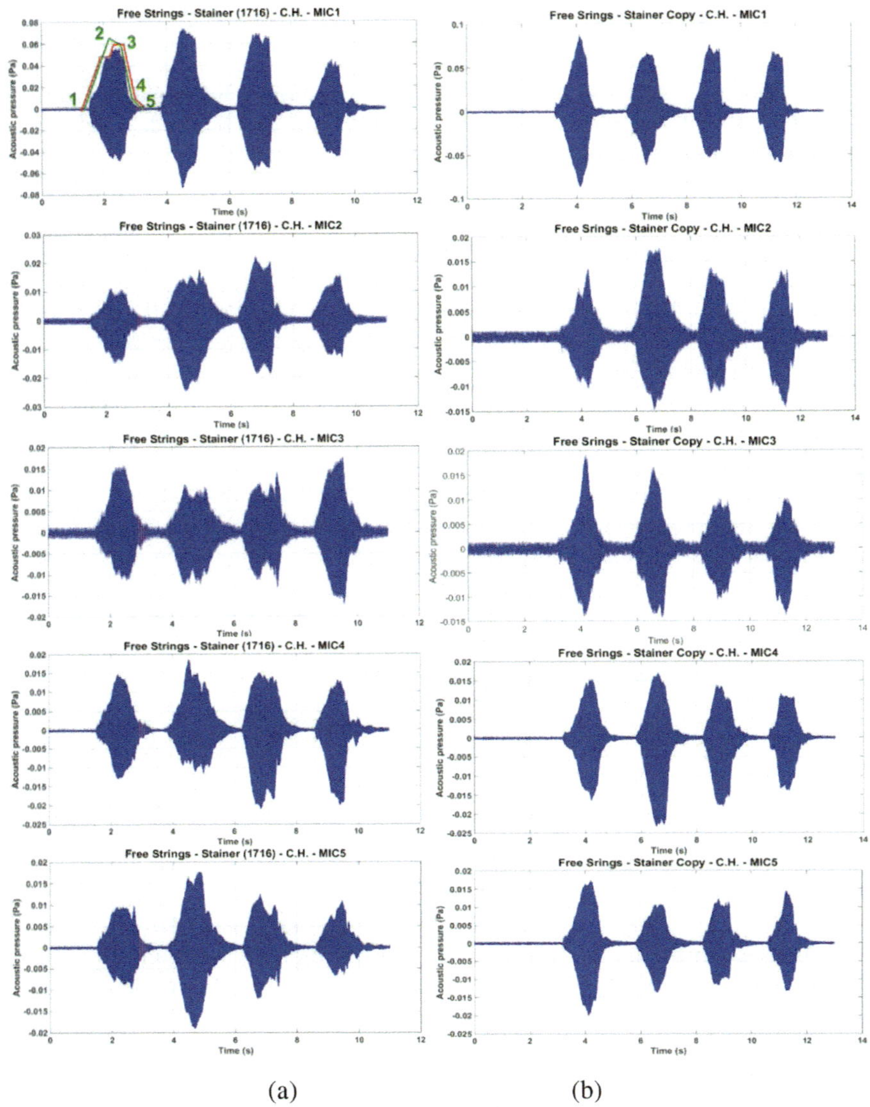

Fig. 13.4 Comparisons regarding the ADSR envelopes for fundamental notes, for the five microphones in the concert hall: **a** Stainer 1716 violin; **b** Stainer copy

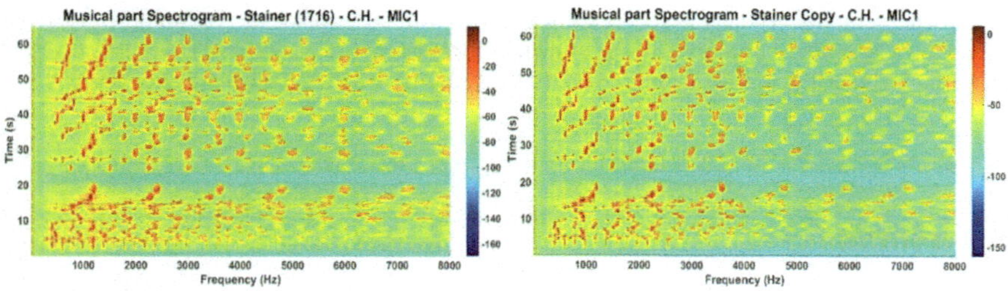

(a) (b)

Fig. 13.5 Comparisons between spectrograms in case of microphone 1 (Mic1): **a** Stainer 1716 violin; **b** Stainer copy

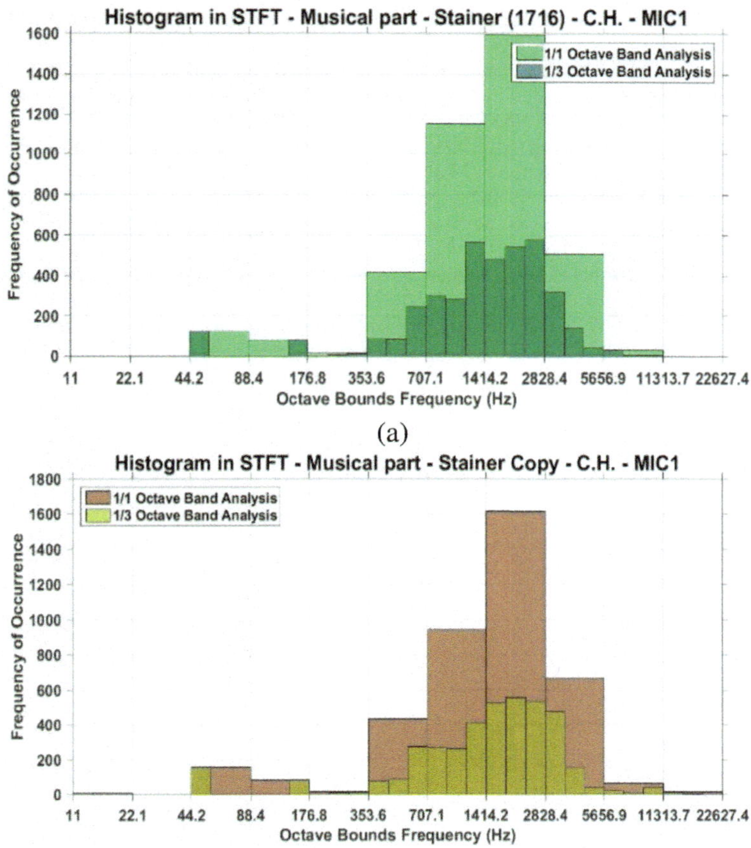

(a)

(b)

Fig. 13.6 Frequency modulus histogram analyzed on the music sequence, in the concert hall, microphone 1: **a** Stainer 1716 violin; **b** Stainer copy

Acknowledgements The audio recording was made with the support of AP Studio Brașov. This research was supported by a grant of the Ministry of Research, Innovation and Digitization, CNCS/CCCDI—UEFISCDI, project number 61PCE/2022, PN-III-P4-PCE2021-0885, ACADIA.

References

1. G. Bissinger, Structural acoustics of good and bad violins. J. Acoust. Soc. Am. **2008**(124), 1764–1773 (2008)
2. S. Le Conte, S. Le Moyne, F. Ollivier, Modal analysis comparison of two violins made by A. Stradivari, in *Proceedings of the Acoustics Nantes Conference*, Nantes, France, 23–27 Apr 2012, hal-00811010 (2012)
3. M.D. Stanciu, M.Ș.A. Mihalcica, *Dinamica Viorii* (Universității Transilvania din Brașov, 2022). (in Romanian)
4. M.D. Stanciu, M. Mihălcică, F. Dinulică, A.M. Nauncef, R. Purdoiu, R. Lăcătus, G.V. Gliga, X-Ray imaging and computed tomography for the identification of geometry and construction elements in the structure of old violins. Materials **14**, 5926 (2021)
5. F. Leccese, et al., The bowed string instruments: acoustic characterization of unique pieces from the Italian lutherie. IOP Conf. Ser. Mater. Sci. Eng. **364**, 012022 (2018)
6. M.D. Stanciu, F. Dinulica, V. Bucur, G.V. Gliga, S.M. Nastac, M. Campean, Changing the vibrational behavior of the wooden thin arched plates—the maestro violins experimental study case. Thin-Walled Struct. **174**, 109042 (2022)
7. M. Schleske, Contemporary violin making. Analysis of design, materials, varnish and normal modes. Catgut Acoust. Soc. J. **4**, 50–65 (2002)
8. S.H. Bae, J.R. Cho, W. Jeong, Free and transient responses of linear complex stiffness system by Hilbert transform and convolution integral. Smart Struct. Syst.Struct. Syst. **17**, 753–771 (2016)

Chapter 14
On Experimental Evaluation of Tortuosity for Cellulose-Based Highly Porous Composites Used Within Noise Insulation Applications

Mihai Seciureanu, Maria-Violeta Guiman, Silviu-Marian Nastac, Petronela Nechita, Carmen-Nicoleta Debeleac, and Gigel-Florin Capatana

Abstract This study deals with the area of noise insulation using innovative porous composites, and starts from the practical necessity of material characteristics within the acoustics simulation. The authors present a method to experimental identification of tortuosity characteristic for porous materials, taking into account that this parameter has a major importance in empirical formulations of actual numerical models widely used within acoustics simulations. The laboratory tests were performed for a set of innovative highly porous composite materials based on foam-forming processes, from cellulose fibers and biopolymers. This type of composites structures provides a competitive alternative to existing petroleum-based foams materials, currently and widely used in technical applications, and especially on sound insulation. The preliminary results indicate that these new porous composites are able to provide noise insulation performances at comparative level to the commercial products large-scale used nowadays.

14.1 Introduction

The porous composites have widely applications within the area of noise protective solutions. Recent studies highlight an innovative way to produce highly porous materials based on cellulose fibers [1–8]. The foam-forming process represents a

M. Seciureanu · S.-M. Nastac (✉) · P. Nechita · C.-N. Debeleac · G.-F. Capatana
Faculty of Engineering and Agronomy in Braila, "Dunarea de Jos" University of Galati, Galati, Romania
e-mail: snastac@ugal.ro

M.-V. Guiman · S.-M. Nastac
Faculty of Mechanical Engineering, "Transilvania" University of Brasov, Brasov, Romania
e-mail: violeta.guiman@unitbv.ro

N. Herisanu and V. Marinca (eds.), *Acoustics and Vibration of Mechanical Structures—AVMS-2023*, Springer Proceedings in Physics 302,
https://doi.org/10.1007/978-3-031-48087-4_14

relative simple way to obtain porous composites with low density. Hereby, the cellulose fibers and other additional components are mixed with foam instead of water such as within the conventionally methods. Three dimensional structure of the foam prevents the flocculation of fibers and enables the production of lightweight and high porosity composites with useful properties for soundproofing applications [9–14]. The effect of different parameters of foam-formed composites (e.g. air content, thickness, porosity, consistency) was reported in many research reports [15–22].

Within their previous studies, the authors had produced the foam-formed cellulose composites by mixing a commercial surfactant with cellulose fibers (from virgin pulp and/or recovered paper) at high shear velocity (2000 rot/min) to entrain air, dewatering on Buchner funnel using low vacuum, and air drying in non-restrained conditions [9–12]. It was obtained various samples, with 36–38 kg/m^3 density, and geometrical characteristics that usually respect the requirements of sample holder of the impedance tube used in acoustical properties investigations.

Besides the basic experimental acoustical investigations, the authors were performed additional computational analyses in order to estimate the sound insulation capabilities of these composites for some configurations encountered in common practice. These analyses had required specific values for a lot of parameters (e.g. airflow resistivity, porosity, tortuosity, viscous/thermal characteristic length) according to the theoretical and/or empirical approaching expressions, widely available in specialized literature [22–25].

Computational investigations regarding noise insulation characteristics were performed using various types of foam-formed cellulose composites obtained both from bleached hardwood cellulose fibers (usually with 100–200 μm length and 15–35 μm diameter), and the recycled cellulose fibers from recovered paper [23].

The main reasons that had led to the authors interest on these bio-composites based on lignocellulosic fibers foams are the lowest density and, respectively, the highest porosity comparing with available petroleum-based worldwide used foam materials, and the simple procedure of obtaining, completely using environmental friendly materials, highly recyclable and biodegradable. Because of their highly porous structure, with predominant open pores, results an improved efficiency in sound and thermal insulation due to increasing of the energy damping capability. This ability, together with a relative lack on information regarding both manufacturing and applications of these lignocellulosic bio-foams, have also attracted and supplied a great interest, highlighting the novelty of this research.

Taking into account permanently interests and constantly investigations within the area of bio-composites from cellulose fibers foams, intended to soundproofing applications, the authors have attempted to evaluate the practical values of specific parameters such as porosity, tortuosity, airflow resistivity etc. Within this paper, the authors present preliminary results on experimental evaluation of tortuosity parameter, based on acoustic measurement method. This method was adopted because of its simplest practical procedure, in the same time with highly accuracy for provided results. In addition, it become easily to evaluate the porosity parameter based on simple empirical expression, which will be presented into next section.

14.2 Theoretical Background

14.2.1 Preliminary Background

The goal of this research consists by practical evaluation of tortuosity parameter involved in most of available mathematical poroacoustics models, thus enabling accurate simulations related to the soundproofing capability of fiber foam-based composites. Computational analysis facilitates the behavioral estimation of these innovative materials, on various practical configurations related to the noise insulation application, in terms of normal sound incidence and frequency. More specifically, it was considered the specific surface impedance, the absorption and the reflection coefficients, and the sound transmission loss characteristics.

Hereby, within their analyses, the authors had considered empirical and multi-parameters poroacoustics models (both microstructural and phenomenological models) such as Delany–Bazley (DB), Biot, Miki, Johnson–Champoux–Allard (JCA), Johnson–Champoux–Allard–Pride–Lafarge (JCAPL). Microstructural schematization, developed according to classical DB model, is intended for simulation the behavior of highly porous materials, having enough accuracy for large scale of materials, but providing unconformable results for very high or too low frequencies [23, 25]. But then, the phenomenological models (JCA, JCAPL) present the ability to simulate porous materials behavior accounting their random geometry. Thus, one of the most used schematization, the JCA porous matrix model can be defined by the complex dynamic rigid density $\rho(\omega)$ in Eq. (14.1) [23]

$$\rho(\omega) = \frac{\tau_\infty \rho_f}{\varepsilon_p}\left(1 + \frac{\sigma \varepsilon_p}{i\omega\rho_f\tau_\infty}\sqrt{1 + \frac{4i\omega\tau_\infty^2\mu\rho_f}{\sigma^2\Lambda^2\varepsilon_p^2}}\right), \tag{14.1}$$

respectively the equivalent bulk modulus $K(\omega)$ in Eq. (14.2) [23]

$$K(\omega) = \frac{\gamma\, p_A}{\varepsilon_p}\left[\gamma - (\gamma - 1)\left(1 + \frac{8\mu}{i\omega\, \Lambda'^2 P_r\rho_f}\sqrt{1 + \frac{i\omega\, \Lambda'^2 P_r\rho_f}{16\mu}}\right)^{-1}\right]^{-1}, \tag{14.2}$$

where τ_∞ denotes the tortuosity factor, ρ_f is the fluid density, ε_p is the porosity, σ is the airflow resistivity, μ is the dynamic viscosity, p_A is the quiescent pressure, γ is the ratio of specific heats, Λ is the viscous characteristic length, Λ' is the thermal characteristic length, and P_r represents the Prandtl number. In addition, the JCAPL model had introduced few corrections to the dynamic density and bulk modulus respectively at low frequencies, through the viscous and thermal behavior, corrections that were not captured by the initial JCA model [23, 25].

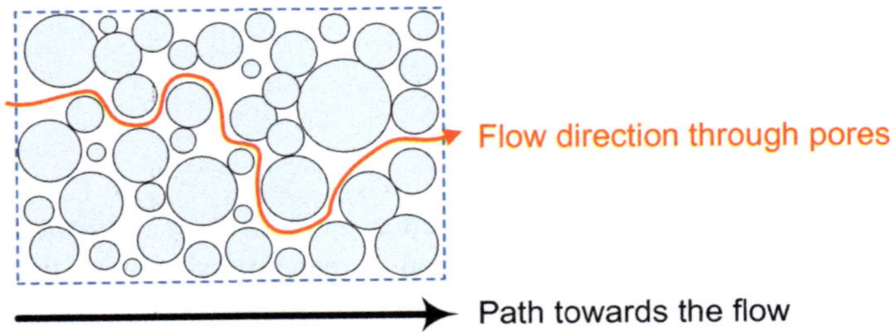

Fig. 14.1 An example of 2D porous structure with tortuosity schematization

14.2.2 Tortuosity Evaluation

Within these phenomenological models, tortuosity denotes the transport property and represents an important non-acoustical parameter that enables the high-frequency acoustical behavior of a porous composite layer. In fact, this parameter accounts for the twistiness of the pores inside the material, being a measure of the fiber deviation from the normal way—please see the planar (2D) schematization in Fig. 14.1.

Thus, the tortuosity parameter can be approximated by the ratio of the actual path (h_e) traveled by a high frequency sound through the porous sample to the sample thickness d, with Eq. (14.3)

$$\tau_\infty \cong \frac{h_e}{d}. \tag{14.3}$$

Within his researches, Allard had assumed that, for high frequencies, the thickness of viscous boundary layer is still considerably smaller than the characteristic pore dimension [23]. Hereby, the phase velocity c_b of the sound wave into the porous material approaches

$$c_b \cong \frac{c_0}{\sqrt{\tau_\infty}}. \tag{14.4}$$

In addition, taken into account the parameter of material porosity, tortuosity can be evaluated using of one of the most known expression (also known as Bruggeman relation) in Eq. (14.5)

$$\tau_\infty = \frac{1}{\sqrt{\varepsilon_p}}, \tag{14.5}$$

which enables the evaluation of porosity based on the tortuosity values, experimentally gained.

14.3 Experimental Investigations and Results

Practically, the tortuosity can be experimentally evaluated by the phase speed of a high frequency wave passing through the sample, comparing the time delay between the incident and the transmitted wave fronts respectively. This supposes the acoustical method, which basically starts from Eq. (14.4).

A simple schematic diagram of this method was depicted in Fig. 14.2. Experimental laboratory setup was presented in Fig. 14.3, where images respectively depict a general view, a detail of sample holder with mounted microphone, and the digital equipment (function waveform generator and DAQ boards system). A high frequency tweeter, mounted on a side of the sample holder tube, was used in order to enable sound waves up to frequency 40 kHz.

The sample holder has to allow synchronous evaluation of incident and transmitted waves nearby the both sides of the inspected material. The setup has to ensure aligned centers of the microphones, sample and wave generator respectively.

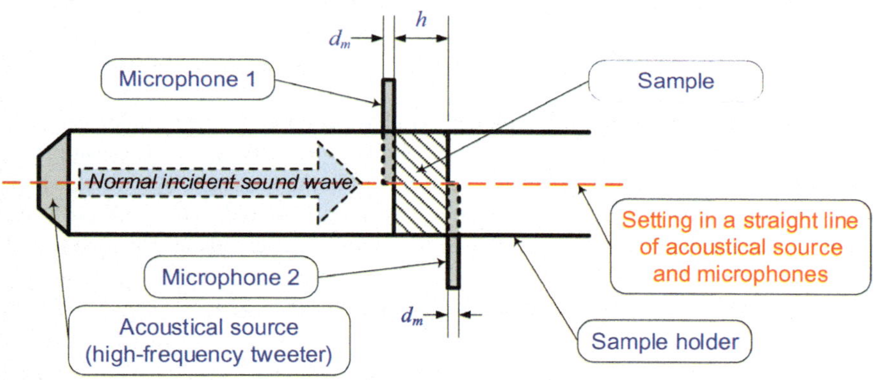

Fig. 14.2 Schematic diagram of the experimental setup for tortuosity evaluation

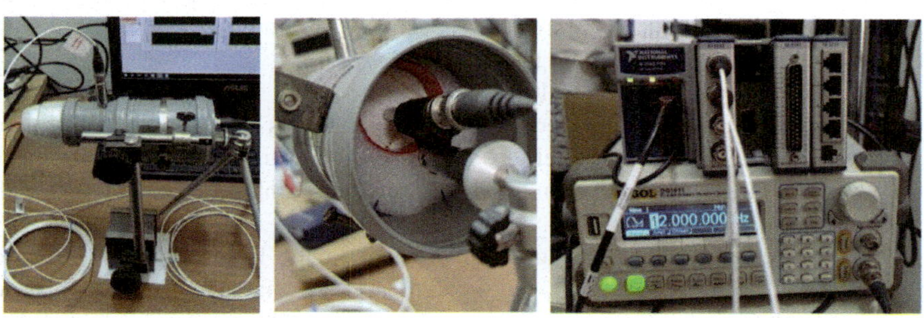

Fig. 14.3 General and detailed views of laboratory setup used for tortuosity evaluation

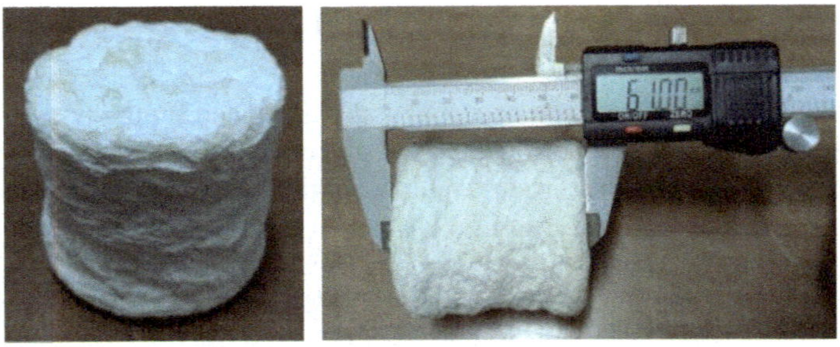

Fig. 14.4 Highly porous composite sample used within the experimental investigations

Taking into account the speed of sound c_0, microphones diameter d_m, sample thickness d and the time delay Δt, experimentally evaluated between incident and transmitted wave fronts, and replacing in Eq. (14.4), the tortuosity results using Eq. (14.6)

$$\tau_\infty = \left(\frac{c_0 \, \Delta t - d_m}{d} \right)^2 . \tag{14.6}$$

During the experimental tests, the authors have used acoustical transducers with $d_m = 6.35 \times 10^{-3}$ m and 25 kHz incident sound wave frequency. The samples of highly porous cellulose-based composite used within the tests have cylindrical shape, with 50 mm diameter. Samples thickness was initially stated to 62 mm, but it was slightly altered during air drying process, thus that it was evaluated for each sample separately (see details in Fig. 14.4). In order to calibrate the experimental setup in respect to each sample, Eq. (14.6) was reconfigured as Eq. (14.7)

$$\sqrt{\tau_\infty} = \frac{d + d_m}{d} \frac{\Delta t}{\Delta t_0} - \frac{d_m}{d}, \tag{14.7}$$

which represents the expression practically used by the authors in tortuosity evaluation, and where Δt and Δt_0 represent respectively the time delays evaluated with and without mounted sample into the holder.

The estimated tortuosity results as an average value for entirely set of tests performed on each sample. Experimentally investigations were conducted on five samples from the same type of cellulose fiber-based porous composite, with thickness into the range [61.0, 62.4] mm, and supposed a number of ten tests for each sample.

Final values of the average tortuosity related to the five samples result into the range [2.099506, 2.426504]. Taking into account that entirely investigation was performed for the same composite structure, the authors have concluded that tortuosity parameter presents average value 2.263005.

14.4 Concluding Remarks

The authors had adopted an acoustical method for tortuosity evaluation of highly porous composites cellulose fiber-based, which presents the main advantage of simplicity for both the setup and the operational procedure. In the same time, this method is very sensitive to the sound wave frequency and to the digital acquisition rate. Nevertheless, this method allows a relative simple way in order to estimate both the tortuosity, and additionally, using Eq. (14.5), the material porosity, entirely based on a unique experimental configuration. Supposing the empirical character of Bruggeman relation, beyond its large scale utilization, it have to be mentioned that porosity values may require a supplementary validation through specific experimental methods, in order to assure high level of accuracy for further simulations.

Taking into account that this kind of porous composite presents a certain novelty character, the literature available datum related to specific parameters sporadically come into sight, having a particularly character according to the material particular structure. Hereby, the future research will be focused on evaluation of specific geometrical and transport properties, in addition to acoustics characteristics, for the foam-formed composites based on cellulose fibers.

During experimental tests it was observed that the second microphone (which was intended to acquire the transmitted wave) presents high sensitivity to surroundings noise, thus that the laboratory investigations have to be performed into an anechoic enclosure (strictly including the experimental setup).

Acknowledgements This research was funded by "Dunarea de Jos" University of Galati, Romania, Grant number RF 9418/2023.

References

1. T. Hjelt et al., Foam forming of fiber products: a review. J. Dispersion Sci. Technol. **43**(10), 1462–1497 (2022)
2. C.N. Wang, J.H. Torng, Experimental study of the absorption characteristics of some porous fibrous materials. Appl. Acoust. **62**, 447–459 (2001)
3. M.J. Valkonen et al., Preparation of fully bio-based sound absorbers from waste wood and pulp fibers by foam forming. BioResources **18**(2), 2657–2669 (2023)
4. A.E. Ketola et al., Changing the structural and mechanical anisotropy of foam-formed cellulose materials by affecting bubble-fiber interaction with surfactant. ACS Appl. Polymer Mater. **4**(10), 7685–7698 (2022)
5. M.D. Stanciu, et al., Research regarding acoustical properties of recycled composites, in *Proceedings of the 8th International DAAAM Baltic Conference Industrial Engineering*, ed. by T. Otto (Tallinn University of Technology, Estonia, 2012), pp. 741–746
6. M.N. Baba, F. Dogaru, M.V. Guiman, Low velocity impact response of laminate rectangular plates made of carbon fiber reinforced plastics. Procedia Manuf. **46**, 95–102 (2020)
7. H. Teodorescu-Draghicescu, et al., Some advanced symmetric composite laminates subjected to off-axis loading systems. A stiffness evaluation, in *Proceedings of the 13th International*

Conference on Modern Technologies, Quality and Innovation ModTech (POLITEHNIUM Publishing House of Technical University "Gheorghe Asachi" Iasi, Romania, 2009), pp. 647–650

8. H. Teodorescu et al., Mechanical behaviour of pre-tensioned glass fiber reinforced composite tubes subjected to internal pressure. WSEAS Trans. Appl. Theor. Mech. **2**(2), 29 (2007)
9. P. Nechita, S. Nastac, Foam-formed cellulose composite materials with potential applications in sound insulation. J. Compos. Mater. **52**(6), 747–754 (2018)
10. P. Nechita, S.M. Nastac, Overview on foam forming cellulose materials for cushioning packaging applications. Polymers **14**(10), 1963 (2022)
11. S. Nastac et al., The acoustic performance of expanded perlite composites reinforced with rapeseed waste and natural polymers. Sustainability **14**(1), 103 (2022)
12. S. Nastac, C. Debeleac, P. Nechita, Assessments on shock absorption properties of foam-formed low density cellulose composites. Acta Technica Napocensis Ser. Appl. Math. Mech. Eng. **60**(4), 565–572 (2017)
13. S. Nastac, et al., On acoustic panels with high performances for pollutant individual working place insulation, in *Proceedings of 8th International DAAAM Baltic Conference Industrial Engineering*, ed. T. Otto (Tallinn University of Technology, Estonia, 2012), pp. 69–73
14. M.D. Stanciu, A. Savin, S.M. Nastac, Mechanical and surface properties of lignocellulosic fibres reinforced composites. Stroj. Vestn **64**, 698–705 (2018)
15. E.M. Samsudin, L.H. Ismail, A.A. Kadir, A review on physical factors influencing absorption performance of fibrous sound absorption material from natural fibers. J. Eng. Appl. Sci. **11**(6), 3703–3711 (2016)
16. H.T. Luu, C. Perrot, R. Panneton, Influence of porosity, fiber radius and fiber orientation on the transport and acoustic properties of random fiber structures. Acta Acust. United AC **103**, 1050–1063 (2017)
17. X. Zhu et al., Recent advances in the sound insulation properties of bio-based materials. BioResources **9**(1), 1764–1786 (2014)
18. L. Egab, X. Wang, M. Fard, Acoustical characterisation of porous sound absorbing materials: a review. Int. J. Vehicle Noise Vibration **10**, 129–149 (2014)
19. K.V. Horoshenkov, A review of acoustical methods for porous material characterisation. IJAV **22**, 92–103 (2017)
20. I.C. Rosca, et al., Wave propagation through ultrasonic horns-Simulation and testing. Ann. Faculty Eng. Hunedoara Int. J. Eng. **14**(4), 33 (2016)
21. J. Cucharero et al., Sound absorption properties of wood-based pulp fibre foams. Cellulose **28**(7), 4267–4279 (2021)
22. K. Dong, X. Wang, J. Yang, Predicting model of sound absorbing properties of cellulose-based porous materials prepared by foam forming. J. Nat. Fibers **19**(14), 9614–9623 (2022)
23. C. Debeleac, P. Nechita, S. Nastac, Computational investigations on soundproof applications of foam-formed cellulose materials. Polymers **11**(7), 1223 (2019)
24. M.E. Delany, E.N. Bazley, Acoustical properties of fibrous absorbent materials. Appl. Acoust. **3**(2), 105–116 (1970)
25. Z. Chen et al., Microstructural characteristics and elastic modulus of porous solids. Acta Mater. Mater. **89**, 268–277 (2015)

Chapter 15
Heart Rate and Surface Electromyography Analysis to Assess Physical Activity Using a Virtual-Reality Exergame

Alin Totorean, Linda Lancere, Brian Horsak, Mark Simonlehner, Dan Ioan Stoia, Mihaela Crisan-Vida, Diogo Moco, Rita Fernandes, Attila Gere, Yasmine Sterckx, Abdul Zulkarnain, Norbert Gal-Nadasan, and Andreea Stoia

Abstract Physical inactivity is considered to be one of the main problems leading to several conditions, such as cardiovascular diseases or diabetes. To combat physical inactivity there have been developed nowadays tools that offer attractive environments and encourage users to practice sport. Virtual Reality is one of the modern tools introducing users in a virtual environment but allowing them to perform physical exercises and playing different sports also at home. The paper presents a case report of a person performing physical exercises (e.g. squats, deadlifts) incorporated in a newly developed VR exergame. The level of physical activity is assessed using heart rate and surface electromyography as relevant physiological parameters. The Positive and Negative Affect Scale questionnaire is used to evaluate the subject's feelings and emotions before and after using the VR exergame.

A. Totorean (✉) · D. I. Stoia · M. Crisan-Vida · N. Gal-Nadasan · A. Stoia
Politehnica University of Timisoara, Piaţa Victoriei No. 2, 300006 Timisoara, Romania
e-mail: alin.totorean@upt.ro

L. Lancere
Vidzeme University of Applied Sciences, Valmiera, Latvia

B. Horsak · M. Simonlehner
St. Polten University of Applied Sciences, St. Pölten, Austria

D. Moco · R. Fernandes
Polytechnic Institute of Setubal, Setúbal, Portugal

A. Gere · A. Zulkarnain
Hungarian University of Agriculture and Life Sciences, Gödöllő, Hungary

Y. Sterckx
UC Leuven-Limburg University of Applied Sciences, Leuven, Belgium

N. Herisanu and V. Marinca (eds.), *Acoustics and Vibration of Mechanical Structures—AVMS-2023*, Springer Proceedings in Physics 302,
https://doi.org/10.1007/978-3-031-48087-4_15

15.1 Introduction

Virtual Reality (VR) became more and more popular worldwide being used nowadays in different fields of activity like medicine [1], sports [2], teaching [3], training [4] or industry [5]. Although most VR-based exergame approaches are designed for a solitary experience without contact with other individuals, there have been recently attempts to developed opportunities for virtual socializing with other VR users.

The use of VR in medical rehabilitation has proven to be efficient, especially for patients with stroke [6], cerebral palsy [7] or patients having the Parkinson disease [8]. Studying the effects of using Virtual Reality environments on human movement in various populations is an ongoing research topic [9, 10].

Physical inactivity and sedentary lifestyle have become a current concern in healthy individuals who do not usually perform physical exercises, leading to important medical conditions, such as diabetes or cardiovascular diseases. As an alternative to traditional sports environments, there have been developed solutions using Virtual Reality tools where users could practice sport and low to high intensity physical exercises. As a solution for sedentary people, VR exergames could be more and more attractive especially because they can be played at home, with no specific requirements except the VR technical setup. To assess the physical exercise level, specific physiological parameters should be measured with high interest on electromyography and heart rate.

The aim of this paper is to present the efficiency of a new VR exergame in case of a sedentary subject by analyzing the electromyography signals and heart rate variation while playing the VR exergame, together with subjective data using the The Positive and Negative Affect Scale (PANAS) questionnaire [11] to assess the feelings and emotions before and after performing the physical exercises.

15.2 Materials and Methods

A Virtual-Reality exergame was designed with the aim of encouraging sedentary individuals to enhance their level of physical activity. The physical exercises were implemented in the exergame according to the medical guidelines and consists in exercises for the upper and lower limbs (e.g., squats, jumps, intensive movements of the upper limbs to destroy bubbles with hands (Fig. 15.1).

The exergame used in this study was developed by Brian Horsak and Mark Simonlehner at the Center for Digital Health & Social Innovation from St. Polten University of Applied Sciences, Austria.

The equipments used in the experimental measurements were:

- A VR Meta Quest 2 headset, with the resolution of 1832×1920 pixels and refresh rate of 120 Hz;
- Four surface electromyography (EMG) sensors (BiosignalPlux) were used. The sensors were wired to a four-channel hub that transmitted the signals to a PC using

Fig. 15.1 Player's view while playing the exergame. The player's hands are tracked and displayed in the VR in real-time. In this game players need to pop soap bubbles that spawn randomly from the bottom of the field of view. The more bubbles are popped the faster and more difficult the game gets

Bluetooth connection. The sampling rate was set to 1000 Hz, and all the EMG signals were acquired simultaneously;
- The Polar belt H10 was used for heart rate (HR) measurement. The belt was placed around the subject's chest using a chest-belt. A dedicated software application was developed and used for HR acquisition. HR data was also sent in real-time to the VR headset, so that the intensity of the VR exergame is adjusted using a bio-feedback algorithm, based on the HR variation.

The experimental setup used in the study is shown in Fig. 15.2.

The subject presented as case report in this paper provided written informed consent and met all the inclusion and exclusion criteria of the study, being sedentary, having the age between 18 and 65 years, the body mass index lower than 29.9 and no medical condition, as shown in Table 15.1.

Four muscle groups were considered for EMG analysis:

- Vastus Medialis;
- Vastus Lateralis;
- Gastrocnemius Medialis;
- Gluteus Medius.

The SENIAM guidelines: Welcome to SEMNIAM were followed in all EMG procedures [12].

The exergame consisted of a warm-up session (5 min) and the physical exercise session (20 min). The EMG and HR measurements were recorded during these sessions.

Objective measurements were performed using the EMG and HR systems to assess the body reaction to different physical exercises. Subjective data was recorded asking the subject to fill the Positive and Negative Affect Scale (PANAS) specific questionnaire to evaluate the feelings and emotions [11] the subject had before and

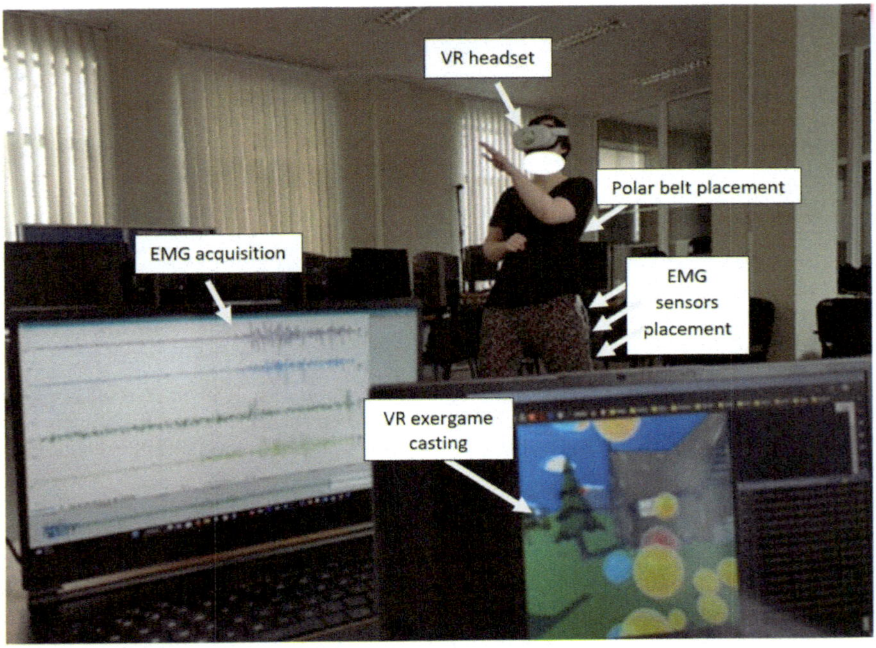

Fig. 15.2 Experimental set up for sensors placement, physical exercise and data acquisition

after the VR session. The results of this questionnaire also give useful information related to the subject's wellbeing status before and after practicing physical exercises, putting in evidence the efficiency of the VR exergame on the wellbeing improvement.

The experimental measurements from this paper were conducted by Linda Lancere in the laboratory of Virtual Reality and Smart Technologies from the Vidzeme University of Applied Sciences, Latvia.

15.3 Results

15.3.1 Surface Electromyography

Surface electromyography (EMG) signals associated to muscle activity between the time-steps 350–446 s of the exergame playing period are shown in Fig. 15.3. The results show successively muscle activation, being more intensive especially for Vastus Medialis and Vastus Lateralis.

The loweest EMG activation signal was associated with the Gastrocnemius Medialis during the analysis period.

Table 15.1 Subject's characteristics related to the inclusion and exclusion criteria

Criteria	Subject's data
Gender	Female
Age	21
Height [cm]	173
Weight [kg]	80.9
Body mass index [kg/m^2]	27
Does the participant have a sedentary lifestyle?	Yes
Does your pain typically exceed 5 points from 10-point scale during exercises, lying on your abdomen, back, side, or in an upright position?	No
When exercising does your pain typically increases by 2 or more points on a 10-point scale?	No
Have you had moderate or severe brain injury?	No
Have you had lower or upper extremity, spinal or other injuries resulting in surgery or limiting your functions longer than 6 weeks or any other injuries that would limit your participation in this study?	No
Have you ever been diagnosed with central nervous system disorder or disease like stroke, cerebral palsy, spinal cord injury or other condition or disease that affects your brain or spinal cord?	No
Are you diagnosed with diabetes mellitus type 1 or type 2 with daily insulin therapy?	No
Have you ever been diagnosed with high pulmonary and/or cardiovascular risk conditions like cardiac stenosis, pulmonary embolism or others?	No
Is the function of your lower and upper extremities limited because of an injury or neurological/musculoskeletal disorder?	No
Do you have a metallic implant or other metallic material containing part after orthopedic or other similar procedures (except dental implant)?	No
Do you have any uncorrected visual or hearing impairments that limits or could limit the ability to see feedback or comply with instructions given during testing?	No
Are you pregnant?	No

15.3.2 Heart Rate

Heart rate variation during the warm-up and VR exergame sessions is shown in Fig. 15.4. HR ranged between 95 and 167 bpm. The highest values of HR are associated to the VR exergame session, indicating that the subject achieved a good level of physical exercise.

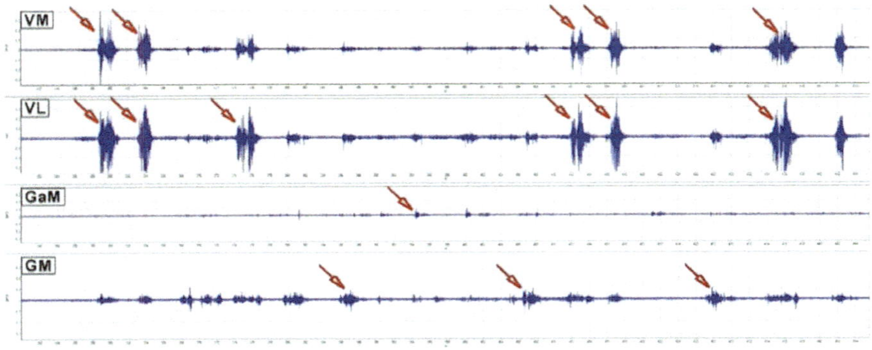

Fig. 15.3 Surface electromyography signals associated to the muscle groups: Vastus Medialis (VM), Vastus Lateralis (VL), Gastrocnemius Medialis (GaM) and Gluteus Medius (GM). X-axis represent the time in seconds, and y-axis represent the electric voltage for muscle activity, measured in mV. Relevant muscle activations are marked with red arrows

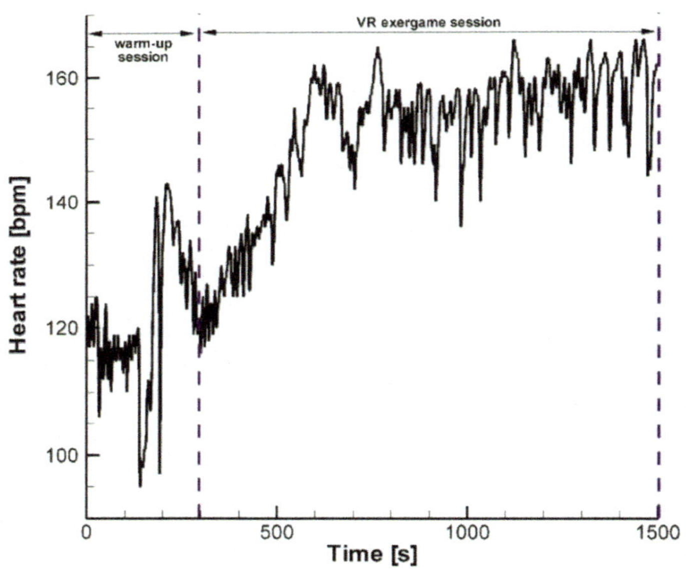

Fig. 15.4 Heart rate variation during the warm-up and VR exergame sessions

15.3.3 Subjective Data: PANAS Questionnaire

The subject was asked to fill the PANAS questionnaire [11] before and after the VR session, marking the feeling and emotions with scores from 1 (very slight or not at all) to 5 (extremely). Higher values for positive score indicate higher level of positive

Table 15.2 PANAS questionnaire before and after the VR session, indicating different feelings, emotions and levels of perception (1—very slight or not at all, 2—a little, 3—moderately, 4—quite a bit, 5—extremely)

Indicate the extent you have felt this way over the past week	Before VR session	After VR session
Interested	3	4
Distressed	4	1
Excited	2	3
Upset	2	1
Strong	2	2
Guilty	3	1
Scared	1	1
Hostile	1	1
Enthusiastic	3	4
Proud	3	4
Irritable	2	1
Alert	2	2
Ashamed	1	1
Inspired	4	3
Nervous	3	2
Determined	3	4
Attentive	3	3
Sum (positive)	27	32
Sum (Negative)	21	13

The sum positive represent the total score for positive affect marked with bold and the sum negative is the total score for negative affect marked with italic

affect, whereas lower negative score is associated to a lower level of negative affect. It can be noticed that the total score for positive affect improved after playing the exergame, from 27 to 32, whereas the negative affect decreased from 21 before the VR session to 13 after the game performance, as shown in Table 15.2.

15.4 Conclusion

Physical inactivity is considered to be one of the main problems leading to several conditions, such as cardiovascular diseases or diabetes. VR solutions are very efficient to combat physical inactivity offering attractive environments and encourage users to practice sport.

The exergame developed in this project allows users to perform physical exercises in accordance with medical guidelines. The level of physical activity increased for the subject presented, as the muscle groups are successively activated and heart rate varies during the exercises from a rest value of 95 bpm to a maximum value of 167 bpm.

The analysis of PANAS score before and after the VR session showed the increase of positive affect and the decrease of negative affect after playing the exergame indicating that this type of exergame could improve subject's wellbeing.

The results presented in this paper are part of a multi-center study that has as main aim the increase of physical activity level in sedentary people using a new VR exergame and analyzing the muscle activity level and heart rate variation during the exergame playing period.

Acknowledgements The work from this paper is supported by the project no. 101004069, Engaged and Entrepreneurial European University as Driver for European Smart and Sustainable Regions, E3UDRES2, co-funded by the Erasmus+ Programme of the European Union.

References

1. O. Stamm, R. Dahms, U. Muller-Werdan, Virtual reality in pain therapy: a requirements analysis for older adults with chronic back pain. J. Neuroeng. Rehabil. **17**, 129 (2020)
2. O.R.L. Farley, K. Spencer, L. Baudinet, Virtual reality in sports coaching, skill acquisition and application to surfing: a review. J. Human Sport Exercise **15**(3), 535–548 (2020)
3. X. Shao, Q. Yuan, D. Qian, Z. Ye, G. Chen, K. le Zhuang, X. Jiang, Y. Jin, D. Qiang, Virtual reality technology for teaching neurosurgery of skull base tumor. Med. Educ. **20**, 3 (2020)
4. K.R. Vaghela, A. Trockels, J. Lee, K. Akhtar, Is the virtual reality fundamentals of arthroscopic surgery training program a valid platform for resident arthroscopy training? Clin. Orthop. Relat. Res. **480**, 807–815 (2022)
5. S. Choi, K. Jung, S.D. Noh, Virtual reality applications in manufacturing industries: past research, present findings, and future directions. Concurr. Eng. **23**(1), 40–63 (2015)
6. M.E. Moan, E.K. Vonstad, X. Su, B. Vereijken, M. Solbjor, N. Skjaeret-Maroni, Experiences of stroke survivors and clinicians with a fully immersive virtual reality treadmill exergame for stroke rehabilitation: a qualitative pilot study. Front. Aging Neurosci. **13**, 735251 (2021)
7. J.M. de Oliveira, R.C.G. Fernandes, C.S. Pinto, P.R. Pinheiro, S. Ribeiro, V.H.C. de Albuquerque, Novel virtual environment for alternative treatment of children with cerebral palsy. Comput. Intell. Neurosci. **2016**, e8984379 (2016)
8. C.-S. Chuang, Y.-W. Chen, B.-Y. Zeng, C.-M. Hung, Y.-K. Tu, Y.-C. Tai, Y.-C. Wu, C.-W. Hsu, W.-T. Lei, S.-L. Wu, B. Stubbs, A.F. Carvalho, K.-P. Su, T.-Y. Chen, P.-T. Tseng, M.-K. Wu, C.-H. Tsai, Effects of modern technology (exergame and virtual reality) assisted rehabilitation vs conventional rehabilitation in patients with Parkinson's disease: a network meta-analysis of randomised controlled trials. Physiotherapy **117**, 35–42 (2022)
9. B. Horsak, M. Simonlehner, L. Shoffer, B. Dumpart, A. Jalaeefar, M. Husinsky, Overground walking in a fully immersive virtual reality: a comprehensive study on the effects on full-body walking biomechanics. Front. Bioeng. Biotechnol. **9**, 780314 (2021)
10. E.-C. Park, S.-G. Kim, C.-W. Lee, The effects of virtual reality game exercise on balance and gait of the elderly. J. Phys. Ther. Sci. **27**, 1157–1159 (2015)
11. D. Watson, L.A. Clark, A. Tellegen, Development and validation of brief measures of positive and negative affect: the PANAS scales. J. Pers. Soc. Psychol. **54**(6), 1063 (1988)
12. H.J. Hermens, B. Freriks, R. Merletti, D. Stegeman, J. Blok, G. Rau, C. Disselhorst-Klug, G. Hägg, European recommendations for surface electromyography. Roessingh Res. Dev. **8**(2), 13–54 (1999)

Chapter 16
Upper Limb Motion Analysis During a Virtual Reality Fitness Game

Dan Ioan Stoia, Mihaela Galescu-Blidariu, Andreea Stoia, and Alin Totorean

Abstract Virtual reality (VR) games became more popular everyday due to the realistic and engaging experience of the users. The immersive environment of the game allows players to interact with it using their own body movements and therefore creating a deep feeling of being part of the game. The virtual interactive world creates a very good scenario for homme training, rehabilitation and even sport. In order to quantify the intensity of the exercises, the body kinematics and kinetics during VR game, motion analysis systems can be used to track the body movements and muscle activation. The purpose of this study was to determine the accelerations of the dominant upper limb and the muscle activations during a VR fitness game exercise. The game is focused on the activity of the upper limb and comprises of objects punching, uppercutting, squatting and alternating combination of those. The results are estimating quantitatively the body dynamics required for accomplishing a good score in the game.

16.1 Introduction

Virtual reality (VR) is a simulated environment that allow users to interact with the game or exercise using their own body. The technology emerges due to the progress in computational speed and memory, graphics power, physical tracking, interaction

D. I. Stoia (✉) · M. Galescu-Blidariu · A. Stoia · A. Totorean
Politehnica University Timisoara, Piaţa Victoriei Nr. 2, 300006 Timişoara, Jud. Timiş, România
e-mail: dan.stoia@upt.ro

M. Galescu-Blidariu
e-mail: mihaela.galescu@yahoo.ro

A. Stoia
e-mail: andreea.stoia2@upt.ro

A. Totorean
e-mail: alin.totorean@upt.ro

N. Herisanu and V. Marinca (eds.), *Acoustics and Vibration of Mechanical Structures—AVMS-2023*, Springer Proceedings in Physics 302,
https://doi.org/10.1007/978-3-031-48087-4_16

technologies (various types of sensors) and artificial intelligence. All these developments have facilitated affordable virtual reality setups that can operate on a standard personal computer [1]. The VR technology consist of highly interactive computer-generated simulations that monitor the user's movements and substitute or add feedbacks to sensors in order to produce the feeling of complete involvement of the user within the simulation [2]. As consequence, playing games in VR will lead to an increase level in dopamine due to bold colors, risk-taking, reward-anticipation, and exploratory behavior mechanisms [3]. This technology helps to prevent the negative effects caused by excessive digital gaming (recognized as a disorder by the World Health Organization). Prolonged periods of sedentary behavior in front of screens have been linked to higher incidences of obesity and hypertension, both being significant risk factors for cardiovascular morbidity [4].

Over the years, the VR technology has developed several fitness games in order to increase the physical activity at home, without the need of the gym or practical courses. Among other things, a regular fitness routine decreases risk of high blood pressure, encourages retention of muscle mass and it is known for its psychological benefits such as reducing and managing stress, anxiety and also depression. Favorable results have been seen by using VR-based treatments for various disorders with significant efficacy in treating fears, weight issues, persistent discomfort, and eating/drinking [5–7].

Assessing the kinematic, kinetic and electromyographic parameters associated with human movements were extensively studied since the developments of motion analysis systems. The studies showed the capacity of motion analysis to give results in both prevention and rehabilitation of various illness [8–10]. However fewer studies involve VR as stimulation environment for motion analysis [11–13].

The purpose of this study was to quantify the movement of the dominant upper limb during a VR game exercise, in order to identify the effect of the VR on the acceleration and muscle activation of the upper limb.

16.2 Materials and Methods

The equipment involved in the study were the VR Meta Quest 2 (Oculus Quest 2) headset having an image resolution of 1832×1920 pixels and a refresh rate of 72 Hz and the MuscleBAN wearable devices (Fig. 16.1).

The MuscleBAN measures the electromyography (EMG) in the range of $\pm$ 1.14 mV, acceleration in 3 directions (ACC) in the range of $\pm$ 4 g and the magnetic field (MAG) in the range of $\pm$ 1200 μT. The sampling rate of the device can be set between 1 and 1000 Hz, and in this case was to 100 Hz. The uncertainty of measurement of this device is not provided by the producer nevertheless it will be determined using laboratory measurements.

The subject involved in the study was a 21 years Female, 1.74 m height, 67.7 kg and 22.9 body mass index, athletic constitution. She engaged in VR game 3 times

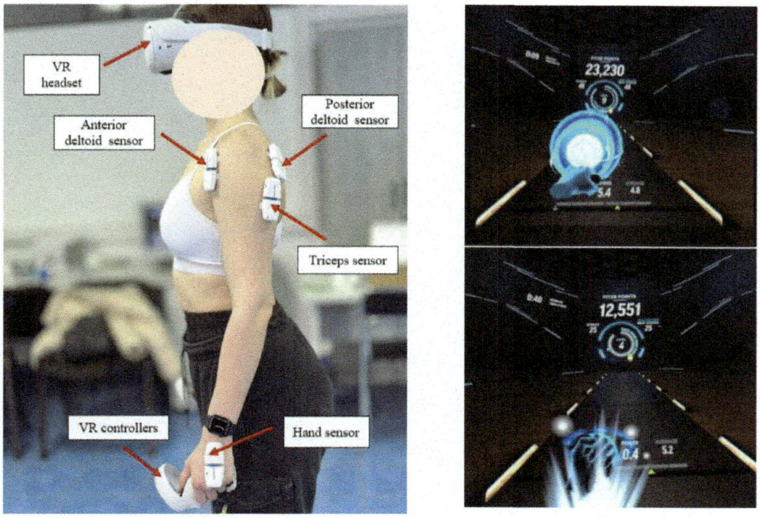

(a) (b)

Fig. 16.1 Sensor positioning (**a**) and game action (**b**)

and prior of that in 10 min warm-up training. The accommodation with the virtual reality environment was done through a fitness round of two minutes.

Four sensors were used in order to record EMG and acceleration signals. The placement of the sensors was: on the back of the hand, on the triceps muscle, on anterior deltoid muscle and on posterior deltoid muscle, as they can be observed in the Fig. 1a.

The VR game consist of several type of different movements: punching (four left hand followed by four right hand), uppercutting (two left had followed by two right hand), squatting (four times followed by a combination of right-left punches and uppercuts) and alternating combinations of those (Fig. 1b).

16.3 Results and Discussions

The electromyographic activity and acceleration of the hand were recorded for the entire VR game exercise. The EMG signal was processed using OpenSignals software for identifying the muscle activations. Both amplitude (mV) and duration (s) of every muscle activation were recorded. The low pass filtering of the raw signal allows for clear identification of activation peaks that can be observed in the Fig. 16.2.

The channels 2, 3 and 4 are accelerations on X, Y and Z axes and while channel 1 is the EMG signal. High acceleration and deceleration amplitudes can be observed for the hand in X axis, this axis corresponding to the horizontal direction (sagittal plane).

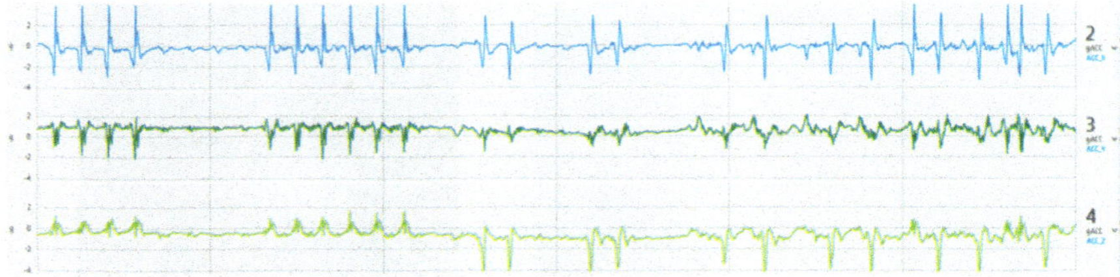

(a) Back of the hand

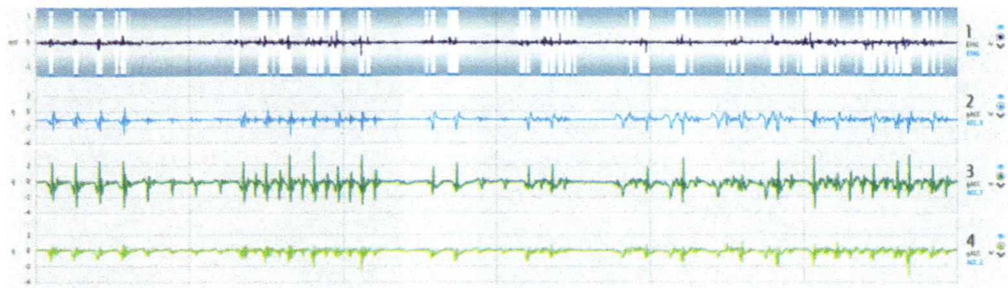

(b) Triceps muscle

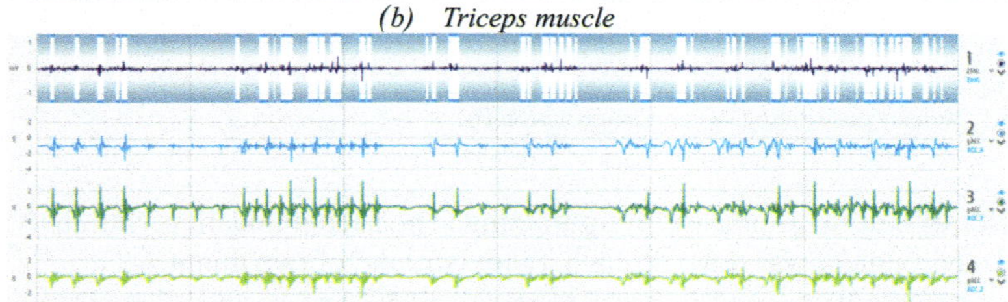

(c) Anterior deltoid muscle

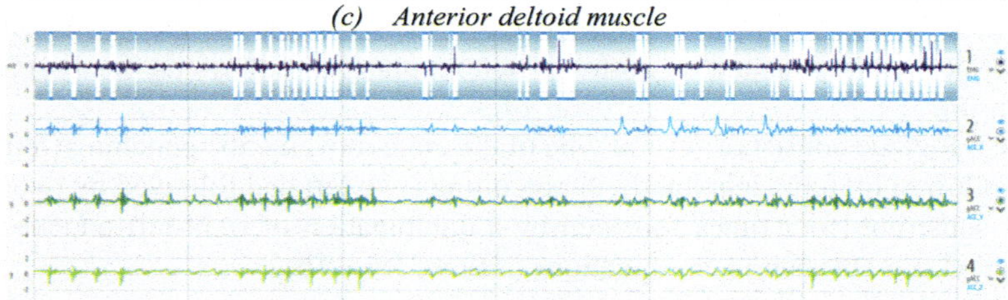

(d) Posterior deltoid muscle

Fig. 16.2 Time series of electromyographic and acceleration signals acording to sensor placement

In the same time, the acceleration peaks are sustained by the high activation of anterior and posterior deltoid muscles, as it can be observed in Fig. 16.2c, d. The accelerations at the shoulder level are much lower than for the hand because the movement amplitude is much more reduced in this site. The Z direction of acceleration corresponds to the movements in transversal plane of the body which are rather reduced.

In the case of uppercutting movements, the EMG signals are higher in amplitude than for punching and the acceleration larger, leading us to the conclusion that VR stimulation on this case produces higher effects on dynamics of the hand.

In Fig. 16.3 maximum accelerations of the hand, triceps and anterior and posterior deltoids are presented in accordance to the axis of recording. The positive values (orange boxes) are the accelerations when executing the movements while the negative accelerations (blue boxes) are the accelerations corresponding to the returning movement. It can be identified that, in order to execute the next movement (keeping the tempo of the VR game), the body uses almost the same dynamic while returning.

Overall, the results obtained by our measurements are consistent with the literature. Similar kind of surface electromyography sensors and experimental procedures were used by many authors [14–17] for determining the electric response of various body segments. Nevertheless, none of them use the EMG and accelerometer sensors on human subjects acting in VR.

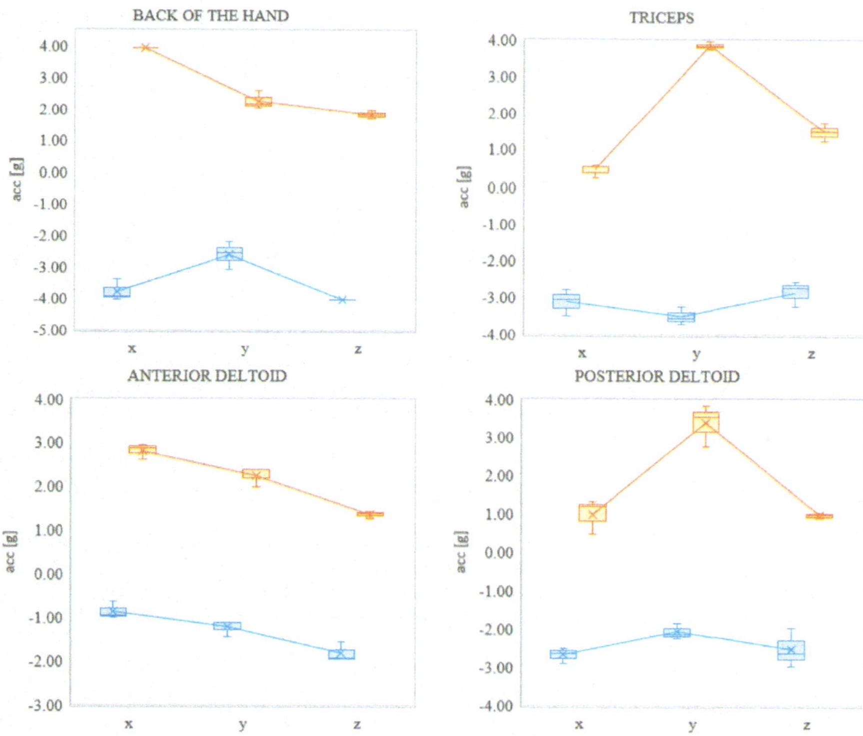

Fig. 16.3 Accelerations on hand, triceps, anterior deltoid and posterior deltoid according to the sensor axis

16.4 Conclusions

The paper presents the movement parameters of the dominant upper limb during a VR game exercise that implies punching, uppercutting, squatting and combination of these. The recorded parameters were the accelerations and the electromyographic signals for hand, triceps muscle, anterior deltoid muscle and posterior deltoid muscle.

The results reveal in a quantitative manner the effect of virtual reality environment on the acceleration of the upper limb and electromyographic impulse of the muscles.

The best muscle activation and accelerations were recorder for the uppercutting movements rather than punching. Another observation was that in order to execute the next movement, for keeping the tempo of the VR game, the body uses almost the same dynamic in returning that in determinate action, so the training effect is double.

So, the motion analysis can be used as a tool for measuring the effect of the VR environment in order to design or optimize virtual fitness games for a better fitting to a certain group of subjects or a certain exercise.

References

1. A. Rizzo, B. Lange, E. Suma, M. Bolas, Virtual reality and interactive digital game technology: new tools to address obesity and diabetes. J. Diabetes Sci. Technol. **5**(2), 256–264 (2011)
2. S. Ali, X. Wang, P. Li, Y. Jung, L. Bi, J. Kim, Y. Chen, D. Feng, N. Magnenat Thalmann, J. Wang, B. Sheng, A systematic review: virtual-reality-based techniques for human exercises and health improvement. Front. Public Health **11**, 1143947, 01–14 (2023)
3. E. Matar, J.M. Shine, S.L. Naismith, S.J. Lewis, Virtual reality walking and dopamine: opening new doorways to understanding freezing of gait in Parkinson's disease. J. Neurol. Sci. **344**(1–2), 182–185 (2014)
4. B. Feodoroff, I. Konstantinidis, I. Froböse, Effects of full body exergaming in virtual reality on cardiovascular and muscular parameters: cross-sectional experiment. JMIR Ser. Games **7**(3), e-12324 (2019)
5. N. Zeng, Z. Pope, Z. Gao, Virtual reality exercise for anxiety and depression: a preliminary review of current research in an emerging field. J. Clin. Med. **7**(3), 42 (2018)
6. F.A. Djawas, V.R. Prasasti, R. Pahlawi, M. Noviana, A.D. Pratama, The Effectiveness of virtual reality exercises to reduce fall risk in Parkinson's disease: a literature review. Proceedings **83**, 33 (2022)
7. B. Brea-Gómez, I. Torres-Sánchez, A. Ortiz-Rubio, A. Calvache-Mateo, I. Cabrera-Martos, L. López-López, M.C. Valenza, Virtual reality in the treatment of adults with chronic low back pain: a systematic review and meta-analysis of randomized clinical trials. Int J. Environ. Res. Public Health **18**(22) (2021)
8. L. Lancere, Technological solutions for low back pain physical therapy real-time monitoring with feedback. Baltic J. Mod. Comput. **9**(4), 436–452 (2021)
9. D.I. Stoia, M. Toth-Tascau, Influence of treadmill velocity on joint angles of lower limbs during human gait, in *2011 E-Health and Bioengineering Conference (EHB)* (Iasi, Romania, 2011), pp. 1–4
10. N.J. Jarque-Bou, A. Scano, M. Atzori et al., Kinematic synergies of hand grasps: a comprehensive study on a large publicly available dataset. J NeuroEngineering Rehabil **16**, 63 (2019)
11. L. Lancere, M. Jürgen, H. Gapeyeva, Mixed reality and sensor real-time feedback to increase muscle engagement during deep core exercising. Virtual Real. (2023)

12. M. Żuk, M. Wojtków, M. Popek, J. Mazur, K. Bulińska. Three-dimensional gait analysis using a virtual reality tracking system. Measurement **188** (2022)
13. R. Maskeliūnas, R. Damaševičius, T. Blažauskas, C. Canbulut, A. Adomavičienė, J. Griškevičius, BiomacVR: a virtual reality-based system for precise human posture and motion analysis in rehabilitation exercises using depth sensors. Electronics **12**, 339 (2023)
14. K. Rogers, A. Caravan, J. Scheffey, K. Boddy, Surface electromyographic analysis of the serratus anterior during 1 bench press variations (2019)
15. K. Lee, S. Kim, H.O. Choi, J. Lee, Y. Nam, Analyzing electrocardiogram signals obtained from a nymi band to detect atrial fibrillation. Multimedia Tools Appl. **79**(23), 15985–15999 (2020)
16. L. Lancere, Technological solutions for low back pain physical therapy real-time monitoring with feedback. Baltic J. Mod. Comput. **9**(4), 436–452 (2021)
17. Internet resource accessed on 05.06.2023. https://milefysiotherapie.nl/meet-en-trainingsapparatuur/physioplux-muscleban/

Chapter 17
Jitter and Shimmer Parameters in the Identification of Vocal Tract Pathologies. Current State of Research

Laura Jurca and Cosmina Vigaru

Abstract Even though parts of a speech signal might seem stationary, there are always small fluctuations in it, as vocal fold oscillation is not exactly periodic. These fluctuations in frequency and amplitude outline two of the most important parameters that characterize the voice signal, namely jitter and shimmer. Both features characterize different aspects concerning particular types of voices and are largely used to describe pathological voices. This paper aims to emphasize the significance of the jitter and shimmer parameters in the identification of the pathologies and disorders that can occur in the vocal tract. The focus of this study is on the previous works developed by different researchers over the world, on the analysis of voiced sounds with the help of jitter and shimmer parameters.

17.1 Introduction

Throughout time, many clinicians and scientists have developed numerous methods for the analysis of voiced sounds, in a way that they could find, understand and describe the characteristics of normal voices. Although it was a challenging, it further lead to the assessment of the pathological field regarding the vocal tract.

For the assessment of the pathologies of the vocal tract, over the years were used various techniques, from listening to a patient's voice production by an expert jury and using computerized software which requires acoustic, aerodynamic measures through complex medical equipment like laryngoscopy, stroboscopy, electroglottography, etc. to the development of acoustic analysis programs similar to audiometric testing in schools and industry [1].

Among the most important characteristics in describing voice production and pathology detection are jitter and shimmer parameters, which have been extensively used for the identification of voice pathologies as they are caused by irregular vocal

L. Jurca (✉) · C. Vigaru
University Politehnica Timisoara, 300222 Timisoara, Romania
e-mail: maria.jurca@upt.ro

N. Herisanu and V. Marinca (eds.), *Acoustics and Vibration of Mechanical Structures—AVMS-2023*, Springer Proceedings in Physics 302,
https://doi.org/10.1007/978-3-031-48087-4_17

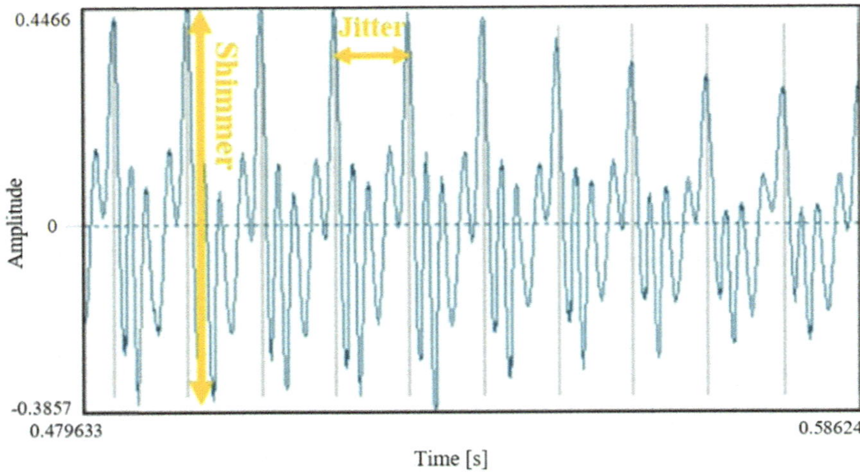

Fig. 17.1 Jitter and shimmer variations for a recorded sound

fold vibration [2]. Even though significant levels of these two parameters appear in voice pathologies, natural speech also contains some level of jitter and shimmer.

The jitter parameter refers to variations in signal frequency, representing the measure of cycle-to-cycle variations of the fundamental glottal period, whereas shimmer represents variations in signal amplitude, respectively the cycle-to-cycle variations of the fundamental glottal period amplitudes [3], as depicted in Fig. 17.1. The jitter is primarily affected by the lack of control of the vocal cords' vibration. The shimmer varies with the reduced glottal resistance and mass lesions on the vocal cords [4]. High values of jitter and shimmer are often related to pathological voices. Both jitter and shimmer are most clearly detected from long, sustained vowels [2].

Most commonly used jitter values can be determined by the average absolute difference between consecutive periods and is known as absolute jitter. There is also a relative jitter that can be obtained by dividing the absolute jitter by the average period. The absolute and relative jitter cand be determined using the following equations:

$$Jitter(absolute) = \frac{1}{N-1}\sum_{i=1}^{N-1} \|T_i - T_{i+1}\| \tag{17.1}$$

$$Jitter(relative) = \frac{\frac{1}{N-1}\sum_{i=1}^{N-1} \|T_i - T_{i+1}\|}{\frac{1}{N}\sum_{i=1}^{N-1} T_i} \tag{17.2}$$

where T_i are the extracted F_0 period lengths and N is the number of extracted F_0 periods [2]. Shimmer parameter can also be divided into absolute and relative shimmer and can be obtained with the help of the equations:

$$Shimmer(dB) = \frac{1}{N-1}\sum_{i=1}^{N-1} \|20\log(A_{i+1}/A_i)\| \quad (17.3)$$

$$Shimmer(relative) = \frac{\frac{1}{N-1}\sum_{i=1}^{N-1}\|A_i - A_{i+1}\|}{\frac{1}{N}\sum_{i=1}^{N-1} A_i} \quad (17.4)$$

where A_i are the extracted peak-to-peak amplitude data and N is the number of extracted fundamental frequency periods.

Absolute shimmer refers to the average absolute base—10 logarithm of the difference between the amplitudes of consecutive periods multiplied by 20 and relative shimmer represents the average absolute difference between the amplitudes of consecutive periods divided by the average amplitude [2].

17.2 Current State of Research

The studies included in this paper were selected based on the selection of participants and study methods, as each participant from selected groups (groups with no known phonetic disorders in contrast with groups with different vocal disorders) was asked to do the same action, namely to phonate sustained vowels for which Jitter and Shimmer were identified. Other research studies were not taken into consideration as a lot more parameters were considered and they did not present conclusive results regarding Jitter and Shimmer parameters.

Detecting various pathologies of the voice has been a constant pursuit for scientists and doctors all over the world. This is why the research done so far on the variations in frequency (jitter) and amplitude (shimmer) of the vocal fold vibrations is of the utmost importance for the evaluation of the speech mechanism, quality and detection of possible voice disorders.

Many different studies have been developed, with reference to jitter and shimmer, based on multiple papers of different authors some of which reported that shimmer is a more accurate index of perceived vocal noise than jitter, or suggested that shimmer is better discriminator of pathology than jitter, but also including the fact that jitter displayed significantly different results during morning, afternoon and evening, etc. [5].

One important approach regarding jitter and shimmer parameters in detecting voice pathologies was introduced by Parley W. Newman, Richard W. Harris and Laurence M. Hilton. Their study, "Vocal jitter and shimmer in stuttering", aims to test and highlight the presence of significantly different jitter and shimmer during vowel production in the fluent phonations of stutterers compared to nonstuttering matched controls [6].

The tests were run on two groups of subjects matched according to sex and age, stutterers and nonstutterers. The groups consisted in 14 people each (2 females and 12 males), with a mean age of 24 years. Subjects were instructed to phonate and sustain each vowel as steadily as possible for at least 5 s at a comfortable loudness level [5] so that the scientists could obtain jitter and shimmer measures. The recording of the vowels was made through a microphone and measures for jitter and shimmer were obtained through a miniature accelerometer linked to a vibro-meter [6].

The recorded vowels (i, e, u, a) were produced in the same sequence by every subject in a random order and repeated 9 times each. The analysis of jitter (in percent) and shimmer (in dB) parameters of the recorded sounds was made by a computer program named SEARP, using just the mid 3 segments of signals of the recordings.

For stutterers the mean jitter for all vowels was 0.625% with a standard deviation of 0.260. For nonstutterers the mean jitter for all vowels was 0.525% with a standard deviation of 0.185. For each vowel the mean percent of jitter was larger for the stutterers than the nonstutterers. The same was true for standard deviations [6].

The mean shimmer measure for stutterers for all vowels combined was 0.269 dB with a standard deviation of 0.114. The mean shimmer value for nonstutterers was 0.173 dB with a standard deviation of 0.08. For each vowel, the mean shimmer measures were larger for stutterers than nonstutterers. The same was true for standard deviations.

Even though in jitter they did not find the differences to be statistically significant, the important difference discovered between stutterers and nonstutterers was on measures of shimmer. The significant differences between the two groups of subjects can be observed in the determined values for means and standard deviations of jitter and shimmer that are presented in Table 17.1 [6].

The sustained phonation has been proven to be less stable in terms of frequency and intensity for the group of stutterers, as both jitter and shimmer means were higher than the ones for the group of nonstutterers. These results also show that the direction of this difference suggests that stutterers have less stable neuromuscular control over the events regulating the aerodynamics of the laryngeal and respiratory systems during sustained fluent vowel articulations than nonstutterers [6].

Another paper of interest in the field voice pathology detection with the help of jitter and shimmer parameters is [5]. Here, Hwa-Young Pyo and Hyun-Sub Sim, also used a microphone and a computer program for acoustic analysis to record a

Table 17.1 Jitter and shimmer means and standard deviations for stutterers and nonstutterers

	Jitter [%]				Shimmer [dB]			
	Stutterers		Nonstutterers		Stutterers		Nonstutterers	
	Means	Standard deviation	Means	Standard deviation	Means	Standard deviation	Means	Standard deviation
u	0.608	0.211	0.528	0.169	0.269	0.140	0.165	0.074
a	0.612	0.243	0.523	0.199	0.277	0.119	0.177	0.079
i	0.636	0.253	0.521	0.183	0.636	0.090	0.165	0.080
e	0.644	0.322	0.526	0.188	0.270	0.106	0.186	0.089
/u, a, i, e/	0.625	0.260	0.525	0.185	0.269	0.114	0.173	0.080

group of Korean female subjects in order to investigate how shimmer changes with the variation of the fundamental frequency F_0. The /a/, /i/ and /u/ vowels have been chosen to be voiced and sustained by the subjects, with a comfortable loudness and pitch.

Both jitter and shimmer parameters had a direct link with F_0 variation in the same way. As F_0 increased, the two parameters decreased. Even though the mean of jitter also decreased as F_0 increased like that of shimmer, most of subjects with the exception of 2 persons did not follow the pattern. So, the authors focused on shimmer [5].

On the whole, from the 60 subjects, 30 showed that the order of F_0 increase is the same with that of shimmer decrease when the link between F_0 and shimmer is taken into consideration. From the 30 subjects, 28 of them had a lower fundamental frequency for the vowel /a/ than /i/ and /u/, which led to a higher shimmer for the vowel /a/ than for the other two vowels. For the other two subjects, the fundamental frequency for the vowel /u/ was found in between the fundamental frequencies of the other two vowels.

In contrast, the case in which F_0 increases and shimmer also increases could only be found in 3 of the 60 subjects. These were the cases in which the fundamental frequency for the vowel /u/ was lower than for the vowels /i/ and /a/ in this particular order, F_0 for /u/ was lower than for the vowels /i/ and /a/ and F_0 for /i/ was lower than for the vowels /a/ and /u/.

Furthermore, there was a number of 27 cases for which the scientists could not find any pattern in the relationships between F_0 and shimmer change.

After also reviewing the results of the analysis of each vowel separately, Hwa-Young Pyo and Hyun-Sub Sim stated that as F_0 increases, both jitter and shimmer decrease, leading to the conclusion that increasing the fundamental frequency, voice quality can be improved.

Other researchers, such as Haydee F. Wertzner, Solange Schreiber and Luciana Amaro have taken into consideration the topic of jitter and shimmer parameters in children with phonological disorders. The objective of the study was to verify vocal characteristics related to the intensity and fundamental frequency F_0 and their disturbance indexes—jitter and shimmer—in children with phonological disorders [7]. The study was conducted on 40 children, divided in groups of 20 (one group of children without disorders—control group GC—and another one of children with pathology—group with phonological disorder).

For the recording and acoustic analysis (through vocal parameters such as jitter and shimmer) of the vowels /a/, /e/ and /i/ was used the Computer Speech Lab program. Subjects remained seated and were instructed to produce vowels at comfortable intensity and height [7], three times each vowel. To properly analyze the recording and neutralize the effect of vocal attach, they discarded the first part of the recording.

To shimmer and the level of significance of 5%, researchers of this paper concluded that there was no evidence that the means of GTF were different from GC ($p = 0.191$). For jitter, the test did not show differences between GC and GTF ($p = 0.865$) [7].

The comparison between the control group GC and the group with phonological disorders GTF, in terms of jitter and shimmer showed no statistically significant differences. As far as each separate vowel is concerned, the vowel /i/ had a mean of jitter lower than for the other two vowels and vowel /a/ had a mean of shimmer higher than for vowels /i/ and /e/.

In [8], M. Farru´s and J. Hernando use jitter and shimmer parameters in a speaker verification and identification experiments, as both parameters characterize aspects concerning particular or pathological voices.

In contrast with many other studies that analyze only sustained vowel in order to determine jitter and shimmer, authors in [8] use all the voiced speech segments in their current experiments. Praat voice analysis is the software from which jitter and shimmer parameters were extracted. This program reports different kinds of measurements for jitter and shimmer [8]. For jitter, these measurements are absolute and relative jitter, jitter (rap) that represents the Relative Average Perturbation, and jitter (ppq5), the five-point Period Perturbation Quotient. For shimmer authors use shimmer (dB) and relative shimmer, along with shimmer (apq3) which represents the three-point Amplitude Perturbation Quotient, shimmer (apq5), the five-point Amplitude Perturbation Quotient and the 11-point Amplitude Perturbation Quotient, shimmer (apq11).

A nine-feature vector was extracted for an acoustic system based on the nine jitter and shimmer measurements described in the previous paragraph [8], that were extracted with Praat software.

In the verification experiments, M. Farru´s and J. Hernando, state that the absolute measurements for both jitter and shimmer can be useful in recognition of the speaker and that the only measurements that give a reliable result are the absolute measurements of jitter and shimmer, but also the apq3 shimmer.

The identification experiments of paper [8] are based on the recording of sets of 100 and 50 speaker, all gender balanced, have led to the same result as the verification experiments, respectively, that the absolute values of jitter and shimmer are the most reliable when it comes to the analysis of the voice.

In the study "Algorithm for jitter and shimmer measurement in pathologic voices" [9], researchers have used the Saarbrucken Voice Database, that contains recorded segment of speech from people with different pathologies of the voice and also healthy voices. From these recordings, the authors extracted 9 record files of the vowels /a/, /i/ and /u/ for each of the 14 people that had Reinke's Edema.

These recordings were used to create an algorithm to determine the jitter and shimmer parameters according to the equation of each jitter and shimmer measure.

This could be done only after the fundamental issue of the algorithm was solved, namely to find the maximum value and the positions of the glottal periods [9].

To determine the accuracy of the algorithm, the results of a synthesized signal using a 4 formants model (with predefined jitter and shimmer values) [9] were compared to analytical calculations using the equations for both parameters or to values extracted from Praat software.

The algorithm showed reliable results in comparison to Praat and analytical calculations and is now able to deal with almost any damaged voice, with the exception of severely damaged ones.

In the study of Lathadevi, Suresh Pundalikappa Guggarigoudar, a comparison in terms of voice parameters was made between normal and abnormal voices. The study was done on a group of 43 people with healthy voices and another 43 people between 18 and 55 years presenting with hoarseness and lesions of larynx like vocal nodules, vocal polyps, chronic laryngitis, voice abuse, early glottis carcinoma etc., were selected during the study [1]. The patients were asked to phonate the vowel /a/ for a few seconds and the parameters were analyzed with the Praat software.

The parameters determined were compared to the values of the normal voice database of the Department of ENT, BLDE (Deemed to be University) Shri BM Patil Medical College, Hospital and Research Centre, Vijayapura, Karnataka, India.

The comparison of the parameters between normal and abnormal male voices showed that shimmer values were different in relation to a level of significance of $p < 0.05$, but not significant (p was 0.414), but the value of jitter was of great importance (p-value was 0.026). For the comparison between female normal and abnormal voices, the parameters are the opposite of the ones for males, meaning that shimmer parameter had significant values (p value was 0.035).

The method of comparison can provide a non-invasive and objective tool to identify and document abnormal voices [1], but authors state that it would be recommended for each region to have voice samples as both human characteristics and programs differ from one to another.

Regarding the analysis of the voice with the help of jitter and shimmer, other researchers from India, along with researchers from Denmark have developed a paper in which they used a database of 17 subjects who had stuttering disorder and 25 nonstutterers. The database was collected in the Marathi language and jitter and shimmer parameters were extracted from Praat scripts.

The authors determined that value for the mean fundamental frequency for the stutterers was 344 Hz, whereas for the nonstutterers was determined to be 254 Hz.

After making the comparison between the two groups, stutterers and nonstutterers, in respect to fundamental frequency, Jitter and Shimmer, it led to the fact that the major difference between the stutterers and nonstutterers was found in the value of absolute jitter, 8.7632% in stuttered speech, in comparison to 0.0001% in nonstutterer speech. For each word, the mean percentage of jitter was larger for the stuttered than non-stuttered [4]. Shimmer values were not of great importance, as the difference between the values was small, 0.3819 dB for stutterers and 0.2787 dB for nonstutterers.

Another important approach regarding voice analysis through acoustic parameters for the evaluation of voice quality in patients with voice disorders such as laryngitis, laryngospasm, laryngeal tumors, spasmodic dysphonia, and vocal cord paralysis [10] involves 4 groups of subjects, divided as follows: one group formed of 29 subjects with no known pathology in respect to the vocal tract and 3 groups of patients with pathologies that were separated into one group of 55 patients with vocal cord polyps (VCP), another 35 patients for the early-stage laryngeal carcinoma (ELC) and one group consisting in 17 patients with mutational falsetto (MF).

The participants were asked to emit a sustained vowel /a/, starting with a steady pronunciation for 0.5–1 s. Subsequently, the participants were asked to abruptly raise

Table 17.2 Summary table

Paper	Phonetic disorders	Methods	Parameters of interest	Conclusions
[1]	• Hoarseness • Lesions of larynx	• Abnormal voices recorded • Normal voices collected from database • Parameters extracted from Praat	• Jitter • Shimmer	• Characterization of voice into either normal or abnormal voices
[4]	• Stuttering	• Recording of stutterers and non-stutterers • Parameters extracted from Praat	• Jitter • Shimmer • Fundamental frequency	• Difference between Stuttered and Non-Stuttered speech analysis. The most important difference was observed in terms of Jitter, while Shimmer did not present a significant value
[6]	• Stuttering	• One group of stutterers and one group of non-stutterers were recorded • Parameters analyzed with program SEARP	• Jitter • Shimmer	• Stutterers have less stable control of respiratory-laryngeal dynamics in sustained phonation
[5]	• Normal voice	• One group of participants recorded. Computerized Speech Lab was used for acoustic analysis	• Jitter • Shimmer • Fundamental frequency	• By increasing the fundamental frequency, voice quality can be improved
[7]	• Phonological disorders	• Recording of one group of children with phonological disorders	• Jitter • Shimmer • Fundamental frequency	• The frequency of the vowel /e/ was smaller in the Phonological Disorder Group • The averages of Jitter and Shimmer showed no difference

(continued)

Table 17.2 (continued)

Paper	Phonetic disorders	Methods	Parameters of interest	Conclusions
[8]	• Normal voice	• Recording of sets of 100 and 50 speaker, all gen-der balanced	• Jitter • Shimmer	• The absolute values of jitter and shimmer are the most reliable when it comes to the analysis of the voice
[9]	• Reinke's disease	• Both normal and pathological voices collected from Saarbrucken Voice Database • Created an algorithm to determine the jitter and shimmer parameters Extract Jitter and Shimmer from Praat	• Jitter • Shimmer	• The algorithm showed reliable results in comparison to Praat • The algorithm can deal with almost any damaged voice
[10]	• Vocal cord polyps • Early-stage laryngeal carcinoma • Mutational falsetto	• Both normal voices and voices with disorders were recorded • The sample recorded was analyzed in Praat	• Jitter • Shimmer • Fundamental frequency	• All parameters are of great importance in the detection of voice disorders

the pitch and keep it stable for 0.5–1 s, then abruptly raise the pitch again and maintain stable pronunciation until the pitch reached the upper limit of the natural vocal range [10]. The comparison between the standard deviation of the fundamental frequency with each group separately lead to different significant results. Therefore, the authors found that between the control group and ELC group, as well as between VCP group and the control group, jitter and shimmer were different, whereas between the MF and the control group there were no significant differences in respect to jitter and shimmer for the sustained vowel /a/.

Another conclusion stated was that there could not be observed any significant differences in jitter or shimmer between different voice samples, not even after the patients abruptly raised the pitch. Last but not least, the instability of the frequency of the voice and weakness in controlling the vocal folds was also proven as there was a substantial variation of the standard deviation of the F0 in patients with mutational falsetto.

To conclude, researchers Gelin Li, Qian Hou, Chi Zhang, Zhen Jiang, Shusheng Gong stated that jitter and shimmer are amongst the important parameters for the evaluation of pitch variation, which is of great significance in the diagnosis of MF and voice disorders [10].

Table 17.2 presents a summary of the scientific works presented in this paper, showing the phonetic disorders, main methods, parameters, and short conclusions for each paper.

17.3 Conclusions

Jitter and shimmer parameters are emphasized in many studies in having a great contribution in detecting different voice pathologies, whether they are determined through analytical calculations or directly from specialized programs or software. The determination of both parameters is, in most studies, done on vowels sustained for a longer period of time (a few seconds for each vowel). The longer subjects voice the vowel, the more visible and easier variations in frequency and amplitude are to determine.

The determination of jitter and shimmer parameters for pathological voices is a subject of great importance that will continuously be developed through different techniques as the need for simple and effective pathology detection methods increases.

References

1. H.T. Lathadevi, S.P. Guggarigoudar, Objective acoustic analysis and comparison of normal and abnormal voices. J. Clin. Diagn. Res. **12**, 1–4 (2018)
2. Introduction to speech processing. https://speechprocessingbook.aalto.fi/Representations/Jitter_and_shimmer.html. Last accessed 13 May 2023
3. J.P. Teixeira, A. Gonçalves, Accuracy of jitter and shimmer measurements. Procedia Technol. **16**, 1190–1199 (2014)
4. S.D. Waghmarea, R.R. Deshmukh, S.N. Kayte, Analysis of fundamental frequency, jitter and shimmer in stuttered and non-stuttered speech of Marathi language, in *Proceedings of International Conference on Communication and Information Processing (ICCIP)* (2019)
5. H.-Y. Pyo, H.-S. Sim, Shimmer change according to fundamental frequency variation of Korean normal adults. Speech Sci. **10**(1), 143–150 (2003)
6. P.W. Newman, R.W. Harris, L.M. Hilton, Vocal jitter and shimmer in stuttering. J. Fluency Disord. **14**(2), 87–95 (1989)
7. H.F. Wertzner, S. Schreiber, L. Amaro, Analysis of fundamental frequency, jitter, shimmer and vocal intensity in children with phonological disorders, Brazilian J. Otorhinolaryngol. **71**(5), 582–588 (2005)
8. J.H. Farru's, Using jitter and shimmer in speaker verification. IET Signal Process. **3**(4), 247–257 (2009)
9. J.P. Teixeira, A. Gonçalves, Algorithm for jitter and shimmer measurement in pathologic voices. Procedia Comput. Sci. **100**, 271–279 (2016)
10. G. Li, Q. Hou, C. Zhang, Z. Jiang, S. Gong, Acoustic parameters for the evaluation of voice quality in patients with voice disorders. Ann. Palliative Med. **10**(1), 130–136 (2021)

Chapter 18
Analysis of Violinist Kinematics During Musical Rehearsals

Mircea Mihalcica, Alina Maria Nauncef, Maria-Violeta Guiman, Ștefania Bîrsan, Silviu-Marian Nastac, and Mariana Domnica Stanciu

Abstract Improving the musical technique while playing the violin is an objective of violinists, for which they allocate a significant amount of study time. In addition to the artistic aspect, it is important for violinists to be aware of the interpretation technique so that they have better performances. The paper presents a case study regarding the kinematics of the motions of an experienced violinist during musical rehearsals. A musical passage with a duration of 40 s, part of the Prelude from Partita III for solo violin (Bach), was performed by the violinist and the activity was recorded. Colored markers were applied to the joints of the hand, elbow, and right shoulder of the active arm of the violinist. Their position in time was captured with the help of a video recording system, and the signals were processed with specialized programs. The applicability of the study consists both in the analysis of the interpretation technique from a kinematic point of view, but also in the capitalization of this information in the teaching–learning activities within the music faculties.

18.1 Introduction

The kinematics of violinists during music rehearsals have been analyzed in several studies to highlight the effect of posture on their health and performance. During musical performance, various factors contribute to achieving artistic expression, but

M. Mihalcica (✉) · M.-V. Guiman · Ș. Bîrsan · S.-M. Nastac · M. D. Stanciu (✉)
Department of Mechanical Engineering, Transilvania University of Brașov, B-dul Eroilor, 29, 500036 Brasov, România
e-mail: mihalcica.mircea@unitbv.ro

M. D. Stanciu
e-mail: mariana.stanciu@unitbv.ro

A. M. Nauncef
Faculty of Music, Transilvania University of Brașov, 500036 Brasov, România

S.-M. Nastac
Faculty of Engineering and Agronomy, Braila, "Dunarea de Jos" University of Galati, Galati, Romania

N. Herisanu and V. Marinca (eds.), *Acoustics and Vibration of Mechanical Structures—AVMS-2023*, Springer Proceedings in Physics 302,
https://doi.org/10.1007/978-3-031-48087-4_18

also to the emergence of health problems [1–3]. Risk factors include age, sex, joint mobility, technique, body posture, and other anthropometric variables (neck length, finger length, arm length, etc.) as highlighted in [4–6]. The most common musculoskeletal problems of the upper body reported by violinists refer to problems of the neck, chin, and left shoulder, as well as injuries/pain of the upper limbs, affecting between 50 and 88% of violinists [1–3]. On the other hand, involvement in the act of interpretation produces physiological changes such as an increase in heart rate during a concert, where the tension felt by violinists is similar to that of elite athletes, and an approximately 60–70% greater increase in heart rate frequency is recorded during concerts [6]. The muscles involved in the kinematics of the violinist during playing include the left superior trapezius muscle, which ensures the lifting of the shoulder to fix the violin between the shoulder and chin, the right sternocleidomastoid muscle (SCM), responsible for left cervical rotation and chin depression, the scapular retractors, especially the rhomboids and middle trapezius muscles, with a role in ensuring scapular stability and supporting the arm that holds the violin. According to [7–10], the position and static loading during musical performances lead to fatigue of the left major rhomboid muscle, usually manifested by pain. Thus, identifying motor patterns related to the coordination of neural and musculoskeletal systems based on the analysis of the kinematics of violinists' movements during music rehearsals can be useful in understanding and developing interpretation skills/techniques for both students and teachers, and on the other hand, can reduce the unpleasant effects of body posture during interpretation [11–14].

In this regard, this study analyzed recordings of an experienced violinist from the Brasov Philharmonic during music rehearsals. Using motion capture (MOCAP), the trajectories of the hand, elbow, and shoulder in Cartesian coordinates, as well as the angle formed by the arm and forearm during the performance of a musical sequence, were evaluated.

18.2 Material and Methods

18.2.1 Data Collection

The recording of the experienced violinist's movements during music rehearsals was done using a GHS mirrorless video camera, filming at 180 frames per second, at Full HD resolution (1980 × 1080 pixels) from the side, perpendicular to the plane in which the violinist primarily moves their hand during the rehearsal. As a result, data on the temporal position of markers installed at the main joints of the arm-forearm pair were collected: the hand joint, the elbow joint, and the shoulder joint, as seen in Fig. 18.1. With the expressed consent, the subject—a violinist with over 25 years of experience—agreed to participate in this study.

The musical sequence was chosen by the violinist, and the recorded passage is part of the Prelude from Partita III for solo violin by J. S. Bach. The beginning of the

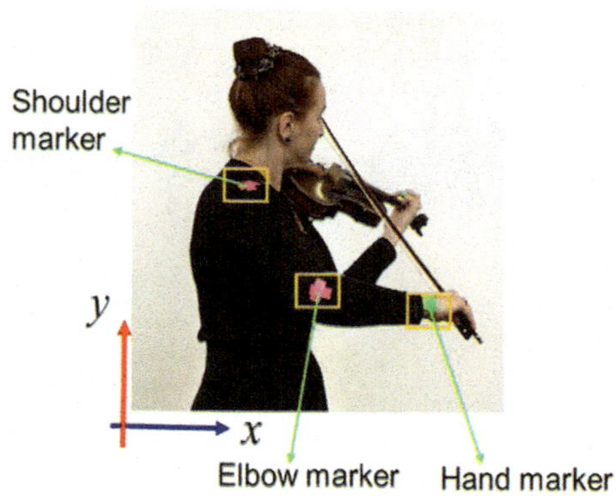

Fig. 18.1 The position of the tree markers installed at the hand, elbow and shoulder

Prelude is brilliant, soloistic, with a rhythmic formula. From a technical standpoint, the first measures are played with a large bow, in the upper part of the bow, with a very clear and precise string attack and a fast bow speed. These technical aspects lead to the production of a generous, soloistic, and brilliant sound. All of these technical details are achieved through a large mobility of the entire arm, which will perform opening movements at the elbow level to facilitate the bow reaching from the middle to the tip. The arm movements from bottom to top must be accompanied by a slight jerk of each stroke, to achieve the separation of each note (with an eighth note value). Figure 18.2 presents the time analysis and spectrogram of the recorded musical sequence.

18.2.2 Data Analysis

Using the Kinovea application, the X and Y coordinate projections in the camera plane of the 3 markers were extracted for a representative sequence corresponding to 13 s of repetition, chosen between second 3 and second 16. Although the Z coordinate was not taken into account, the analysis of the movement in the projection plane was done in the same way for all the recordings, comparing the obtained data. So, the X and Y axes belong to a plane fixed in space. Each marker resulted in a number of 2341 frames, resulting in a distance between frames of approximately 0.0056 s, expressed in pixels. Additionally, to establish potential correlations between movement and acoustics, sound was recorded using 4 AKG Cardioid Condenser Small Diaphragm P170 microphones and a Zoom h6 audio recorder. Using the trajectory tracking tools in Kinovea, the three markers installed at the main joints of the right upper limb were set as regions of interest. The data was exported in tabular format as X, Y coordinate pairs for each marker and finally taken and processed with MATLAB to determine a series of relevant parameters.

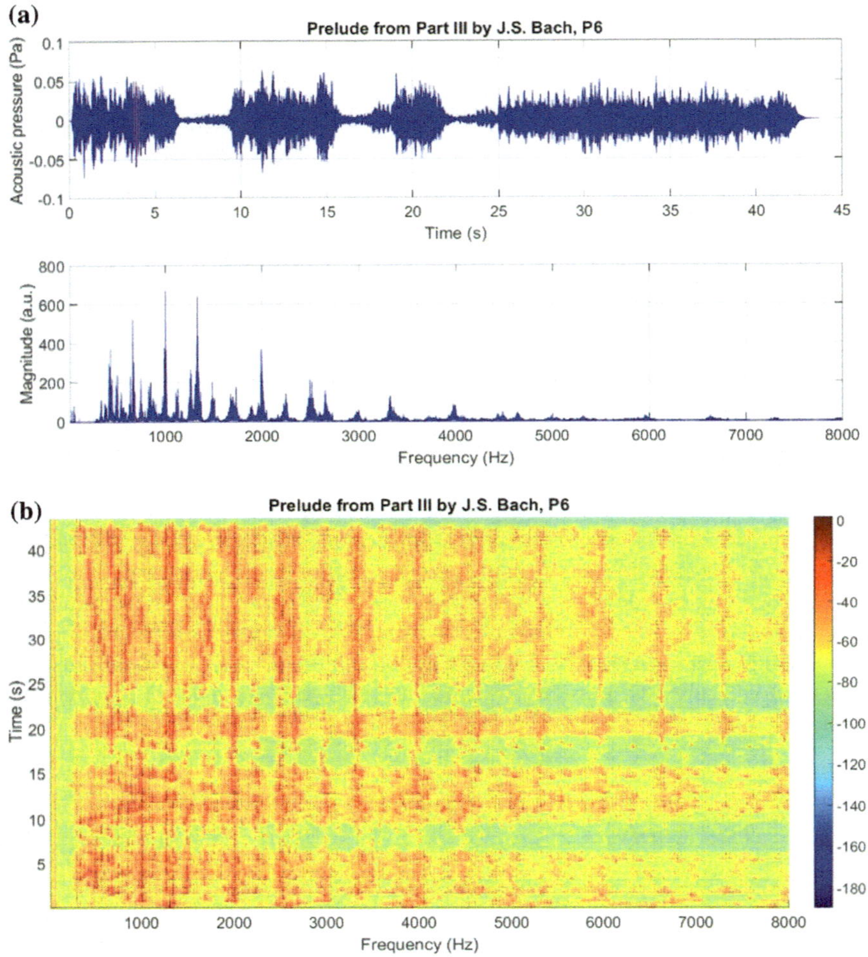

Fig. 18.2 The configuration of the recorded signal: **a** time analysis; **b** spectrogram

18.3 Results and Discussion

18.3.1 Hand Trajectories

The analysis of the hand joint marker was carried out over 7 repetitions. In order to have a relevant image, the representation of the marker displacement was chosen separately on the X and Y axes (Fig. 18.3). From the comparative analysis of the hand marker trajectory on the X axis, it can be observed that there are areas with relatively similar movement patterns, corresponding to different sequences that make up the musical fragment, but there are also differences between executions, which can reveal possible trends that already exist or may occur over time as the repetition progresses and fatigue sets in for the artist. For example, the similarity between executions 6 and 7 (the last ones performed) can be easily observed, as well as the differences in execution between them and the rest of the repetitions (Fig. 18.3a). The reasons may be related to fatigue that occurs at the limb level and how the artist tries to compensate for it. Although the amplitude of the movement on the Y axis is in the same range of values as that obtained for the corresponding parameter analyzed on the X axis, in this case, the graphs are much closer to each other, as the musician tends to control the bow much better in a more controlled area on this axis. However, some repetitions still stand out as outliers (Fig. 18.3b).

An important parameter in motion pattern analysis is the maximum amplitude of movement, as highlighted in studies [10–14]. Thus, the maximum amplitude value of the movement was extracted for the x and y axes for the seven repetitions (the difference between the maximum and minimum position), a parameter that can be visualized in Fig. 18.4. The way in which this parameter changes for the seven repetitions can be observed, as well as an interesting evolution over time: if the musician started with a high amplitude of hand movement, a decrease can be observed with repetition 4 (where fatigue sets in), and then the musician compensates, with the maximum amplitude increasing again in the last two repetitions (Fig. 18.4a).

Although the amplitude of movement on the y-axis is in the same range as that obtained for the respective parameter analyzed on the x-axis, in this case, the graphs are much closer to each other, as the musician tends to control the bow in a much better controlled area on this axis (Fig. 18.4b).

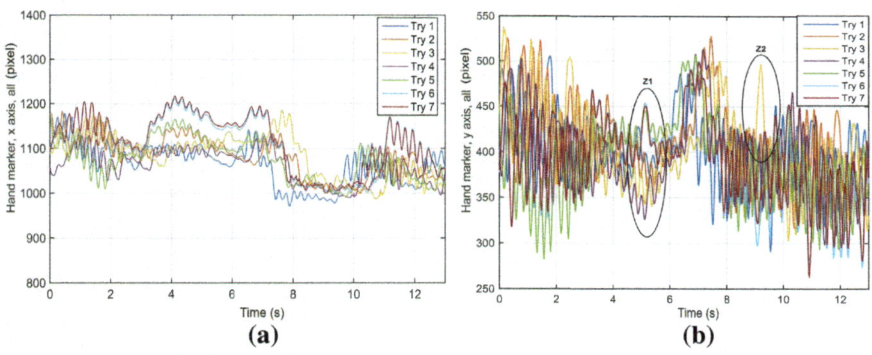

Fig. 18.3 The trajectory of the hand marker for the seven repetitions: **a** on the x-axis; **b** on the y-axis

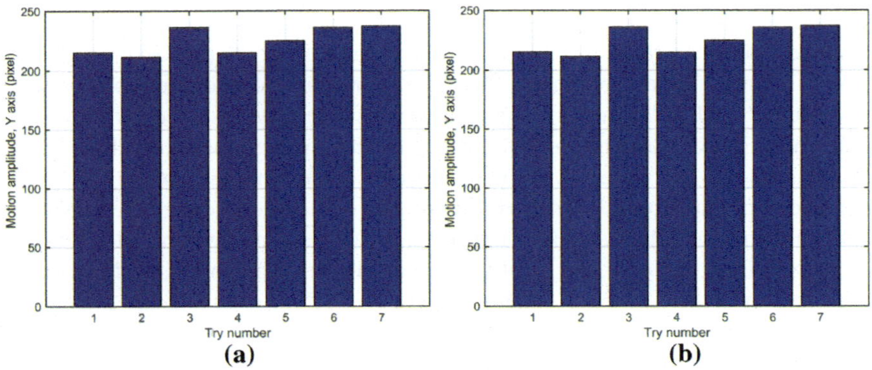

Fig. 18.4 The maximum amplitude of the hand: **a** on the x axis; **b** on the y axis

18.3.2 The Elbow Trajectory

The analysis of the elbow motion shows significant differences between executions, as evidenced by the overlaid trajectories on the x-axis graph (Fig. 18.5a). However, as expected, the last two repetitions (6th and 7th) are very close to each other even in this case (Fig. 18.5b). Throughout the entire performance, the right upper limb is highly solicited, as it has to perform multiple types of string plane changing movements. Thus, the minimum value of the maximum amplitude of movement on the x-axis was obtained in the 4th repetition, while the maximum value was obtained in the 6th repetition, the difference being greater than in the case of the values for the hand joint (Fig. 18.6a). The minimum value of the maximum amplitude of movement on the y-axis was obtained in the 5th repetition, while the maximum value was obtained in the 3rd repetition, the difference also being greater than in the case of the values for the hand joint (Fig. 18.6b).

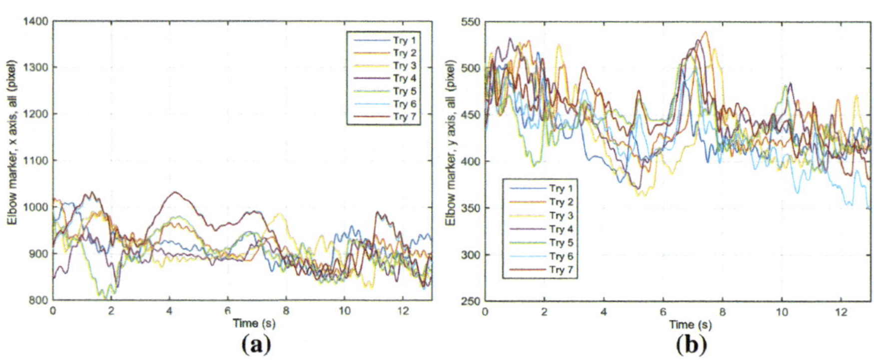

Fig. 18.5 The trajectory of the elbow marker, seven repetitions: **a** on the x-axis; **b** on the y-axis

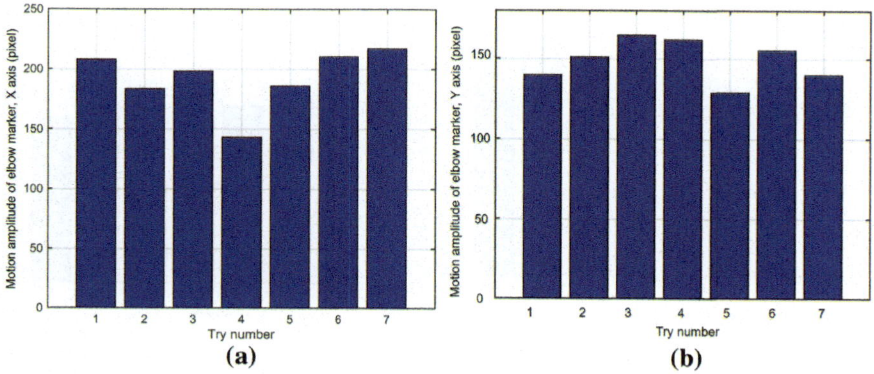

Fig. 18.6 The maximum amplitude of the elbow: **a** on the x axis; **b** on the y axis

18.3.3 The Shoulder Trajectory

Initially, the analysis in terms of the shoulder marker was based on the idea that the shoulder is a relatively stable reference point, with more interest focused on the movement along the x-axis—a very visible movement on the x-axis can represent the tendency to shift weight from one foot to the other (Fig. 18.7). For the shoulder, the values of the maximum amplitudes on both the x and y axes were the lowest compared to the other analyzed joints: the minimum value of the maximum amplitude of the movement on the x-axis was obtained in the 4th repetition, while the maximum value was obtained in the 7th repetition. The minimum value of the maximum amplitude of the movement on the y-axis was obtained in the 5th repetition, while the maximum value was obtained in the 4th repetition (Fig. 18.8).

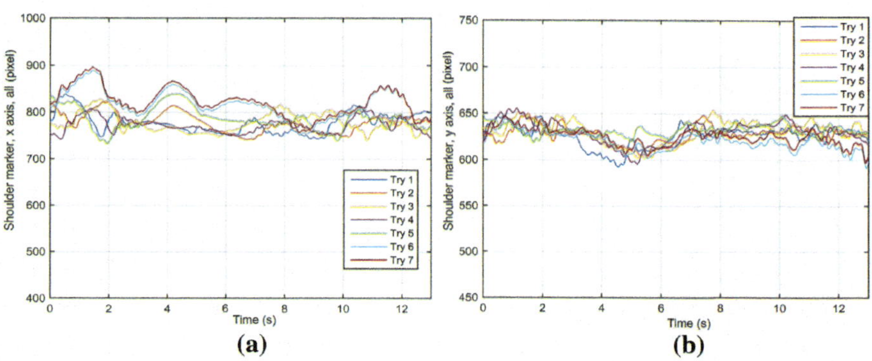

Fig. 18.7 The trajectory of the shoulder marker, seven repetitions: **a** on the x-axis; **b** on the y-axis

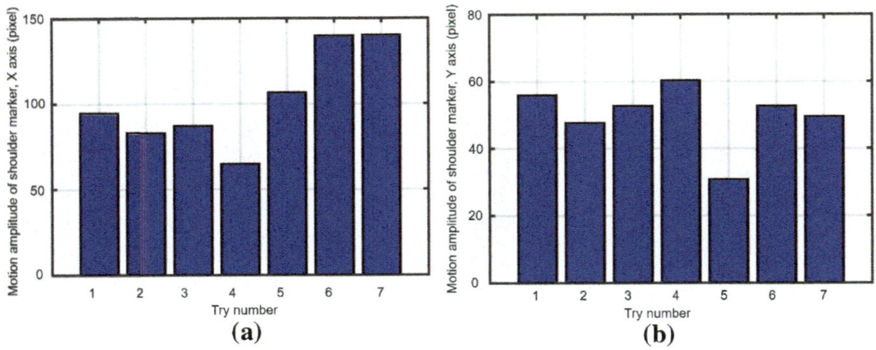

Fig. 18.8 The maximum amplitude of the shoulder **a** on the x axis; **b** on the y axis

18.3.4 The Angle Between Arm and Forearm

A different perspective analysis takes into account the angle created between the arm and forearm during the movement. By using the coordinates of the markers at the wrist, elbow, and shoulder, a triangle can be defined and the angle can be calculated using the cosine theorem. Thus, the elbow angle has a minimum variation of 43.88°, found in the 5th repetition, and a maximum variation of 88.35°, found in the 3rd repetition. It should be noted that when mild fatigue set in, the variation in the elbow angle decreased, with repetitions 4 and 5 having the lowest values. The interpretation of these parameters is that, as mild fatigue sets in, the musician tends to open the elbow less and perform less extensive movements. Once the need to compensate for fatigue is realized, the values of the parameters return to the initial state values (Fig. 18.9).

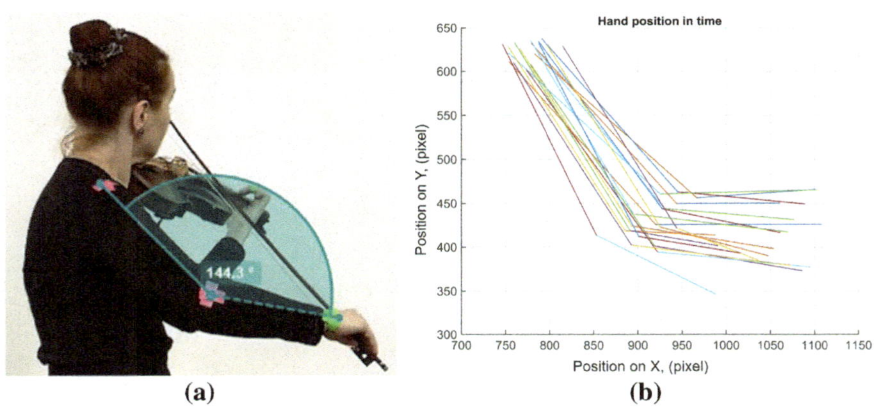

Fig. 18.9 Angle between arm and forearm: **a** representation on the actual model; **b** successive angles of the right upper limb over time

18.4 Conclusions

Having the positions and angles in time makes it easy to obtain other parameters such as velocities, accelerations etc. by simple numerical differentiation. This study highlights the complexity of movements that violinists perform during musical performances, with analysis of the kinematics of movement being an important aspect in the prevention and/or treatment of upper musculoskeletal system (MSKP) problems, as well as in extracting useful movement patterns for practicing interpretation skills and techniques for both music students and for improving the performance of each violinist. In future studies, the kinematic results will be correlated with the sound recording data on different types of musical sequences.

Acknowledgements This research was supported by a grant of the Ministry of Research, Innovation and Digitization, UEFISCDI, project number 61PCE/2022, PN-III-P4-PCE2021-0885, ACADIA. The video–audio recording was made with the support of AP Studio Brașov.

References

1. S.L. Hiner, K.D. Brandt, B.P. Katz, R.N. French, T.J. Beczkiewicz, Performance-related medical problems among premier violinists. Medical Probl. Performing Artists, 67–71 (1987)
2. C. Zaza, C. Charles, A. Muszynski, The meaning of playing-related musculoskeletal disorders to classical musicians. Soc Sci Med **47**, 2013–2023 (1998)
3. B. Ackermann, R. Adams, E. Marshall, The effect of scapula taping on electromyographic activity and musical performance in professional violinists. Aust. J. Physiother. **48**(3), 197–203 (2002)
4. A.M. Abreu-Ramos, W.F. Micheo, Lifetime prevalence of upper-body musculoskeletal problems in a professional-level symphony orchestra: age, gender, and instrument-specific results. Med. Probl. Perform. Artists **22**(3), 97–104 (2007)
5. C. Inesta Mena, *Demanda fisiologica en musicos profesionales [dissertation]* (Functional Biology Department. University of Oviedo, Asturias, Spain, 2006)
6. S. Hallam, The development of metacognition in musicians: implications for education. Br. J. Music Educ. **18**(1), 27–39 (2001)
7. J.W. Davidson, Visual perception of performance manner in the movements of solo musicians. Psychol. Music **21**(2), 103–113 (1993)
8. V. D'Amato, E. Volta, L. Oneto et al., Understanding violin players' skill level based on motion capture: a data-driven perspective. Cogn. Comput. **12**, 1356–1369 (2020)
9. M. Mihalcica, M. Cristea, *Metode de captură și analiză a datelor în biomecanică* (Universitatii Transilvania din Brasov, 2023)
10. S.M. Nastac, V.G. Gliga, M. Mihalcica, A.M. Nauncef, F. Dinulica, M. Campean, Correlation between acoustic analysis and psycho-acoustic evaluation of violins. Appl. Sci. **12**, 8620 (2022)
11. P. Visentin, G. Shan, E.B. Wasiak, Informing music teaching and learning using movement analysis technology. Int. J. Music. Educ. **26**(1), 73–87 (2008)
12. G. Shan, P. Visentin, A quantitative three-dimensional analysis of arm kinematics in violin performance. Med. Probl. Perform. Artist. **18**(1), 3–10 (2003)

13. M.H. Schwartz, A. Rozumalski, A new method for estimating joint parameters from motion data. J. Biomech. **38**(1), 107–116 (2005)
14. S. Duprey, B. Michaud, E. Gagne, M. Begon, Bow influence on violinist gestures: upper arm angles and shoulder muscle activity. FA2020, Forum Acusticum 2020, Dec 2020, Lyon, France. 3 p. ff10.48465/fa.2020.0696ff. ffhal-03130299f (2020)

Chapter 19
Design and Construction of a Mechatronic System for Measuring Oxygen Concentration in Hospital Institutions

Alexandru-Constantin Tulică, Ionel Şerban, and Constantin-Adelin Grigorescu

Abstract This article shows how a monitor for measuring oxygen concentration in hospital facilities was designed and built. The device components are identified: the hardware components and the software component, in which the source code was inserted. An oxygen concentration measurement system detects gas leakage from a room as a priority, and different detection thresholds can be set, thus the staff is alerted at the very moment of changes in the composition of the ambient air, and measures to eliminate the danger can be carried out in the shortest possible time.

19.1 Introduction

The portable oxygen concentration indicator OX-07, is a lightweight instrument designed for continuous monitoring of oxygen levels. The OX-07 provides an oxygen concentration measurement, and an alarm whenever the oxygen level drops below a preset point.

A low battery alarm alerts the user when power is low. The mounted oxygen sensor easily detaches for mounting on an optional extension cable to provide remote data sampling in tanks or enclosures before entering these or other potentially hazardous areas [1]. The SMART SENSOR gas detector is a type of device capable of detecting certain gases in an area, similar to an alarm system. This type of device is used to detect the presence of gases and command a system so that a process can be stopped automatically. Such a detector can trigger an alarm to the operators in the area where the leak occurs, to leave the room. An O_2 Monitor is important because it can save lives. Such detectors are used to detect combustible, flammable, toxic gases and oxygen concentrations [2].

A.-C. Tulică (✉) · I. Şerban · C.-A. Grigorescu
Transilvania University, Braşov, Romania
e-mail: alexandrutulica@yahoo.com

N. Herisanu and V. Marinca (eds.), *Acoustics and Vibration of Mechanical Structures—AVMS-2023*, Springer Proceedings in Physics 302,
https://doi.org/10.1007/978-3-031-48087-4_19

19.1.1 Electrochemical Oxygen Sensors

Oxygen sensors are utilized to gauge and oversee the level of oxygen concentration inhaled and exhaled by patients connected to ventilators or anesthesia machines. Oxygen sensors are also referred to as FiO_2 sensors or O_2 cells. The fraction of inspired oxygen (FiO_2) represents the concentration of oxygen in the gas mixture. In atmospheric room air, the fraction of inspired oxygen is 21%, indicating that the oxygen concentration in room air is 21% [6]

All electrochemical oxygen sensors operate based on the principles of self-powering, diffusion limitation, and the metal-air battery concept. They consist of three main components: an anode, a semi-solid electrolyte paste, and an air cathode. The sensor cell is enclosed and contains two electrodes: a flat PTFE tape coated with an active catalyst, serving as the cathode, and a lead block acting as the anode. The enclosure is mostly sealed, except for a small capillary that regulates the rate at which oxygen can enter the cell. Oxygen molecules enter the sensor through the capillary, where they encounter the cathode and undergo immediate reduction, forming hydroxyl ions according to the following reaction:

$$O_2 + 2H_2O + 4e^- \rightarrow 4OH^-$$

These hydroxyl ions then migrate through the electrolyte to the lead anode. At the anode, the hydroxyl ions react with the lead, causing its oxidation to lead oxide:

$$2Pb + 4OH- \rightarrow 2PbO + 4e^- + 2HO$$

As these processes occur, an electric current is generated, which can be measured by the instrument. The magnitude of this current is directly proportional to the oxygen concentration, allowing for accurate determination of oxygen levels in gas analysis applications [6].

An oxygen sensor can be used for both single-patient and multiple-patient applications, operating without issues for several hundred thousand hours of oxygen exposure. This duration can be utilized for a single patient or divided among multiple patients during that period. To provide a clearer understanding of the sensor's lifespan, the oxygen hours have been converted into years. Under normal conditions of air at 25 °C and 60% relative humidity, a typical oxygen sensor has a service life of 24 months.

The theoretical designed life or life expectancy of the sensor can be expressed by the product of the oxygen concentration percentage and the time in hours, provided that the sensor is used under constant conditions of 25 °C temperature and 1 atm pressure. However, the expected lifespan of a sensor is significantly influenced by the amount of anode material and the signal output. A higher signal output leads to a shorter lifespan because the anode material is consumed at a faster rate. In reality, the expected lifespan of a sensor takes into account the signal output range [6].

Electrochemical oxygen sensors are self-powered devices with a diffusion-limited, metal-air battery design. They comprise an anode, a semi-solid electrolyte paste, and an air cathode. The sensor cell is enclosed and contains two electrodes: a flat PTFE tape with an active catalyst as the cathode and a lead block as the anode. The enclosure is sealed, except for a small capillary that controls oxygen entry [6].

19.2 Design of the Device for Measuring Oxygen Concentration

The design of the device was done with the help of Creo software. Creo is a class of computer-aided design (2D and 3D) applications that enables product design for manufacturers and students and is developed by PTC. It features powerful and proven functionality with new technologies such as generative design, augmented reality, real-time simulation, additive manufacturing, and IoT to reduce costs and improve product quality [3].

The case of the device is made with the help of 3D printing technology, using a printer from the PRUSA range as an execution model. The printed assembly is composed of four distinct parts, as follows:

- Outer casing (Figs. 19.1 and 19.2): the thickness of the walls is 6 mm, with dimensions of 206 mm × 120 mm × 106 mm. The case has, on the front side, the area where the LCD is integrated, the three status LEDs, a slot for the speaker, and the name of the device, on the back side of the device the wall mounts are attached, on the right side the case has of the hole for the accumulator and the connection to the external ventilation system;

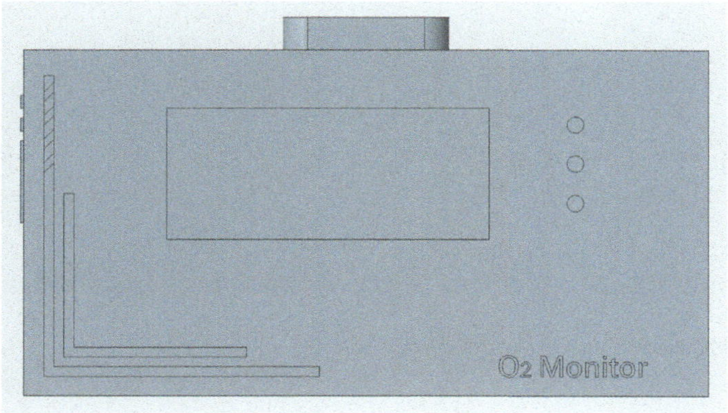

Fig. 19.1 Front view O_2 monitor

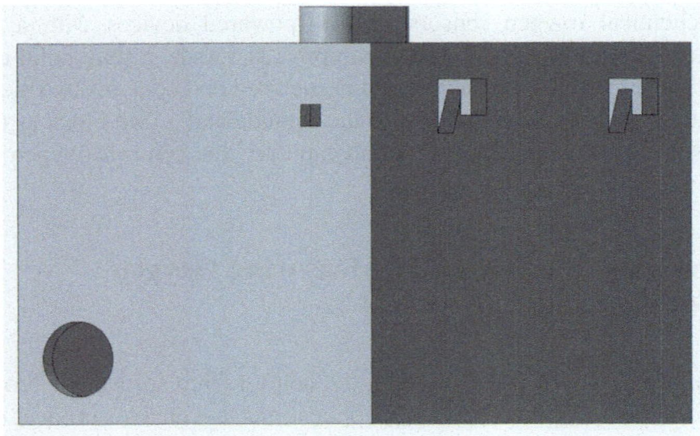

Fig. 19.2 Isometric view of O_2 monitor

- Upper cover (Fig. 19.3): the thickness is 3 mm, with dimensions of 200 mm × 103 mm. The cover has a slot for the sensor, a channel for the lock lever in the closed/open position, and a hole for attaching the handle;
- Fixing handle (Figs. 19.1 and 19.2): helps to operate the device inside and handle the cover;
- Fixing lever (Fig. 19.3): represents the mechanism by which the lid is fixed in the closed/open position.

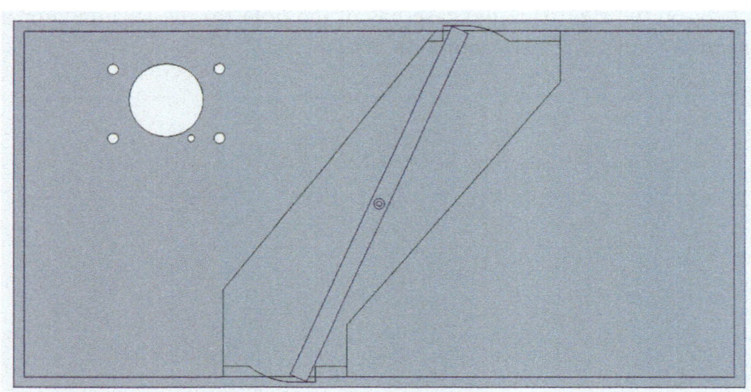

Fig. 19.3 Median section. O_2 monitor cover view

Fig. 19.4 Arduino UNO R3

19.3 Realization of the Device for Measuring Oxygen Concentration—Hardware Components

19.3.1 Arduino Uno R3

Arduino Uno R3 (Fig. 19.4) is a development board based on an ATmega328P microcontroller. It features 14 digital I/O pins (six of which are PWM), six analog inputs, a 16 MHz resonator, USB communication port, electrical connection, and a reset button [4].

19.3.2 O_2 Sensor

The Gravity I2C oxygen sensor (Fig. 19.5) is based on electrochemical principles and can measure ambient O_2 concentration accurately. With high anti-interference ability, high stability, and high sensitivity, this Arduino-compatible oxygen sensor can be widely applied in fields such as portable devices, air quality monitoring devices and industries, mines, warehouses, and other spaces where the air is not ventilated.

This compact oxygen sensor supports I2C output, can be calibrated in the air, and can accurately measure oxygen concentration in the environment. It is compatible with many motherboards like Arduino Uno, ESP32, Raspberry Pi, and so on. Its effective range is 0 ~ 25% Vol, and the resolution can reach 0.15% Vol. It supports a wide range of input voltages: 3.3–5.5 V. In addition, the lifetime is up to 2 years [5].

Fig. 19.5 O_2 sensor

19.3.3 External Fan

The O_2 Monitor device is programmed so that at the time of a gas leak, above the permitted limit, it activates an external ventilation system to ventilate the space in question and reduce the danger to a minimum. The fan (Fig. 19.6) attached to the device works on a voltage between 3 and 5 V and a current of 0.2 A.

Fig. 19.6 External fan

19.4 Realization of the Device for Measuring Oxygen Concentration—Software Components

19.4.1 Arduino IDE-Software

The Arduino Integrated Programming Environment contains a text editor window for typing code, a message box, a navigation bar, a toolbar with shortcuts to the main functions, and a series of menus. Connection is made to Arduino and Genuino hardware to write programs and communicate with them. The programs are implemented using Arduino Software (IDE) (Fig. 19.7) under the name of sketches.

They are designed in the text area and later stored with the characteristic arduino.ino extension. The editor has features like delete/paste and search/replace text. The message box provides a report of the current status while saving, exporting, and displaying, and may contain any errors. In the lower right area of the window,

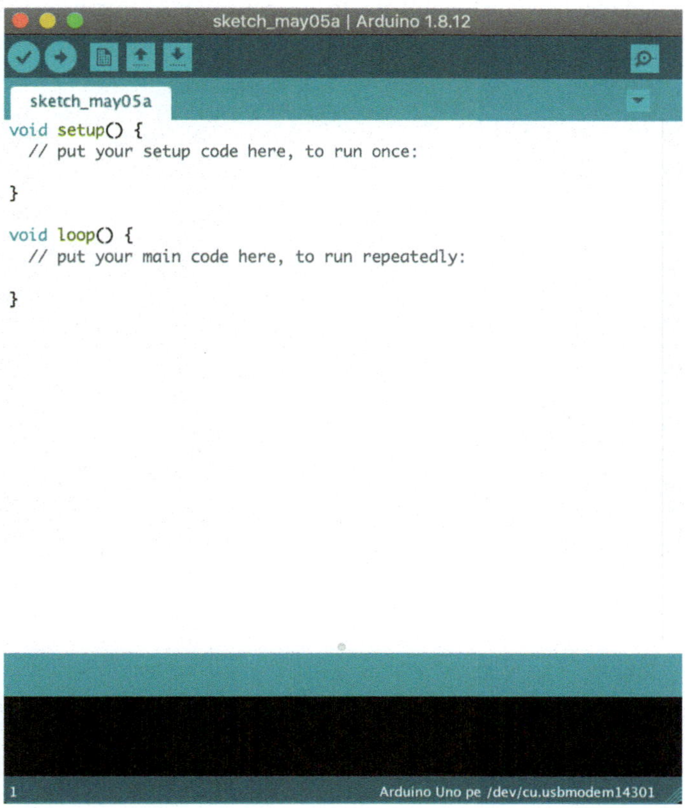

Fig. 19.7 Interface—Arduino IDE

you can see the configured board and the connected serial port. The toolbar functions allow checking, loading programs, creating, opening, and saving sketches and the command that opens the serial monitor [5].

19.5 Realization of the Electrical Scheme of the Oxygen-Measuring Device

The electronic components in a system are the essentials without which a device cannot function. The Oxygen Monitor uses an Arduino board with the role of command and data processing obtained from the peripheral elements, connected to a direct current power supply. The graphic interface of the device consists of a 20 × 4 LCD and an I2C module for transferring data between the board and the LCD. These are connected to the power pins of the main unit and the analog input ports for reading data. The sensor used in data acquisition is connected to the 5 V power port of the Arduino board and implicitly to an analog input for transferring the data obtained from the measurements (Fig. 19.8).

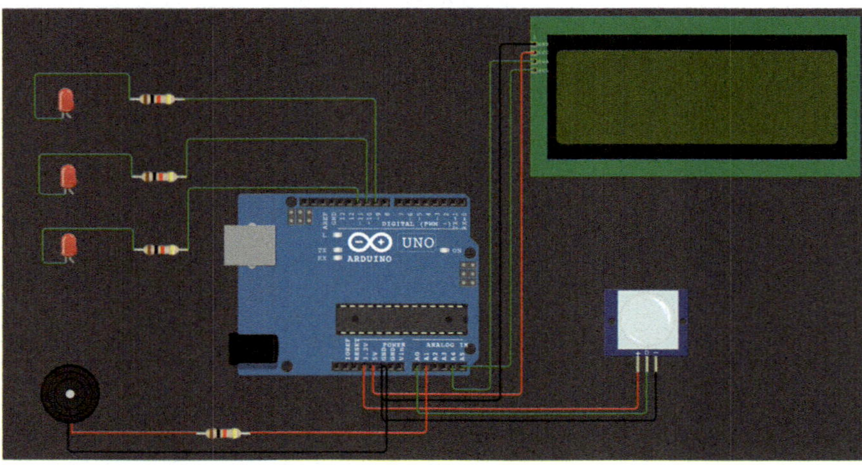

Fig. 19.8 Electrical scheme

Table 19.1 Results—Ambiental temperature

No. crt	Ambiental temperature (°C)	Displayed value (%)
1	21	20.70
2	21	20.83
3	21	20.90
4	21	20.93
5	21	20.93
6	21	20.91
7	21	20.94
8	21	20.94

19.6 Measurement Results

19.6.1 Oxygen Sensor Calibration

To calibrate the oxygen/O_2 sensor, place it outdoors where the oxygen content is 20.9%. Once the sensor data stabilizes, press and hold the calibration button on the sensor for 2 s to complete the calibration process [7].

The results were obtained following multiple measurements, where the ambient temperature was 21 °C for the comparison of recorded data. The measurement was carried out under the same conditions (Table 19.1).

The curve for calibrating the oxygen sensor is depicted when the ambient temperature is 21 °C (Fig. 19.9).

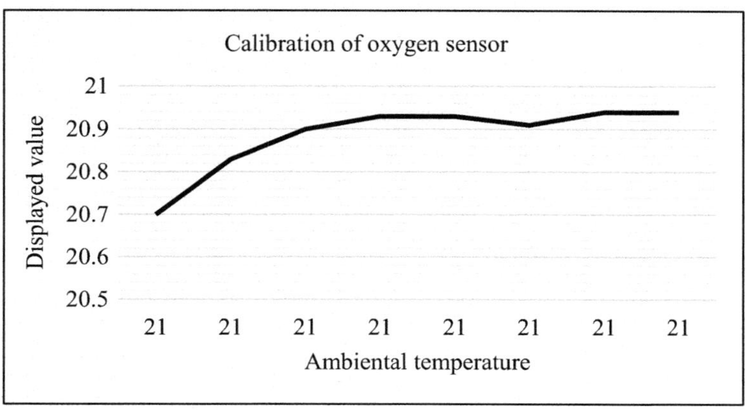

Fig. 19.9 Curve of an oxygen sensor-calibration

19.7 Conclusions

The importance of the existence of such a device within hospital institutions, developed in this paper, contributes greatly to saving human lives through measurement precision, time reliability, and the performance of a fully automated service.

An O_2 Monitor detects with priority the leakage of gases from a room, and different detection thresholds can be set, so the staff is alerted at the very moment when there are changes in the composition of the air in the ambient environment and the measures to eliminate the danger can be performed in the shortest possible time.

Also, the ability of the device to operate a ventilation system considerably facilitates the removal of the fire hazard, and thanks to it, the level of gases in the room is reduced from the very moment the alarm was triggered.

References

1. OX-07 (2023) https://www.rki-gassensor.de/en/products/portable/single-gas/ox-07/. Accessed 1 Sept 2023
2. AR8100 SMART SENSOR Oxygen Detector-Instruction Manual. https://www.comcube.co.th/product/ar8100/?attachment_id=4655&download_file=5d8ae4bc2800b. Accessed 01 Sept 2023
3. Creo CAD Software: Enable the Latest in Design. PTC, 2022. Ptc.com [online]
4. Arduino Uno Rev3 | Arduino Official Store. Store. Arduino. cc [online] (2020). https://store.arduino.cc/arduino-uno-rev3. Accessed 30 March 2023
5. Arduino—Environment. Arduino. cc [online] (2020). https://www.arduino.cc/en/Guide/Environment. Accessed 24 May 2020
6. Oxygen Sensors for Medical Application (2023) https://www.biometriccables.in/blogs/blog/oxygen-sensors-for-medical-application. Accessed 12 July 2023
7. DFRobot Gravity: I2C Oxygen/O_2 Sensor (2023) https://wiki.dfrobot.com/Gravity_I2C_Oxygen_Sensor_SKU_SEN0322#:~:text=Calibration%20Method%3APlace%20the%20oxygen,the%20sensor%20will%20be%20calibrated. Accessed 12 July 2023

Chapter 20
The Effect of Acoustic and Visual Stimuli on Brain Activity

Cornelia Zaharia, Andreea Rohan, Cristina Sandu, Flavius Nițu, Ionel Şerban, and Corneliu Drugă

Abstract This paper aims to analyze individuals' responses to different types of acoustic and visual stimuli. According to state-of-the-art research, it is well known that the human body shows a pattern of reactions as it is exposed to certain external stimuli, but with the advances in technology and research, new directions occur. In the first part of the paper, a critical study is presented regarding the behavior of human subjects to acoustic and visual stimuli. The second part of the paper presents the equipment and method/protocol that was used to investigate. An EEG device (Neuron-Spectrum-65) was used to analyze neurological responses generated by the acoustic and visual stimuli, according to a preset protocol. The results indicate, from the topography of the average amplitudes, for the first subject an average amplitude in the frontal area, for the second subject in the temporal-parietal area, and for the third subject in the occipital area, these are due to the exposure to different types of video media (thriller, comedy, action). The final part of the paper presents the observations and conclusions of this approach to analyzing the behavioral response of human subjects who have been exposed to these stimuli.

20.1 Introduction

The nervous system is a network of neurons whose primary function is to produce, modify, and transfer information across the many organs of the human body. The regulation of essential bodily processes (heartbeat, respiration, digestion), sensations, and movement are only a few of the crucial nervous system activities made possible by this feature.

The conversion of a sensory signal into an electrical signal in the nervous system is the most basic function of a sensory system [1]. The sensory receptor is where this happens; receptor potential refers to the resulting shift in electrical potential.

C. Zaharia (✉) · A. Rohan · C. Sandu · F. Nițu · I. Şerban · C. Drugă
Transilvania University of Braşov, Braşov, Romania
e-mail: zaharia.cornelia99@yahoo.com

N. Herisanu and V. Marinca (eds.), *Acoustics and Vibration of Mechanical Structures—AVMS-2023*, Springer Proceedings in Physics 302,
https://doi.org/10.1007/978-3-031-48087-4_20

The way that each sense's sensory receptor's function varies, receptor specificity being tailored to the kind of stimulus that they are sensing. For instance, several stimuli can activate the receptors for touch, light, and sound. Touch receptors solely respond to pressure or touch; they are not sensitive to light or sound. However, like with olfaction, inputs may be integrated at higher levels in the brain, adding to our perception of taste [2].

The way a person interprets a stimulus is called perception. Even though perception depends on the activation of sensory receptors, perception occurs at the level of the brain rather than the sensory receptor. Through a sensory route, the brain may distinguish between different sensory stimuli because sensory receptors' action potentials flow along neurons specifically tuned to that stimulus [2].

The thalamus and the relevant part of the cortex receive all sensory impulses, except for those from the olfactory system, which are sent through the central nervous system. When a sensory signal leaves the thalamus, it is carried to the region of the cortex that is specifically responsible for processing that sense [2].

The stimulus–response circuit involves three different types of neurons [3]:

- Sensory Neurons: They deliver messages from receptors to the brain.
- Relay Neurons: These neurons relay messages within the brain and spinal cord to aid in decision-making.
- Motor Neuron: To start a reaction, motor neurons transmit information from the central nervous system to effectors (muscles or glands).

The visual system (see Fig. 20.1), which includes the sensory organ (the eye) and central nervous system components (the retina with photoreceptor cells, the optic nerve, the optic tract, and the visual cortex), allows organisms to perceive and process visible light, as well as develop a variety of non-image photo response functions [4].

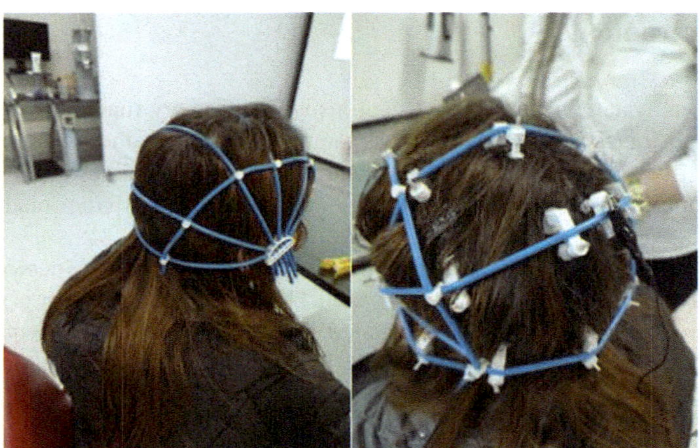

Fig. 20.1 Placement of the grid helmet and bridge electrodes on Subject 1

The sensory system (see Fig. 20.1) for hearing is known as the auditory system. It comprises the auditory components of the sensory system as well as the sensory organs (the ears). The middle-frequency band has higher sound pressure thanks to the outer ear's role in channeling sound waves to the eardrum [5].

In this paper, the main focus was collecting data from the EEG machine during the watching of the video process and then observing the way these stimuli affect the neurological activity of each individual participant.

20.2 Theoretical Aspects of Data Acquisition in Stimuli Reaction Tests

Given the variety of receptors and stimuli, most extrinsic stimuli first generate localized graded potentials in neurons associated with the specific sensory organ or tissue. In the nervous system, both internal and external stimuli can cause two different categories of responses: the first—an excitatory response, normally in the form of an action potential, and the second—an inhibitory response [2].

When a neuron is stimulated by an excitatory impulse, the neuron's dendrites are bound by neurotransmitters that cause the cell to become permeable to a certain type of ion; the type of neurotransmitter determines which ion the neurotransmitter will become permeable too. In excitatory postsynaptic potentials, an excitatory response is generated. This is caused by an excitatory neurotransmitter, which normally binds to glutamate on the dendrites of a neuron, causing an influx of sodium ions through channels located near the binding site [2].

In general, the cooperative properties of neural networks consist of excitatory neurons and inhibitory neurons connected in the form of feedback loops, having an extremely important role in establishing EEG rhythms. The frequency of these oscillations is also influenced by the intrinsic properties of the neuron membrane and the strength of their synaptic interactions. According to this, human neuronal groups with a similar dynamic functional state can be formed, subserving perceptual processes. Furthermore, it has also been demonstrated that the role of synchronized oscillatory EEG activity is to serve as an opening mechanism to the river of neuronal information through the network. The oscillatory activity of the brain has a mechanism that can also regulate changes of state in selected neuronal networks and change the route of information [6].

An EEG device consists mainly of electrodes, amplifiers, an analog-to-digital converter, and a writing unit.

Some studies demonstrated how marketing may benefit from neurophysiology. The authors discuss a particular research case concerning the analysis of a skincare product advertisement: pretests of 2 versions of this TV ad revealed that, although the versions were almost identical, each of them generated significantly different impacts.

To verify this assumption, it was decided to conduct an exploratory study in which the neurophysiological reactions of respondents would be compared to one moment differentiating the two versions of the ad. Electroencephalography (EEG) has the benefit of very high temporal resolution; therefore, changes in brain activity can be monitored second by second [7].

Other studies indicated that the integration and comparison of motion information between visual and auditory modalities involves a specific network of both unimodal and multimodal cortical areas. The parietal cortex, and perhaps the lateral-frontal cortex, appear to be optimally situated to mediate the integration and attentional selection of motion information across modalities. However, interactions between the two modalities can involve both enhancing and suppressing effects, depending on the nature of the stimuli and the task performed by the subject [8].

20.3 Experimental Setup

The device used is an EEG device (Neuron-Spectrum-65) with specific software (Neuron-Spectrum.NET) used for data acquisition. This software acquires data relevant during an EEG examination and it presents various modules that can be used in different types of clinical examinations such as polysomnography and brain function monitoring.

The experimental part of this project started by selecting three participants (*Subject 1—Female/26 years old, Subject 2—Female/28 years old, Subject 3—Male/27 years old*). The grid EEG helmet was attached, after which the bridge electrodes were placed in the positions indicated by the software. The software helped check the correctness of electrode placement.

The grid EEG helmet was attached (see Fig. 20.1) after which the bridge electrodes were placed in the positions indicated by the software. Also, the software helped check the correctness of electrode placement (see Fig. 20.4), this type of bridge electrode is more difficult to place but has the advantage of being reusable.

According to the protocol, the subjects were presented with a collage of clips from different types of media (comedy, thriller, action) with the aim of eliciting a reaction from them. In the case of this experiment, the duration of data acquisition was between 10 and 15 min since the presented collage was approximately 12 min long.

The sampled data is in tabular, graphic, and image form (amplitude topography). In Table 20.1, we will find the position of the electrode from which the values were taken, the average and maximum amplitude for each position, as well as the compression rate of the nerve (Fig. 20.2).

At the end of data collection, the Neuron-Spectrum.NET software has the possibility to create and export data in PDF format. At the same time, the data and images used in the report can be customized and placed on the page according to the user's wishes. In addition, this program compiles and calculates various values, but also

Table 20.1 Data was acquired using Neuro-Spectrum.NET software

Electrode position	Maximum amplitude (μV)			Medium amplitude (μV)			Lempel–Ziv compression ratio of the nerve (%)		
	S1	S2	S3	S1	S2	S3	S1	S2	S3
FP1-R	670	1309	130	29	18	11	50.4	59	64.6
FP2-R	566	851	145	19	17	11	56.2	60.9	68.5
F3-R	258	874	244	15	16	16	62.5	60.9	65.8
F4-R	394	893	261	12	13	17	65	65	65.6
C3-R	302	869	291	11	10	23	70.8	73.2	62.9
C4-R	164	2386	293	8	12	21	77.4	78.8	64.2
P3-R	83	873	314	13	13	21	69.5	71.7	63.9
P4-R	295	874	329	11	12	20	73.1	75.2	64.3
O1-R	123	891	319	14	17	20	66.9	68.9	63.7
O2-R	279	890	332	14	16	20	65.9	70	64
F7-R	808	879	266	14	13	15	62.3	64.7	63.5
F8-R	132	895	261	12	14	14	62.3	61.9	63
T3-R	240	868	295	19	13	17	58.6	62.9	63.4
T4-R	300	862	281	14	12	17	62.8	68.5	64.3
T5-R	269	883	3077	15	16	17	63.6	68.5	62.6
T6-R	314	859	297	13	13	17	64.6	70.1	65..7
FPZ-R	1418	1105	118	23	19	11	53.4	58.4	72
FZ-R	114	861	249	10	10	15	72	72.8	67
CZ-R	168	871	294	1	9	18	85	79.9	64.6
PZ-R	136	882	313	11	12	20	72.4	74.6	64
OZ-R	117	882	328	14	18	21	66.3	69.1	64.1

provides data in graphic format, such as the duration of the user's attention on the material presented to him and the success rate of his concentration.

20.4 Results and Conclusions

In conclusion, it can be seen, both from the numerical and graphical data (see Fig. 20.3), that the three subjects, although they reacted differently to the presented clip, also had stronger reactions. At the same time, the rate of attention and the success of neurofeedback is relatively constant throughout the duration of the experiment.

For the first two subjects, there have been two exact moments where a longer disturbance period can be observed: from 00:05:49 to 00:06:47 for the first subject and from 00:00:01 to 00:04:11 for the second subject.

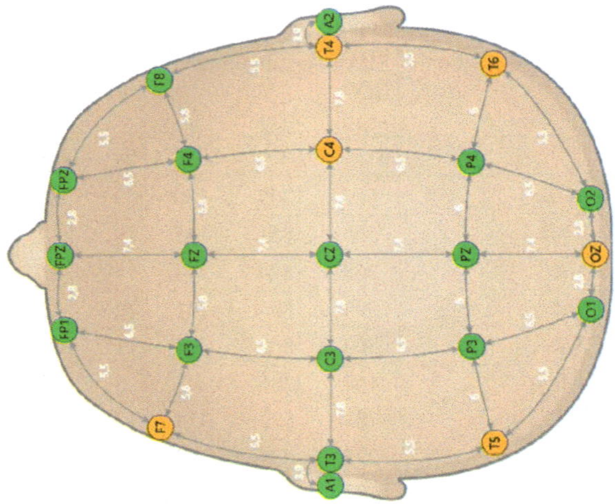

Fig. 20.2 The correctness of the placement of the electrodes as verified by the Software Neuron-Spectrum.NET software

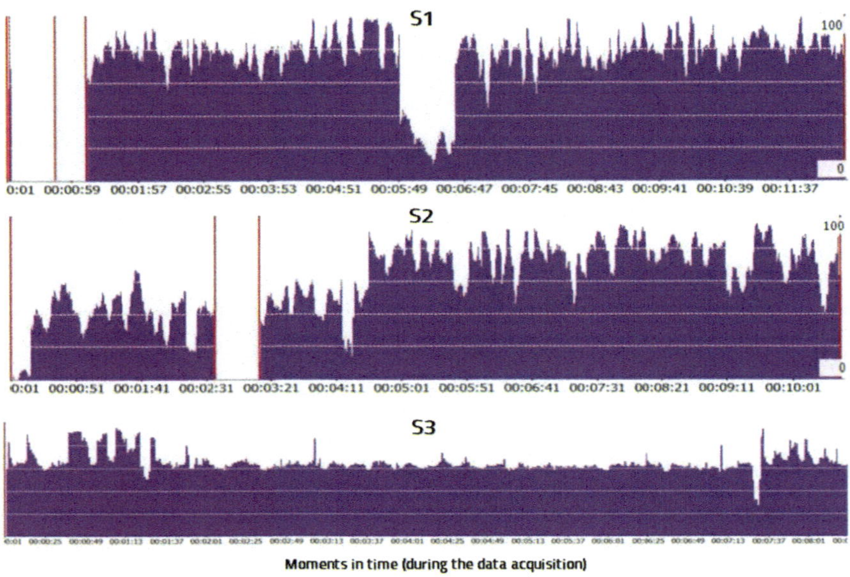

Fig. 20.3 Neurofeedback training success rate for the three subjects

The third subject has presented a very constant attention rate, the only moment where a perturbance can be observed is around the 00:07:37 mark.

It is mentioned that this experiment was not carried out in an environment devoid of external stimulus and because of this, disturbances may occur for the subjects and, implicitly, in the data that were retrieved.

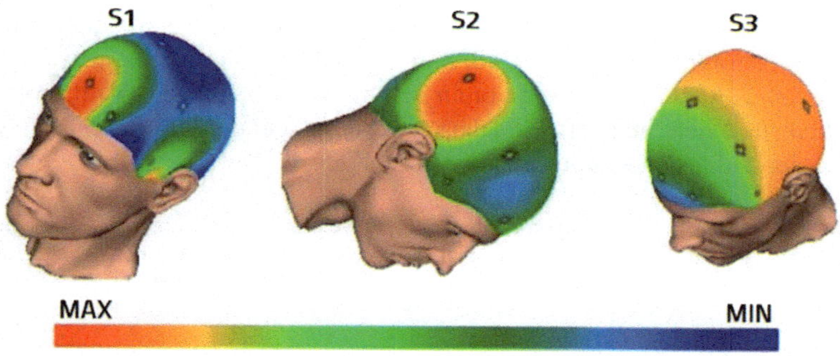

Fig. 20.4 Area of mean amplitudes for the three subjects

The imperfect contact of the electrodes on the skin surface must be considered for perspective on the results. Since this contact cannot be perfect and there will be disturbances during the potential acquisition, the results will be smoothed using the software with the help of various filters.

Another element that can be considered restrictive is the imperfection of the environment in which the experiment was carried out. As it presents external auditory and visual stimuli such as sounds from people/objects around the subject or sudden movements, they can change the result of the experiment.

At the same time, the rate of attention and the success of neurofeedback is relatively constant throughout the duration of the experiment. It is mentioned that this experiment was not carried out in an ideal environment, completely devoid of external stimulus and because of this, disturbances may occur for the subjects and, implicitly, in the data that were retrieved. However, these differences in data have been smoothed by the software using different types of filters and statistics.

Observing the areas of mean amplitudes for the three subjects it can be observed that for Subject 2 the biggest concentration is represented in the auditory area of the brain, which can lead to the conclusion that this subject was mainly focused on the sounds presented much more than the images.

For Subjects 1 and 3 the focal point is in the visual processing area, meaning that these subjects' points of interest were the images presented and the movements within them, more than the sounds they have heard in conjunction with them.

Another important point is that for Subject 3 the area of mean amplitudes is a lot larger than for the other two subjects. This, correlated with the neurofeedback training success chart (see Fig. 20.3) shows that the third subject was the one that maintained its focus on the clips presented for the longest. At the same time, he was also the only one to present the focal point of the area of mean amplitudes at the base of the head, exactly where the visual processing region is.

The results indicate, from the topography of the mean amplitudes, for the first subject a mean amplitude in the frontal area, for the second subject in the temporal-parietal area, and for the third subject in the occipital area (see Fig. 20.4). This experiment sought to demonstrate that different auditory and visual stimuli incite a cognitive response in the participant. Although there are external and internal disturbing factors, it can be concluded that certain visual and auditory stimuli, if strong enough, will influence the brain activity of the subjects. With the help of sophisticated devices and software, data acquisition and quantification can be made much easier and faster without intense user influence.

Acknowledgements In these experiments, we have developed the investigations with equipment from LIAMED's portfolio and the research is part of a project developed at Transilvania University of Brașov.

References

1. H. Blumenfeld, *Neuroanatomy Through Clinical Cases*, 2nd edn. (Oxford University Press, United Kingdom, 2010)
2. C. Molnar, J. Gair, S. Fowler, *Concepts of Biology*, 1st Canadian edn. (BCcampus, BC Open Textbook Project, VIC, 2015)
3. Khanacademy, *Unit 33, Lesson 2: The Neuron and Nervous System*, https://www.khanacademy.org/science/biology/human-biology/neuron-nervous-system/a/overview-of-neuron-structure-and-function. Accessed 18 Apr 2023
4. Visual Processing: Cortical Pathways, https://nba.uth.tmc.edu/neuroscience/m/s2/chapter15.html. Accessed 18 Apr 2023
5. How the Human Eye Sees, https://www.webmd.com/eye-health/amazing-human-eye, Accessed 18 Apr 2023
6. F. Lombardi, S. Pepić, O. Shriki et al., Statistical modeling of adaptive neural networks explains co-existence of avalanches and oscillations in resting human brain. Nat. Comput. Sci. **3**, 254–263 (2023)
7. T.P. Ghuntla, H.B. Mehta, P.A. Gokhale, C.J. Shah, A comparison and importance of auditory and visual reaction time in basketball players. Saudi J. Sports Med. **14**(1), 35–38 (2014)
8. NEURON-SPECTRUM-65, https://neurosoft.com/en/catalog/eeg/neuron-spectrum-61-65. Accessed 18 Apr 2023

Chapter 21
An Analysis of the Vibrations Transmitted to the Crew of a Ferry on the Danube

Mihaela Picu and Constantin Picu

Abstract There are areas over rivers without bridges, in which case the crossing from one bank to the other is done by ferry. This paper, examines the main quantities that characterize the whole body vibrations transmitted to the navigating personnel on a ferry on the Danube. The quantities used were those indicated by Directive 2002/44/EC of the European Parliament on the minimum health and safety requirements regarding the exposure of workers to the risks arising from physical agents (vibration). The Whole Body Vibrations (WBV), under different conditions, were measured on the x, y and z axes. WBV were measured by means of the 01 dB NetdB multichannel system, the SEAT-pad triaxial accelerometer, and the Piezotronics 356A16 PCB triaxial accelerometer. Data analysis was carried out with dBFA Suite—an acquisition and post-processing data software. The data collected during the experiments was processed, using the calculations provided by ISO 2631-1:1997 and 2631-5:2018. Parameters indicated by ISO 2631-1 include: 1. Mean square acceleration (A_w) calculated at floor level; 2. Partial exposure, which is vibration exposure (in m/s^2 A(8)) for each process and calculated based on the vibration size and the duration of exposure; 3. Vibration Dose Value (VDV). This quantity is more sensitive to impulse vibrations, reflecting the total vibration more accurately, than a mere mediation; 4. The length of time after which the exposure value triggering the action (T_{EAV}) is reached and 5. The exposure limit value (T_{ELV}). All the values characteristic of the vibrations transmitted to the navigating personnel are far exceeded.

M. Picu (✉) · C. Picu
"Dunarea de Jos", University of Galati, Galaţi, Romania
e-mail: Mihaela.Picu@ugal.ro

C. Picu
e-mail: Constantin.Picu@ugal.ro

N. Herisanu and V. Marinca (eds.), *Acoustics and Vibration of Mechanical Structures—AVMS-2023*, Springer Proceedings in Physics 302,
https://doi.org/10.1007/978-3-031-48087-4_21

21.1 Introduction

Along rivers are areas without bridges and the crossing from one bank to the other is done by ferry. This is the case for Brăila—Smârdan crossing area, Tulcea county (Figs. 21.1 and 21.2). The ferry operates around the clock (24/7).

Such a crossing poses several problems: the ferry does not sail in bad weather or at regular intervals (one must wait till the ferry is fully occupied before it sets sail). Also, these boats lack guidance equipment for bad weather, which has led to a series of accidents. In addition, whether the ferry is operational depends on the Danube water levels (if the Danube is very low or near the flooding limit, the crossings are stopped). Another important problem is the low ground guard of certain cars, which can make it difficult for them both to board and disembark).

Working on such a boat is particularly difficult; besides the unpredictable weather and the ensuing treacherous river, there is also the economic aspect: the boats are old, transmitting very high vibrations to the working staff, who cannot avoid exposure as there is no vibration-free area to take refuge in during the Danube crossing. The influence of vibrations on the crew has been taken into account by several international supervisory and regulatory organizations (IMO, ABS, Lloyd's Register, BV, DNV, etc.) that have imposed rules since the design of ships.

In the specialized literature, there are few works that analyze the effects of ship vibrations on seafaring personnel, and even fewer that refer to ships on rivers. Most studies address vibrations, their reduction, and their action on vessel structure and components [1]. We have found only a few papers that study the influence of ferry vibrations on rivers on the crew [2, 3].

Ship vibrations are usually measured with vibrometers from different manufacturers, and the level of fatigue induced by them is analyzed with various psychological tests and questionnaires [4].

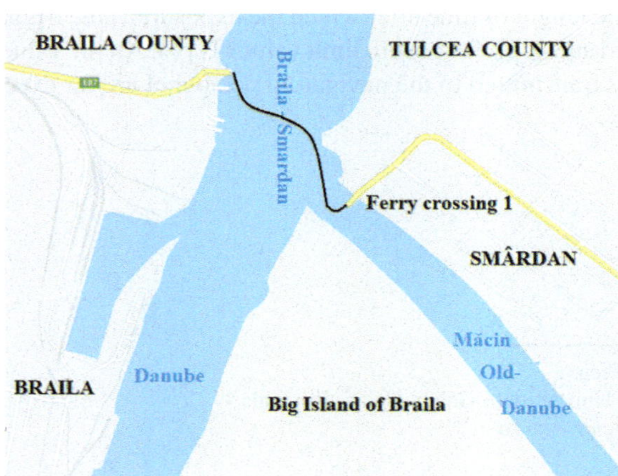

Fig. 21.1 Crossing by ferry from Braila to Smârdan

Fig. 21.2 Crossing the Danube by ferry

21.2 Materials and Methods

In this paper we will analyse the main values that characterize the vibrations transmitted to the whole body of the navigating personnel on a boat of this type; we will refer to the values indicated by DIRECTIVE 2002/44/EC OF THE EUROPEAN PARLIAMENT AND OF THE COUNCIL of 25 June 2002 on the minimum health and safety requirements regarding the exposure of workers to the risks arising from physical agents (vibration) [5].

According to this Directive, the limit values of the vibrations transmitted to man are:"for whole-body vibration: (a) the daily exposure limit value standardised to an eight-hour reference period shall be 1.15 m/s^2 or, at the choice of the Member State concerned, a vibration dose value of 21 $m/s^{1.75}$; (b) the daily exposure action value standardised to an eight-hour reference period shall be 0.5 m/s^2 or, at the choice ofthe Member State concerned, a vibration dose value of 9.1 $m/s^{1.75}$." [5]

There is a great deal of literature on the impact of vibrations on workers in the occupational environment [6–8]. Depending on the type of vibrations, their intensity and frequency, as well as the duration of exposure but also on the area most exposed to the action of the vibrational movements (on the whole body-WBV and on the hand-arm system-HAV), these workers are prone to various types of diseases.

Beard and Griffin [9] analyzed "the discomfort caused by a lateral oscillation and a compensated lateral oscillation on a roll, at frequencies between 0.25 and

1.0 Hz when sitting on a rigid seat and sitting on a cushion, both without a backrest" and concluded that "the frequency dependence of the anticipated discomfort can be improved by adjusting the frequency weight for the acceleration roll at frequencies between 0.25 and 1.0 Hz" [9].

The determinations took place in winter, when the sky was often overcast, the wind blew strongly, and high waves broke and splashed (Fig. 21.3).

The general objective of the paper is the analysis of the influence of vibrations generated by the equipment of the two ferries upon the navigating personnel that ensure the Danube crossing between Brăila and Măcin.

To fully meet the overall objective of the paper, it is necessary to measure and calculate specific parameters of the vibrations transmitted to people, as well as to analyse the discomfort of the navigating personnel.

After the details of the experiments were presented to them and after they gave their consent, ten subjects participated in the experiments in December 2022, each of them having a long experience in the field.

The vibration of the whole body, under different conditions, was measured on the 3 axes (x, y and z) of the centre of the human body. Whole-body vibrations were measured by means of the 01 dB NetdB multichannel system, with the SEAT-pad triaxial accelerometer mounted on the floor (Figs. 21.4 and 21.5). The axes were oriented in the directions specified in EN 1032. The accelerations were calculated using the factors established by ISO 2631 [10, 11]. The calibration of the accelerometers was made with VE-10 Rion.

Data analysis was done with dBFA Suite—Acquisition control and data post-processing software. The collected data during the experiments were processed, using

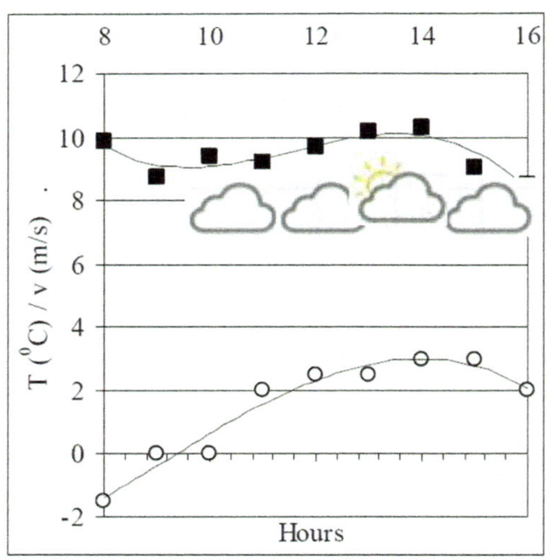

Fig. 21.3 Meteogram for studied period (white circle)—temperature (°C); (black square)—wind speed (m/s)

Fig. 21.4 Multichannel system NetdB 01dB

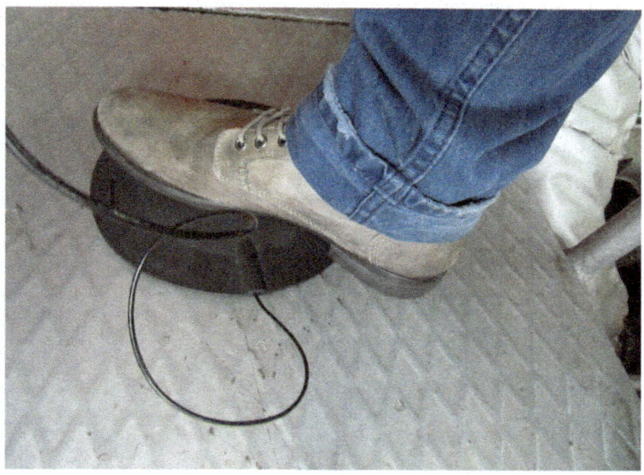

Fig. 21.5 SEAT-pad triaxial accelerometer

the calculations provided by ISO 2631-1/1997 and 2631-5/2018, using "Whole body vibration calculator" [10, 12, 13].

21.3 Results and Discussions

The parameters indicated by ISO 2631-1 that were measured and calculated were: Vibration magnitude, Exposure duration, Partial VDV, Partial exposure, Time to reach EAV (A(8) option), Time to reach ELV (A(8) option only), Total VDV, Total exposure (Tables 21.1, 21.2, 21.3 and 21.4). In all experiments, the exposure time is 30 min.

Although the indicated values in the Directive refer to 8 h of work and our own measurements were made over a period of 30 min (it would have been practically

Table 21.1 Measured and calculated parameters in case of the small ferry moving downstream

Vibration magnitude m/s² r.m.s	Partial VDV m/ s$^{1.75}$	Partial exposure m/s² A(8)	Time to reach EAV (A(8) option)		Time to reach ELV (A(8) option only)		Total VDV m/ s$^{1.75}$	Total exposure m/s² (8)
			h	min	h	min		
5.33334	48.634609	1.333335	0	4.218739	0	22.317131	78.3066	3.35855
5.35354	48.818812	1.338385	0	4.186963	0	22.149035		
5.51247	50.268091	1.378117	0	3.949014	0	20.890288		
5.49523	50.110880	1.373807	0	3.973831	0	21.021571		
5.55141	50.623184	1.387852	0	3.893808	0	20.598249		
5.65421	51.560615	1.413552	0	3.753508	0	19.856058		

Table 21.2 Measured and calculated parameters in the case of the small ferry moving upstream

Vibration magnitude m/s² r.m.s	Partial VDV m/ s$^{1.75}$	Partial exposure m/s² A(8)	Time to reach EAV (A(8) option)		Time to reach ELV (A(8) option only)		Total VDV m/ s$^{1.75}$	Total exposure m/s² (8)
			h	min	h	min		
5.98652	54.590943	1.496632	0	3.3483617	0	17.712833	86.2565	3.70074
6.02587	54.949772	1.506467	0	3.3047737	0	17.482253		
6.10457	54.667438	1.526145	0	3.2201128	0	17.034396		
5.98941	54.617297	1.497362	0	3.3451312	0	17.695744		
6.02598	54.950778	1.506495	0	3.3046530	0	17.481614		
6.12598	55.862676	1.531495	0	3.1976438	0	16.915536		

Table 21.3 Measured and calculated parameters in the case of the large ferry moving downstream

Vibration magnitude m/s² r.m.s	Partial VDV m/ s$^{1.75}$	Partial exposure m/s² A(8)	Time to reach EAV (A(8) option)		Time to reach ELV (A(8) option only)		Total VDV m/ s$^{1.75}$	Total exposure m/s² (8)
			h	min	h	min		
6.68147	60.928167	1.670367	0	2.6880491	0	14.219779	97.1739	4.16826
6.89222	62.850001	1.723055	0	2.5261723	0	13.363451		
6.74665	61.522552	1.686662	0	2.6363610	0	13.946349		
6.66666	60.793125	1.666666	0	2.7000054	0	14.283028		
6.98226	63.671074	1.745565	0	2.4614397	0	13.021016		
6.86533	62.604972	1.716332	0	2.5460006	0	13.468340		

impossible for a person on board the ferry to sit on a pad for 8 h on end), the partial results do not show great changes, so a parallel can be drawn with the standardized values.

Table 21.4 Measured and calculated parameters in the case of the large ferry moving upstream

Vibration magnitude m/s² r.m.s	Partial VDV m/s$^{1.75}$	Partial exposure m/s² A(8)	Time to reach EAV (A(8) option)		Time to reach ELV (A(8) option only)		Total VDV m/s$^{1.75}$	Total exposure m/s² (8)
			h	min	h	min		
7.66258	69.874897	1.915645	0	2.0437661	0	10.811522	108.021	4.63415
7.48922	68.294031	1.872305	0	2.1394792	0	11.317844		
7.51641	68.541976	1.879102	0	2.1240284	0	11.236110		
7.42115	67.673302	1.855287	0	2.1789076	0	11.526421		
7.68996	70.124575	1.922494	0	2.0292384	0	10.734671		
7.62216	69.506308	1.905544	0	2.0654996	0	10.926493		

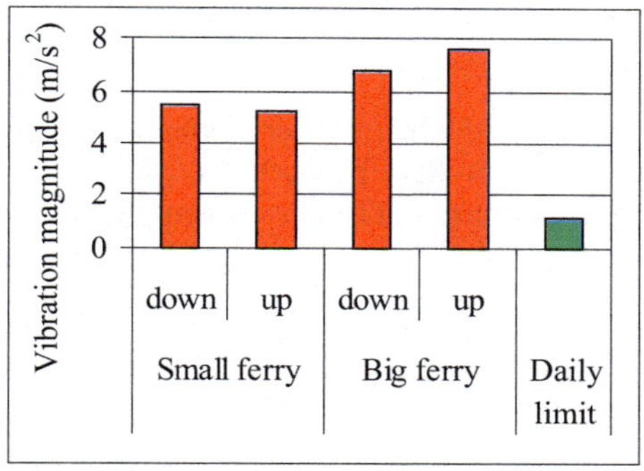

Fig. 21.6 Average vibrations magnitude versus daily exposure limit value standardized, where down = downstream and up = upstream

From Fig. 21.6 it can be seen that the average size of the vibrations is 5.45 times higher than the daily exposure limit value standardized (1.15 m/s²); similarly, Fig. 21.8 shows that the average total VDV is 4.4 times higher than the standardized limit value (21 m/s$^{1.75}$). From Fig. 21.7 it can be seen that the times to reach EAV and ELV, respectively, decrease if it is about the big ferry sailing upstream, while, naturally, the average total exposure increases (Fig. 21.9).

21.4 Conclusions

The ten subjects were asked to rate the discomfort created by the vibrations transmitted by the ferry, according to a 5-point Likert scale [14]. The classification of vibrations is done as follows [15, 16]:

Not detected	1
Weak	2

(continued)

(continued)

Not detected	1
Moderate	3
Strong	4
Too strong	5

From Tables 21.1, 21.2, 21.3 and 21.4 it can be seen how uncomfortable it is to work on such a means of transport. For the most uncomfortable case (big ferry—upstream), the vibration magnitude is 6.58 times higher than the daily limit, and the total VDV is 4.4 times higher. However, the ten subjects, all seniors, find these vibrations barely noticeable (Fig. 21.10).

Considering Stevens's Power Law, whose "general form is [17]:

$$\psi(\mathrm{I}) = \mathrm{k} \cdot \mathrm{I}^{\mathrm{a}} \tag{21.1}$$

where I is the intensity or strength of the stimulus in physical units (energy, weight, pressure, mixture proportions, etc.), $\psi(\mathrm{I})$ is the magnitude of the sensation evoked by the stimulus, a is an exponent that depends on the type of stimulation or sensory modality, and k is a proportionality constant that depends on the units used" [17, 18], by extrapolation the exponents of a can be obtained for the two situations:

- for downstream: $a = 0.8047$,
- for upstream: $a = 0.5521$.

Considering the intensity of the stimulus as total VDV, Stevens's Power Low can be written for each of the 4 analyzed cases:

- for downstream:

$$\text{Small ferry} \quad \psi = \mathrm{k}_1 \times 78.30^{-0.8047} \tag{21.2}$$

$$\text{Big ferry} \quad \psi = \mathrm{k}_2 \times 97.17^{-0.8047} \tag{21.3}$$

- for upstream

$$\text{Small ferry} \quad \psi = \mathrm{k}_3 \times 86.25^{-0.5521} \tag{21.4}$$

$$\text{Big ferry} \quad \psi = \mathrm{k}_4 \times 108.02^{-0.5521} \tag{21.5}$$

The experimental determinations show how dangerous is the work of the navigating personnel on this means of transport. All the vibration characteristic values transmitted to the navigating personnel are well over the accepted limit. Fortunately, once the construction of the bridge over the Danube has been completed, this rudimentary means of transport will be abandoned.

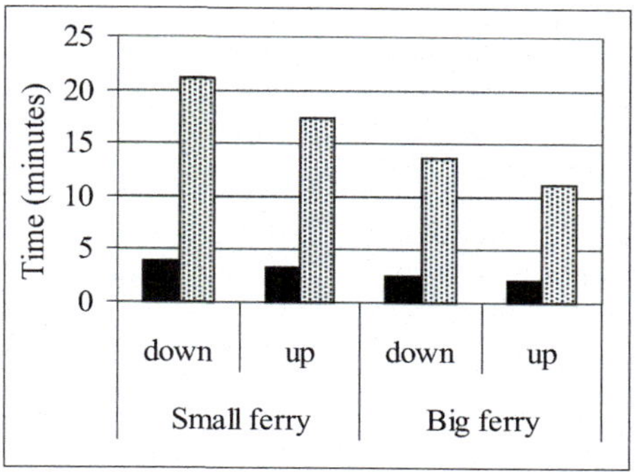

Fig. 21.7 Average times to reach EAV (black bar), respectively ELV (dotted bar), where down = downstream and up = upstream

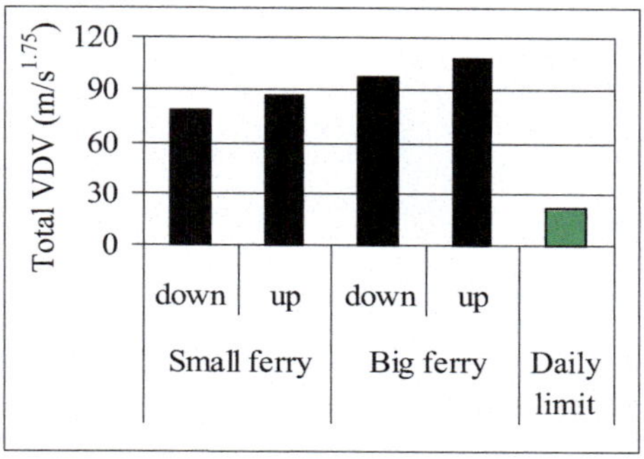

Fig. 21.8 Average total VDV versus daily exposure limit value standardised (21 $m/s^{1,75}$), where down = downstream and up = upstream

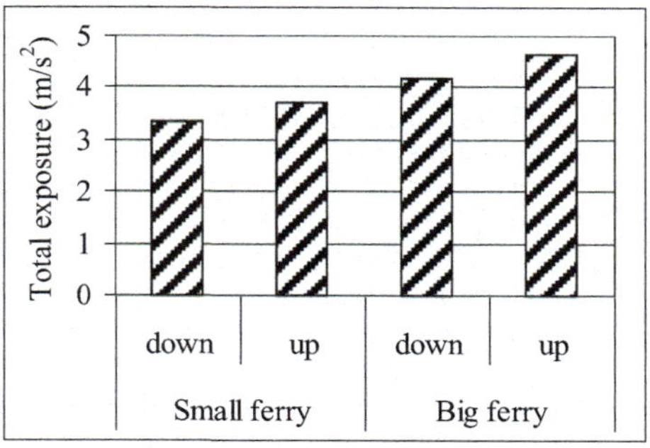

Fig. 21.9 Average total exposure, where down = downstream and up = upstream

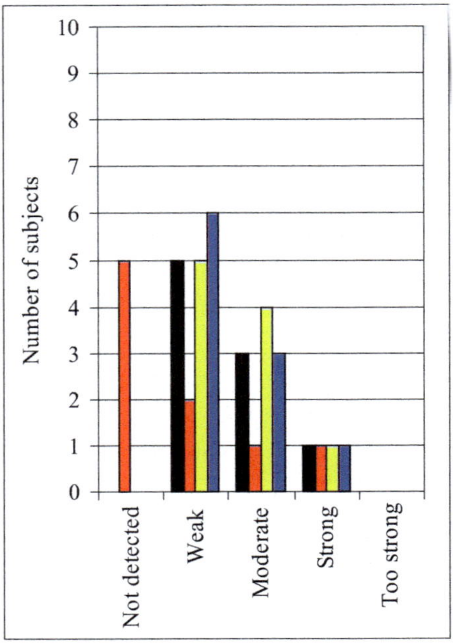

Fig. 21.10 Discomfort evaluation using Likert scale: (black color)—small ferry downstream; (red colour)—small ferry upstream; (yellow colour)—big ferry downstream; (blue colour)—big ferry upstream

References

1. O. Neculet, L. Crudu, Evaluation and control of induced vibrations for a passenger car ferry. Annals of "Dunarea de Jos" University of Galati. Fascicle XI Shipbuild **38**, 213–218 (2015)
2. D.G. Wilder, B.B. Woodworth, J.W. Frymoyer, M.H. Pope, Vibration and the human spine. Spine **7**(3), 243–254 (1982)
3. J. Lloret et al., Environmental and human health impacts of cruise tourism: a review. Mar Pollut. Bull. **173**, 112979 (2021)
4. V. Rumawas, B.E. Asbjørnslett, A content analysis of human factors in ships design. Int. J Mar. Eng. **156**(A3) (2014)
5. Directive 2002/44/EC of the European Parliament and of the Council of 25 June 2002 on the minimum health and safety requirements regarding the exposure of workers to the risks arising from physical agents (vibration), (sixteenth individual Directive within the meaning of Article 16(1) of Directive 89/391/EEC) (2021)
6. L. Picu, E. Rusu, M. Picu, An analysis of how physical and social factors influence the efficiency of crew members of a ship running on the Danube. Springer Proc. Phys. **251**, 185–204 (2021)
7. L. Picu, E.V.C. Rusu, Evaluation of the human discomfort caused by ship vibrations. J. Mar. Sci. Technol. **29**(1), 63–72 (2021)
8. L. Picu, M. Picu, E.V.C. Rusu, An investigation into the health risks associated with the noise and vibrations on board of a boat—a case study on the Danube River. J. Mar. Sci. Eng. **7**(8), 258 (2019)
9. G.F. Beard, M.J. Griffin, Discomfort of seated persons exposed to low frequency lateral and roll oscillation: effect of seat cushion. Appl. Ergon. **45**(6), 1547–1557 (2014)
10. ISO 2631-1:1997, *Mechanical Vibration and Shock. Evaluation of Human Exposure to Whole-Body Vibration. Part 1: General Requirements* (2021)
11. ISO 2631-4:2001, *Mechanical Vibration and Shock. Evaluation of Human Exposure to Whole-Body Vibration. Part 4: Guidelines for the Evaluation of the Effects of Vibration and Rotational Motion on Passenger and Crew Comfort in Fixed-Guideway Transport Systems* (2001).

12. ISO 2631-5:2018, *Mechanical Vibration and Shock. Evaluation of Human Exposure to Whole-Body Vibration. Part 5: Method for Evaluation of Vibration Containing Multiple Shocks* (2022).
13. https://www.hse.gov.uk/vibration/wbv/calculator.htm. Accessed 28 Feb 2023
14. https://en.wikipedia.org/wiki/Likert_scale. Accessed 22 Feb 2023
15. S.S. Stevens, On the psychophysical law. Psychol. Rev. **64**(3), 153–181 (1957)
16. H.V.C. Howarth, M.J. Griffin, The frequency dependence of subjective reaction to vertical and horizontal whole-body vibration at low magnitudes. J. Acoust. Soc. Am. **83**(4), 1406–1413 (1988)
17. https://en.wikipedia.org/wiki/Stevens%27s_power_law, Accessed 22 Feb 2023
18. L. Picu, M. Picu, E.V.C. Rusu, in *18th International Multidisciplinary Scientific GeoConference (SGEM 2018), Conference Proceedings*. Evaluation of Human Exposure to Whole-Body Vibration—Verification Method of Stevens's Power Law, vol. 18, no. 5.2 (Albena, Bulgaria, 2018), pp. 413–420

Chapter 22
Vibration Analysis of Large Eccentric Press Installed on a Concrete Base—A Case Study

Dragan S. Jovanović, Milena Mančić, Miomir Raos, Milan Protić, Marko Mančić, and Milan Banić

Abstract Production processes sometimes cause production machines to be placed outside the technical norms for the way of isolation and foundation of large impact machines. Modern production requires the acquisition of complex machines after the construction of production halls, and then there are no technical possibilities for their proper foundation, so the question arises whether such machines can function safely in the production process. One such instance occurred at the manufacturing plant placed in city of Niš, where a productive eccentric press with a force of around 2500 kN was purchased for punching and cutting metal parts. This paper provides insight into the procedure for checking the impact of mechanical equipment vibrations in working mode in relation to the base on which it is placed, the safety of the operator, the surrounding equipment and the technological process.

22.1 Introduction

Sometimes, the isolation and foundation of huge impact machines [1] require the placement of production equipment outside the bounds of accepted technical practice [2, 3]. After building production halls, modern production necessitates the acquisition of complicated machines, but because there are no technological options for their suitable foundation at that point, the question of whether such machines can operate safely in the production process emerges. One of the ways to check machine safety in relation to technological requirements, operators and working environment is the analysis of machine vibrations in real working conditions [4, 5]. One such case was

D. S. Jovanović · M. Mančić · M. Banić
Faculty of Mechanical Engineering, University of Niš, Aleksandra Medvedeva 14, 18104 Niš, Serbia

M. Mančić (✉) · M. Raos · M. Protić
Faculty of Occupational Safety, University of Niš, Čarnojevića 10a, 18106 Niš, Serbia
e-mail: milena.mancic@znrfak.ni.ac.rs

N. Herisanu and V. Marinca (eds.), *Acoustics and Vibration of Mechanical Structures—AVMS-2023*, Springer Proceedings in Physics 302,
https://doi.org/10.1007/978-3-031-48087-4_22

in the production facility in Niš, where a productive eccentric press for punching and cutting metal parts with a force of ~ 2500 kN was acquired. The machine is placed on a concrete base made with metal pins without a foundation base, with previously performed soil analyzes and foundation with stone aggregate and a layer for normal industrial loads. A legitimate question arises as to whether and how the impact of the machine during the production process affects the bearing capacity of the concrete base [6], the safety of workers in the environment, the surrounding machines and equipment, and the smooth running of the technological process.

Vibrations are transmitted to the foundation with great damping, due to the structure of the press, due to the rubber pads of the press supports, and as a result a standardized measurement of the impact of vibrations on the foundation, as prescribed by the ISO 4150-3 standard, are conducted [5, 7, 8].

22.2 Measurement of the Effect of the Press in the Working Mode of Punching and Cutting

In order to establish the impact of the press operation on the substrate and the working environment, a simulation of the real workload and appropriate measuring equipment were prepared to measure the parameters of the operation and the impact on the working environment and the substrate [6, 9]. Figure 22.1 shows the tested press and its installation in the production facility.

Fig. 22.1 Press with a force of 2500 kN

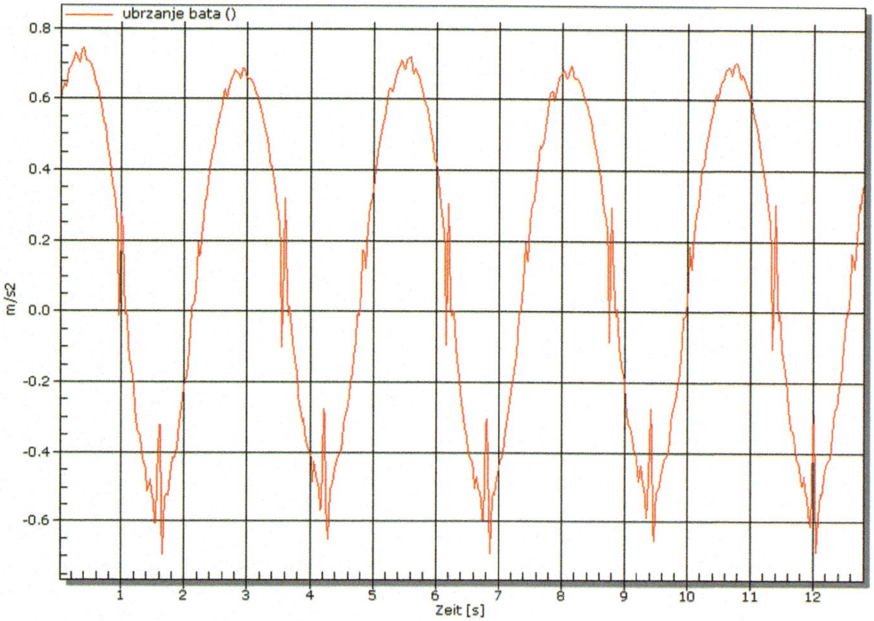

Fig. 22.2 Measurement of press piston acceleration

Measuring equipment consisting of a B12/200 accelerometer and a Quantum MX 830A measuring amplifier with accompanying Catman software, all produced by HBM from Germany, and a Vibroport 80 device with an AS 063 accelerometer from Bruel & Kjaer from Denmark were used for the measurements. The accelerometers for measuring on the press were attached with a magnet, and for measuring the substrate they were placed over the same magnet on a gravity weight of 5 kg.

The initial acceleration flow measured on the press piston is shown in Fig. 22.2.

If it is assumed that the weight of the moving part of the working table of the press is approximately 10 t, the approximate inertial force with which the press bat acts on the supporting structure of the press can be obtained of a little more than 80 t. Part of that force is transferred to the object being processed and the supporting structure of the press, and the most interesting part is the part that is transferred to the table of the press, and that part is shown in Fig. 22.3.

Only a part of the inertial forces is transmitted to the environment, while the technological forces during the operation of the press are transmitted to the support structure of the press and they are not analyzed in this paper, only those forces that cause impacts on the environment of the press. The press is freely supported by rubber pads on four supports on the concrete base of the hall. The measurement of vibrations on the concrete base was carried out using the same methodology, using a gravity weight of 5 kg freely placed on the concrete base, as shown in Fig. 22.4.

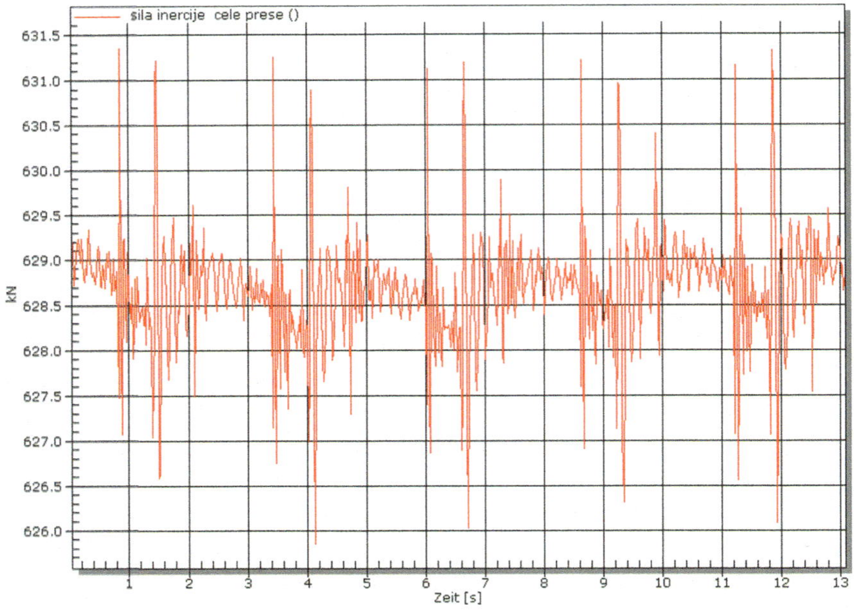

Fig. 22.3 Inertial force on the press table during punching and cutting

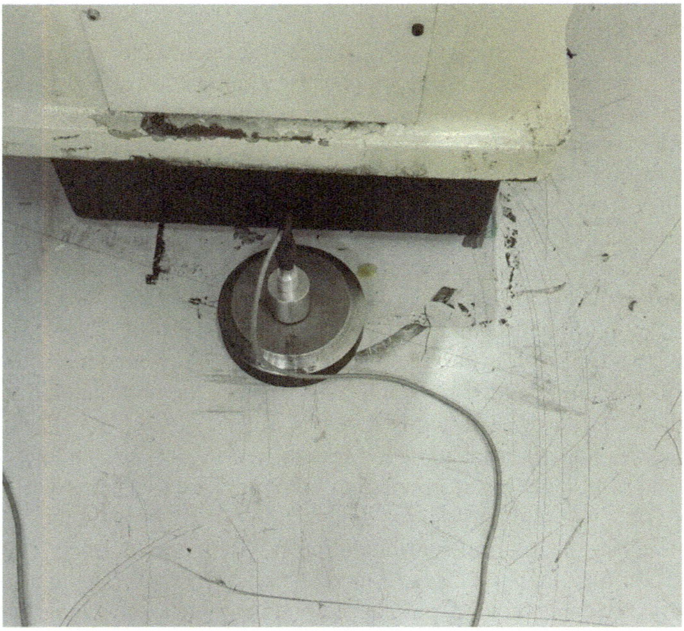

Fig. 22.4 Measurement of acceleration under the support of the press immediately next to the rubber pad

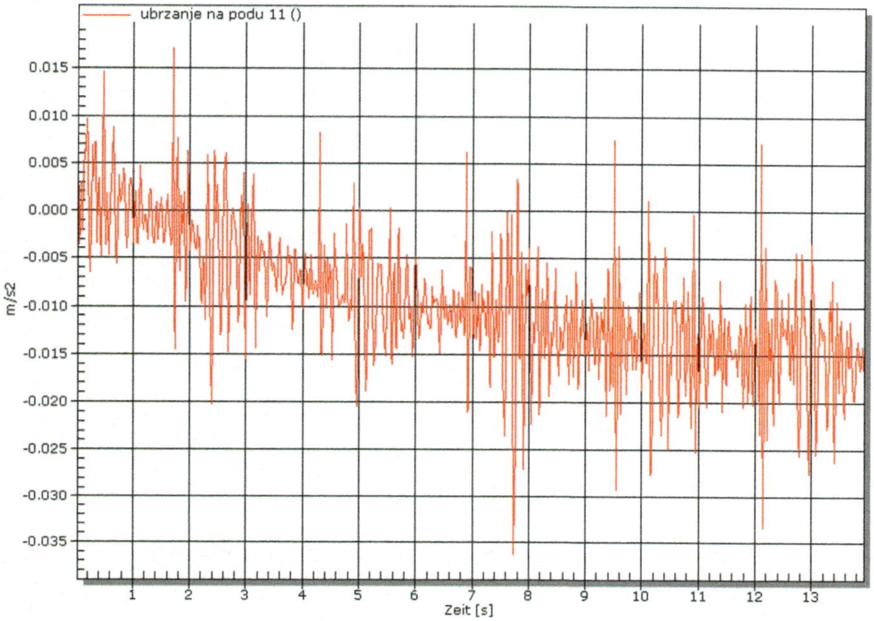

Fig. 22.5 Results of measurement of acceleration next to the support of the press in working mode

Figure 22.5 shows the result of vibration measurement directly next to the press support.

As it was evident from the measurement results that vibrations are transmitted to the foundation with great damping, due to the structure of the press, due to the rubber pads of the press supports, it has been moved to the standardized measurement of the impact of vibrations on the foundation as prescribed by the ISO 4150-3 standard. For this purpose, a raster with measuring points was made around the machine in order to monitor the distribution of vibrations in the space around the machine. The raster and measured vibration velocities are shown in Fig. 22.6.

From the obtained measurement results, it is noticeable that the highest values are obtained on the base, under the very supports of the press, and with the distance in all directions they are significantly weaker, which means that the damping effect of the concrete slab of the hall appears. The measured values were within the limits prescribed by the standard for vibrations of construction objects. However, despite the good measured values, the obtained results were verified by simulation.

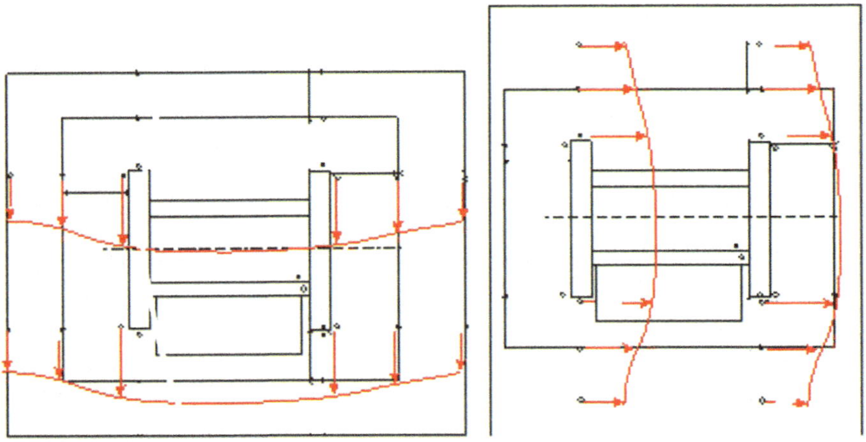

Fig. 22.6 Vibrations on the floor around the machine in a uniform grid

22.3 Simulation of the Stress State of the Reinforced Concrete Slab and Results

In order to carry out the simulation, it was necessary to create a suitable CAD model of the press itself. However, the press arrived with the necessary technological instructions but without technical drawings, so it was necessary to make a model of the press using the appropriate method. For modeling, a photogrammetric method was applied, which discretized the press to the appropriate number of points, as shown in Fig. 22.7.

By creating a CAD model and comparing the basic data with the data from the press documentation, it was established that the model is close enough that the simulation results could be valid. The geometric model was transformed into a discretized model using the automated finite element mesh generation model, which gave a mesh with 275,402 nodes forming 54,361 elements.

As it is shown on Fig. 22.9, there are results of maximum plate deformation and maximum contact pressure during press operation in the vertical direction, and the maximum contact pressure is 3.3923 MPa.

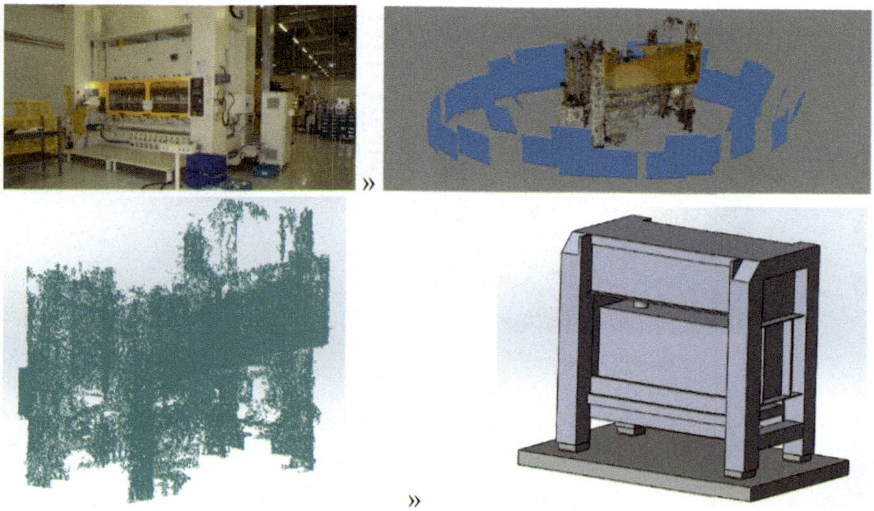

Fig. 22.7 Photogrammetric method for making a press-phase model

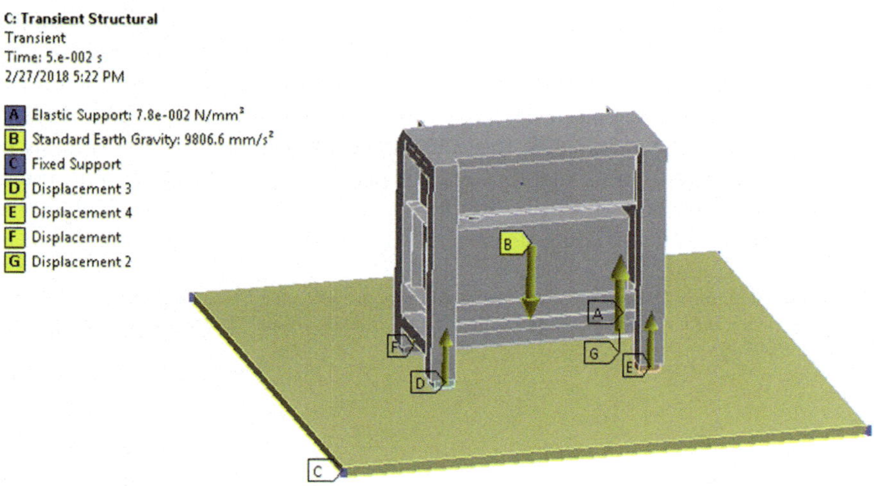

Fig. 22.8 Dynamic constraints in the simulation

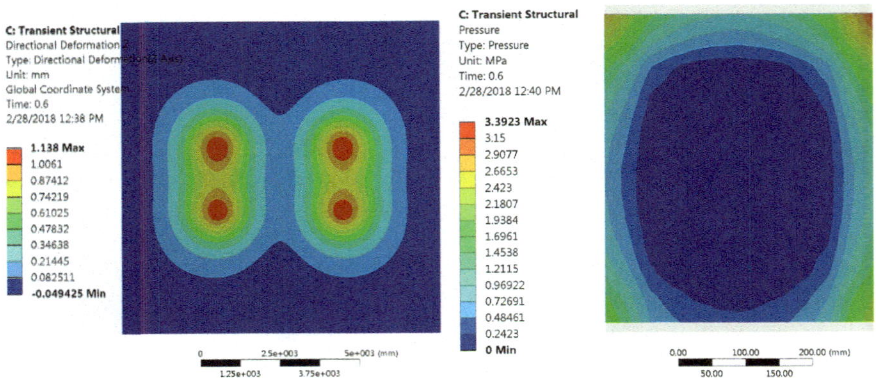

Fig. 22.9 Results of maximum plate deformation and maximum contact pressure during press operation in the vertical direction

22.4 Conclusion

Based on the results of measuring the vibrations of the press itself and its impact on the concrete foundation and the environment, it is shown that despite the faulty foundation of the machine itself, it can work in the production hall. The results were also confirmed by the performed simulation and have been confirmed for several years as long as the press has been operating at full capacity in the production plant. As a conclusion, it can be stated that the conducted measurement and analysis of the vibration conditions of the concrete foundation provide the basis for correct conclusions about the way of placing the machines in the production hall and its possible impact on the building structure and the working environment.

Acknowledgements This research was financially supported by the Ministry of Education, Science and Technological Development of the Republic of Serbia, according to contract No. 451-03-47/2023-01/ 200148.

References

1. I. Chowdhury, S.P. Chowdhury, *Dynamics of Structure and Foundation—A Unified Approach: 2 Applications*, 1st edn. (CRC Press, London, 2008)
2. M. Bharathi, in *Proceedings of Indian Geotechnical Conference*. Modelling of Impact Type Machine Foundations. Paper No. H834. Delhi (2012)
3. A. Amiri, S.N. Moghaddas Tafreshi, A.R. Dawson, Vibration response of machine foundations protected by use of adjacent multi-layer geocells. Geotext. Geomembr. **51**(4), 15–35 (2023)
4. K. Sudhir, in *Modeling and Analysis of Passive Vibration Isolation Systems*. Vibration Isolation System Modeling (Elsevier, 2021), pp. 61–90
5. D. Jovanovic, M. Jovanovic, M. Raos, N. Zivkovic, M. Stankovic, M. Protic, Vibration analysis of insufficiently repaired well pump—a case study. Appl. Mech. Mater. **801**, 207–212 (2015)

6. S. Saran, *Foundations of Impact Machines*, 1st edn. (CRC Press, Boca Raton, 2021)
7. D. Jovanović, M. Jovanović, N. Živković, L. Živković, M. Raos, Belt conveyor drive gear-box problem caused by unpaired gears-a case study. Facta Universitatis, Series Working and Living Environmental Protection **12**(1), 111–121 (2015)
8. ISO 4150-3 Standard, *Vibrations in Buildings/Part 3 Effects on Structures*
9. C. Scheffer, P. Girdhar, *Practical Machinery Vibration Analysis and Predictive Maintenance*, 1st edn. (Elsevier, 2004)
10. D. Jovanović, N. Živković, M. Raos, L. Živković, M. Jovanović, M. Prašćević, Testing of level of vibration and parameters of bearings in industrial fan. Appl. Mech. Mater. **430**, 118–122 (2013)
11. ISO 10816-1, *Mechanical Vibration—Evaluation of Machine Vibration by Measurements on Non-Rotating Parts—Part 1: General Guidelines*
12. S. Jovanović, D. Jovanović, *Dinamičko uravnoteženje kao najvažniji postupak za poboljšanje dinamičkog ponašanja ventilatora* (XXII Konferencija Buka i vibracije, Niš, 2012), pp. 161–164

Chapter 23
Designing an Inertial System for Transforming Rotational Motion into Unidirectional Linear Motion

Attila Gerőcs, Lavinia Culda, Elena Muncut, and Andrei Komjaty

Abstract The present work presents a device with applicability in the field of motor vehicle construction. It is an inertial system that is able to create an inertial force used for the translational movement of the vehicle on which it is mounted without the need to create a kinematic chain from this device to the wheels of the vehicle. The following are addressed in the work: preliminary kinematic scheme, 3D device design in the Solid Works program, mathematical calculations and graphic representations for: displacement, velocity, impulse and inertial force. Finally, the practical realization of the device.

23.1 Introduction

When we talk about a car, one of the specifications we look at is the type of traction the car is equipped with.

The inertial system that we are presenting is a system that, with the help of liquid masses with variable weight in rotational motion, allows the realization of a unidirectional linear movement without the need to use a kinematic chain from this system to the car's wheels on which is mounted, [1–3]. The original element of the system that we will describe in the following, consists in the circulation of a liquid between 8 cylinders placed on a circle of radius R and divided into equal angles, so that the amount of liquid in these cylinders varies from a minimum to a maximum and vice versa at a full rotation. The rotation of the cylinders is done with a constant angular velocity, [4, 5].

A. Gerőcs · L. Culda (✉) · E. Muncut · A. Komjaty
University Aurel Vlaicu of Arad, Revolution 77, Arad, Romania
e-mail: culda.lavinia@yahoo.com

A. Gerőcs · E. Muncut
Babeș-Bolyai University, Cluj-Napoca, Romania

A. Gerőcs · A. Komjaty
University of Szeged, Mars tér 7, Szeged, Hungary

N. Herisanu and V. Marinca (eds.), *Acoustics and Vibration of Mechanical Structures—AVMS-2023*, Springer Proceedings in Physics 302,
https://doi.org/10.1007/978-3-031-48087-4_23

The study is limited to a single cylinder out of the 8 considered. And finally, to get the effect of the 8 cylinders, we multiply the resulting force by 8. The rotational movement of a cylinder can be decomposed into two translational movements perpendicular to each other. In the following we will consider only one of the two translational motions. And the solution for canceling the effect of the second translation is achieved by using two identical systems described above, where the rotation is in opposite directions.

23.2 Description of the Inertial System Proposed to Replace the Classic Transmission

In these cylinders liquid (usually water or hydraulic oil) is introduced in a controlled manner, the amount of which is synchronized with the angular position of the cylinders 1.1; 1.2 … 1.4′. This angular position is synchronized with the linear position of the cylinder′s assembly 9, piston 10 so that the cylinder 1.1 on the left side has the maximum level of liquid, which during the 180° rotation will decrease in a controlled manner when the cylinder 1.1 will take the position cylinder 1.1′, respectively after a further rotation up to 360°, the liquid level will reach the maximum again, in which case the cylinder 1.1 will take the initial position, according with Fig. 23.1.

For the clearer description of the system, in Fig. 23.1, we used the following notations: 1.1 … 1.4’—cylinders in rotation, 2—piston stopper, 3—cylinder support arm, 4- and 5-cylinder feed/empty pipes, 6—cylinder support shaft, 7—fixed support for shaft, bearings and sealing rings, 8—cylinder sealing piston, 9—control cylinder, 10—control piston, 11—piston control guide hole, 12—sliding rod, 13—steering wheel, 14 and 15—bearing, 16—fixed arm, 17—sealing ring, 18—bolt, 19—kinematic chain (ratio – 1:1).

The introduction and removal of the liquid from the cylinders is achieved by means of conduits 4 and 5 with the help of the cylinder 9 and the piston 10, the piston whose movement is made with the help of the rod 12, the bolt 18 mounted on the wheel 13 and the slot 11 mounted on the rod 12. The rotational movement of the wheel 13 is synchronized with the rotational movement of the axis 6 by means of a kinematic chain so that at a complete rotation of the axis 6 the wheel 13 do a complete rotation.

To each assembly of two diametrically opposed cylinders is attached an assembly of piston 10, cylinder 9, rod 12, slot 11, bolt 18, wheel 13, assemblies synchronized in such a way that the maximum position of the liquid level in each cylinder corresponds the same angular position, i.e. the position of cylinder 1.1 according to Fig. 23.1. So more concretely, if we have eight cylinders, 4 assemblies of piston 10, cylinder 9, rod 12, slot 11, wheel 13, bolt 18 are required.

Next, we moved on to the design of the device shown in Fig. 23.2 which consists of eight vertical cylinders 1.1; 1.2; 1.3; 1.4; 1.1′; 1.2′; 1.3′ and 1.4′ assembled by means of eight arms 3 on a vertical axis 6, arms equally spaced at an angle of 45°.

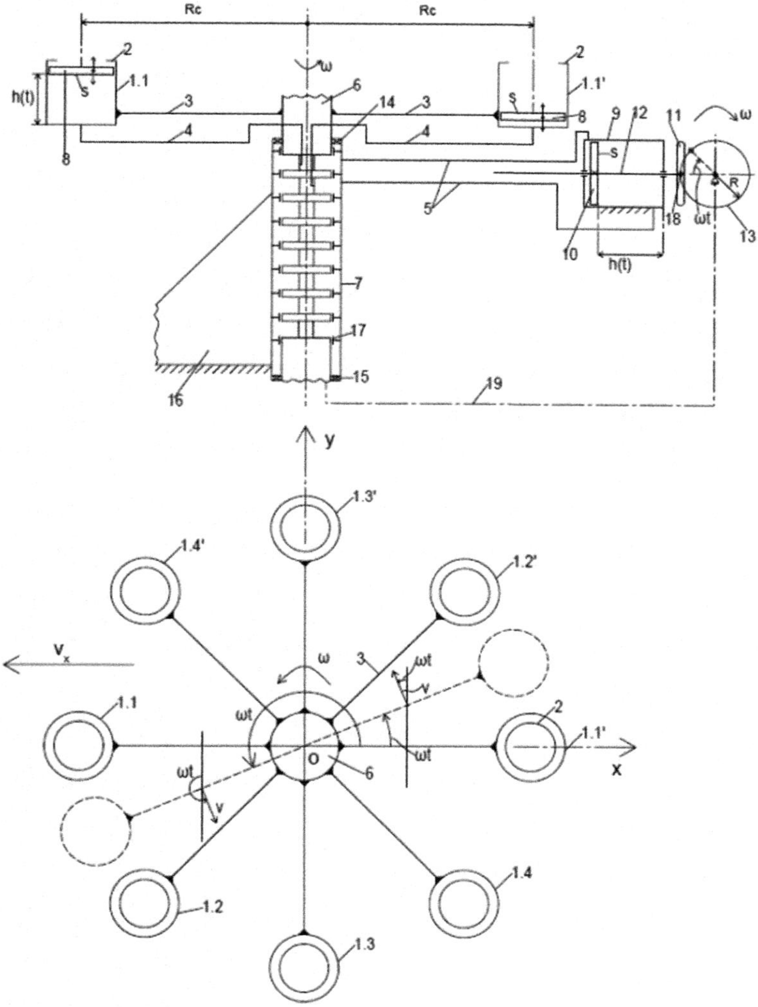

Fig. 23.1 Model of the inertial system in the design phase

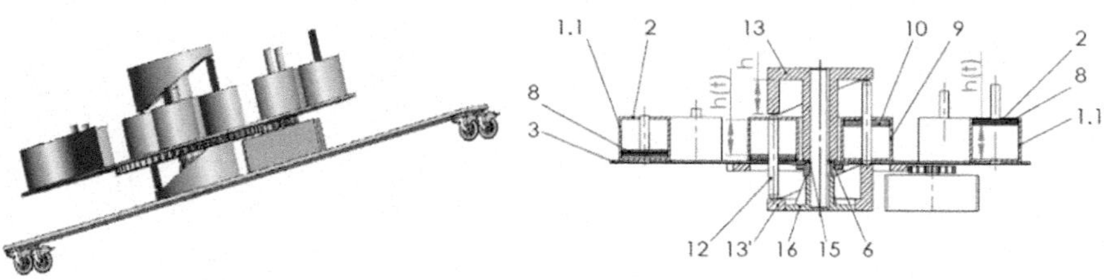

Fig. 23.2 3D drawing of the inertial system viewed from different positions

If we write the displacement law along the Oy direction of the liquid in cylinder 1.1′ it results the Eqs. (23.1):

$$\begin{aligned} y1(t1) &= Rsin(\omega t1); \\ y2(t2) &= Rsin(\omega t2); \\ y3(t3) &= Rsin(\omega t3); \\ y4(t4) &= Rsin(\omega t4). \end{aligned} \tag{23.1}$$

To make the calculation we will use numerical values.

Where: $\omega := 100\,(\mathrm{rad/s})$—the angular speed; $\rho := 1000\,(\mathrm{kg/m^3})$—the density of the liquid in the cylinders; $d_p := 0.2\,(\mathrm{m})$—the diameter of the cylinder; $R := 0.75\,(\mathrm{m})$—the radius of the arm of the cylinders in rotational motion; $h := 0.15\,(\mathrm{m})$—the height of the liquid in the cylinders.

The time t belongs from 0, to and $2\pi/100$ is divided into four intervals, described by the Eqs. (23.2).

$$\begin{aligned} t1 &:= 0, \frac{\pi}{18,000} \cdots \frac{\pi}{200};\ t2 := \frac{\pi}{200}, \frac{\pi}{18,000} + \frac{\pi}{200} \cdots \frac{\pi}{100}; \\ t3 &:= \frac{\pi}{100}, \frac{\pi}{18,000} + \frac{\pi}{100} \cdots \frac{3\pi}{200};\ t4 := \frac{3\pi}{200}, \frac{\pi}{18,000} + \frac{3\pi}{200} \cdots \frac{2\pi}{100}. \end{aligned} \tag{23.2}$$

Next, the mathematical Eqs. (23.3) are presented, for the variation of the liquid volume in cylinders 1.1… 1.4′.

$$\begin{aligned} m1(t1) &:= \left(h - \frac{2h}{\pi}(\omega t1)\right);\ m2(t2) := \left(h - 2h + \frac{2h}{\pi}\omega t2\right)\sin(\omega t2); \\ m3(t3) &:= \left(-h + \frac{2h}{\pi}\omega t3\right)\sin(\omega t3);\ m4(t4) := \left(5h - \frac{2h}{\pi}\omega t4\right)\sin(\omega t4) \end{aligned} \tag{23.3}$$

The equations of the mass variation in the cylinders are shown in Eqs. (23.4) and by the graph in Fig. 23.3:

$$\begin{aligned} M1(t1) &:= \rho\cdot\left(h - \frac{2\cdot h}{\pi}\cdot(\omega\cdot t1)\right)\cdot\left(\frac{\pi\cdot d_p^2}{4}\right);\ M2(t2) \\ &:= \rho\cdot\left(h - 2\cdot h + \frac{2\cdot h}{\pi}\cdot\omega\cdot t2\right)\cdot sin(\omega\cdot t2)\cdot\left(\frac{\pi\cdot d_p^2}{4}\right); \\ M3(t3) &:= \rho\cdot\left(-h + \frac{2\cdot h}{\pi}\cdot\omega\cdot t3\right)\cdot sin(\omega\cdot t3)\cdot\left(\frac{\pi\cdot d_p^2}{4}\right);\ M4(t4) \end{aligned}$$

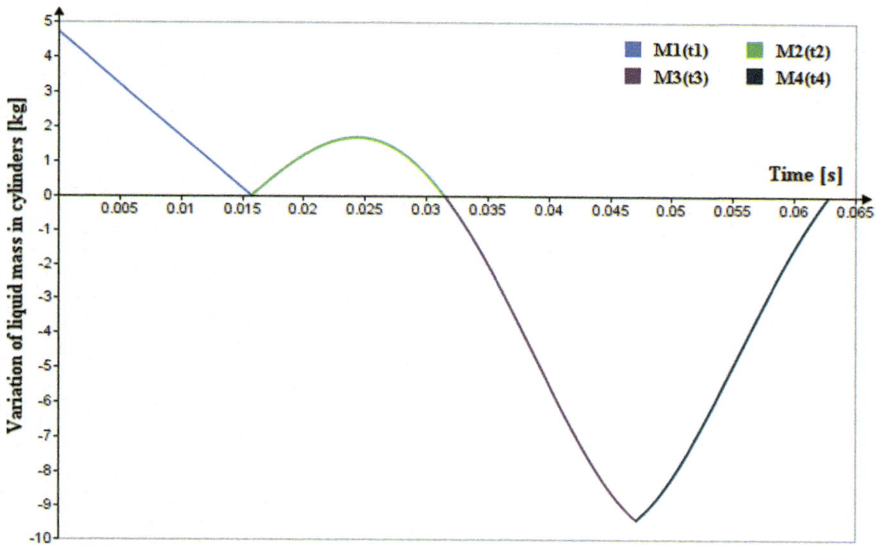

Fig. 23.3 Variation of liquid mass in cylinders as a function of time

$$:= \rho \cdot \left(5 \cdot h - \frac{2 \cdot h}{\pi} \cdot \omega \cdot t4\right) \cdot sin(\omega \cdot t4) \cdot \left(\frac{\pi \cdot d_p^2}{4}\right). \tag{23.4}$$

Deriving the displacement Eqs. (23.1) of the weights in the vertical cylinders leads to obtaining the speed.

To obtain the impulses of the rotating masses, multiply Eqs. (23.4) with speed and get the impulse formulas in the form of Eqs. (23.5) for each time interval and the graph in Fig. 23.4:

$$P1(t1) := M1(t1)v1(t1);\ P1(t1)$$
$$:= \rho\left(h - \frac{2h}{\pi}(\omega t1)\right)\left(\frac{\pi d_p^2}{4}\right)\omega R\cos(\omega t1);$$
$$P2(t2) := M2(t2)v2(t2);\ P2(t2)$$
$$:= \rho\left(h - 2h - \frac{2h}{\pi}\omega t2\right)\sin(\omega t2)\left(\frac{\pi d_p^2}{4}\right)\omega R\cos(\omega t2);$$
$$P3(t3) := M3(t3)v3(t3);\ P3(t3)$$
$$:= \rho\left(-h + \frac{2h}{\pi}\omega t3\right)\sin(\omega t3)\left(\frac{\pi d_p^2}{4}\right)\omega R\cos(\omega t3);$$
$$P4(t4) := M4(t4)v4(t4);\ P4(t4)$$

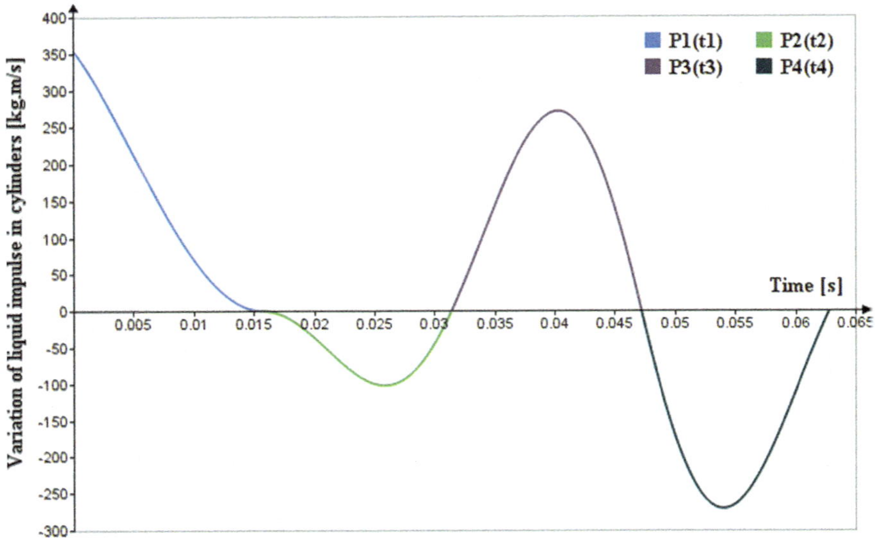

Fig. 23.4 Variation of impulse as a function of time

$$:=\rho\left(5h-\frac{2h}{\pi}\omega t4\right)\sin(\omega t4)\left(\frac{\pi d_p^2}{4}\right)\omega R\cos(\omega t4) \tag{23.5}$$

And if we derive the Eqs. (23.5) we obtain the equations of the inertial forces produced by the vertical cylinders expressed by the Eqs. (23.6–23.9) and the graph in Fig. 23.5:

$$F1(t1):=\frac{\left(-\left(d_p^2\cdot\rho\cdot\omega^2\cdot R\cdot h\cdot\pi\right)+2\cdot d_p^2\cdot\rho\cdot\omega^3\cdot R\cdot h\cdot t1\right)\cdot sin(\omega\cdot t1)}{4}-\frac{2\cdot d_p^2\cdot\rho\cdot\omega^2\cdot R\cdot h\cdot cos(\omega\cdot t1)}{4} \tag{23.6}$$

$$F2(t2):=\frac{\left(d_p^2\cdot\rho\cdot\omega^2\cdot R\cdot h\cdot\pi-2\cdot d_p^2\cdot\rho\cdot\omega^3\cdot R\cdot h\cdot t2\right)\cdot sin(\omega\cdot t2)^2}{4}+\frac{2\cdot d_p^2\cdot\rho\cdot\omega^2\cdot R\cdot h\cdot\cos(\omega\cdot t2)\cdot sin(\omega\cdot t2)}{4}+\frac{\left(2\cdot d_p^2\cdot\rho\cdot\omega^3\cdot R\cdot h\cdot t2-d_p^2\cdot\rho\cdot\omega^2\cdot R\cdot h\cdot\pi\right)\cdot cos(\omega\cdot t2)^2}{4} \tag{23.7}$$

$$F3(t3):=\frac{\left(d_p^2\cdot\rho\cdot\omega^2\cdot R\cdot h\cdot\pi-2\cdot d_p^2\cdot\rho\cdot\omega^3\cdot R\cdot h\cdot t3\right)\cdot sin(\omega\cdot t3)^2}{4}$$

$$+\frac{2\cdot d_p^2\cdot\rho\cdot\omega^2\cdot R\cdot h\cdot\cos(\omega\cdot t3)\cdot sin(\omega\cdot t3)}{4}$$
$$+\frac{\left(2\cdot d_p^2\cdot\rho\cdot\omega^3\cdot R\cdot h\cdot t3-d_p^2\cdot\rho\cdot\omega^2\cdot R\cdot h\cdot\pi\right)\cdot cos(\omega\cdot t3)^2}{4} \tag{23.8}$$

$$F4(t4):=\frac{\left(-\left(5\cdot d_p^2\cdot\rho\cdot\omega^2\cdot R\cdot h\cdot\pi\right)+2\cdot d_p^2\cdot\rho\cdot\omega^3\cdot R\cdot h\cdot t4\right)\cdot sin(\omega\cdot t4)^2}{4}$$
$$-\frac{2\cdot d_p^2\cdot\rho\cdot\omega^2\cdot R\cdot h\cdot\cos(\omega\cdot t4)\cdot sin(\omega\cdot t4)}{4}$$
$$+\frac{\left(5\cdot d_p^2\cdot\rho\cdot\omega^2\cdot R\cdot h\cdot\pi-2\cdot d_p^2\cdot\rho\cdot\omega^3\cdot R\cdot h\cdot t4\right)\cdot cos(\omega\cdot t4)^2}{4} \tag{23.9}$$

$$S(t)=\int_0^{\frac{2\cdot\pi}{100}}F(t)\cdot dt=-353.429[N\cdot s];\ \frac{S(t)}{t}=\frac{353.429}{0.063}=5.61\cdot10^3[N] \tag{23.10}$$

And for 8 cylinders we get: $8\times5.61\times10^3=4.488\times10^4$.

Up to this point, the traction force achieved by a single cylinder has been studied. According to Fig. 23.5, it can be seen that this force during a complete rotation

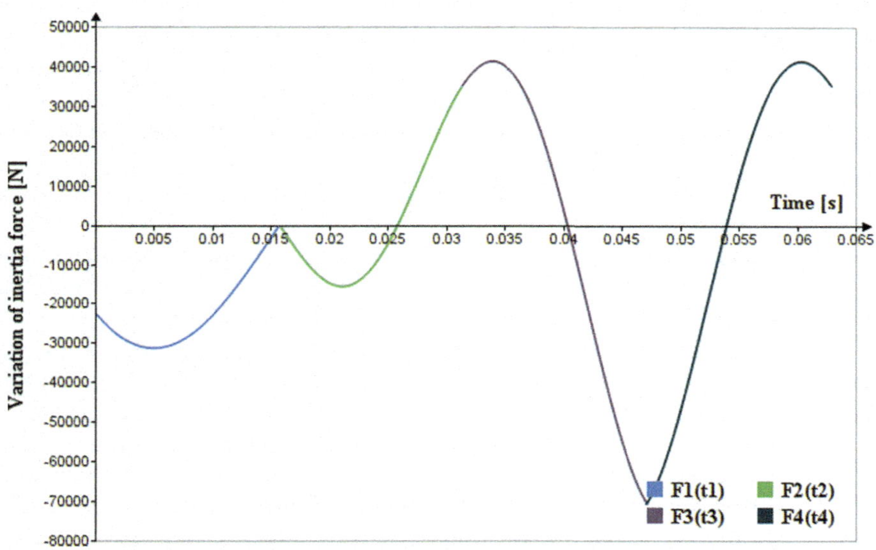

Fig. 23.5 Variation of inertia force as a function of time

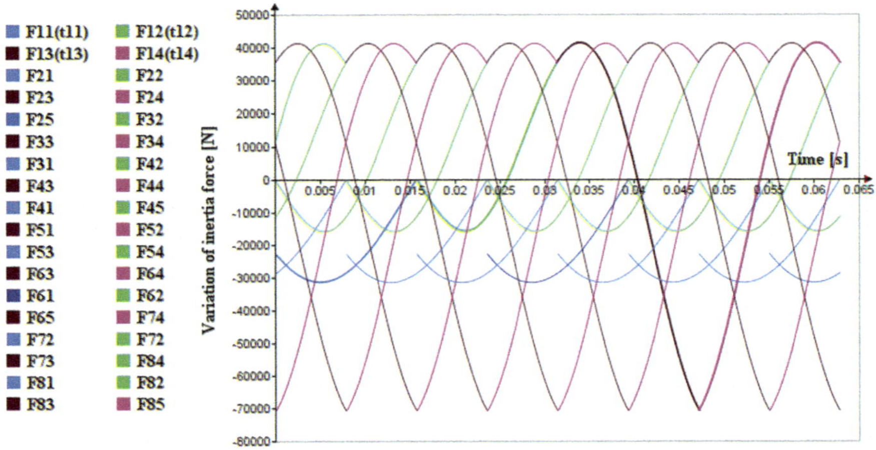

Fig. 23.6 Variation of inertia force given by 8 cylinders as a function of time

is pulsating, taking values from zero to some minimum and maximum values, respectively.

Performing the integration calculation of the formulas of this force covering the entire interval corresponding to a complete rotation, it is found that this value is different from zero (according to the Eq. 23.10).

Extending the calculation for the 8 cylinders with the time gap corresponding to each cylinder, corresponding to each of their momentary positions, the graph according to Fig. 23.6 is obtained.

If we study the phenomenon in more depth, we can realize that the whole phenomenon is repeated when the cylinders are rotated by 45° because at this rotation the next cylinder takes the place of the previous cylinder, thus the phenomenon is repeated.

In the Fig. 23.7 shows the variation of the total force produced by the 8 cylinders in an interval of 45°.

To determine the area for the graph in Fig. 23.7, we entered the function of the total force F(t) and to obtain its average value, we must divide this value by the time interval.

Studying the graph of Fig. 23.7, we can see that the resulting pulsation is no longer so significant, varying between maximum and minimum values lower than in the case of a single cylinder.

This force, if the device is mounted on a vehicle (with the exception of starting or accelerating it) must be equal to the friction force of the vehicle with the ground.

For a better dynamic balancing of the system, it is recommended to double the system, according to which the two semi-devices are rotated in opposite directions to each other, a fact by which any inertial forces that appear perpendicular to the direction of movement cancel each other out.

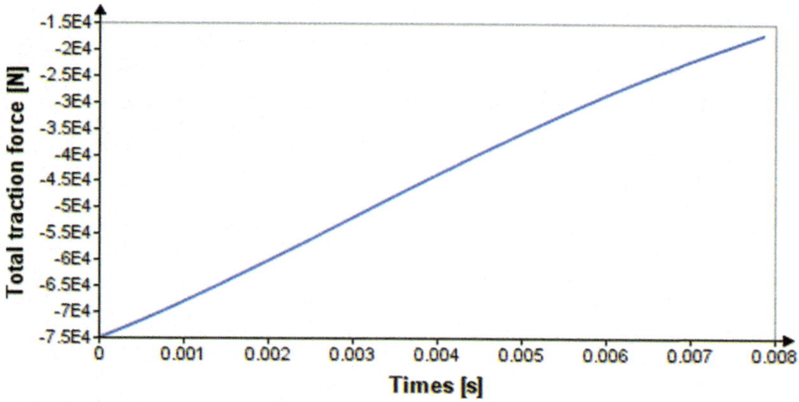

Fig. 23.7 The variation of the total traction force

23.3 Conclusion

The inertial system, the following advantages are obtained: achieving the traction force in a well-established direction; the simplicity of the construction, which reduces manufacturing costs; if we study in more detail the phenomena of accelerations or decelerations of the liquid masses in cylinders, respectively in the supply/drain pipes 4 and 5 and if we use a well-defined flywheel in the kinematic chain 19, we can notice that due to the energy accumulations and their discharge, we only need mechanical work (energy) necessary to overcome the frictional forces, so the system works with a very high efficiency.

Acknowledgements This work was supported by the project "The Development of Advanced and Applicative Research Competencies in the Logic of STEAM + Health"/POCU/993/6/13/153310, project co-financed by the European Social Fund through The Romanian Operational Programme Human Capital 2014-2020.

References

1. L.L. Culda, Portable devices in the fight of sarscov2. Technical University of Cluj-Napoca Acta Technica Napocensis Series: Appl. Math. Mech. Eng. **65**(Speciale I), 81–86 (2022)
2. D. Ciulin, in *Conference Meeting International Joint Conference on Computer, Information, Systems Sciences and Engineering*. System to Produce Mechanical Inertial Force and/or Torque (Bridgeport, CT, 2008), pp. 51–56. https://doi.org/10.1007/978-90-481-3656-8_11
3. A. Gerocs, G. Gillich, D. Nedelcu, Z. Korka, A multibody inertial propulsion drive with symmetrically placed balls rotating on eccentric trajectories. Symmetry **12**(9), (2020). https://doi.org/10.3390/sym12091422
4. E.S. Muncut, G.M. Erdodi, L. Culda, A. Komjati, in *The 16th International Scientific Conference eLearning and Software for Education Bucharest*. Determination of Speed and Acceleration for a

Plane Mechanism with the Instantaneous Rotation Centre Method and Euler's Graph-Analytical Method Using AUTOCAD Mechanical, vol 2 (Bucuresti, 2020), pp. 420–434. https://doi.org/10.12753/2066-026X-20-140
5. I. Biro, B. Csizmadia, G. Krakovits, A. Veha, in *IUTAM Symposium on Dynamic Modeling and Interaction Control in Virtual and Real Environments*. Sensitivity Investigation of Three-Cylinder Model of Human Knee Joint, vol 30. (Budapest, 2010), pp. 177–184. WOS:000394929500020

Chapter 24
Regenerative Braking in the Case of Hybrid Propulsion Systems

Dinel Popa, Sebastian Parlac, Claudia-Mari Popa, and Dragos-Mihai Popa

Abstract In the case of most hybrid vehicles, electric motors used for traction also allow braking by generating electricity, which is then stored in a traction battery. The paper presents the results obtained on an experimental stand that connects three power sources: a heat engine, an electric motor and a generator. A double-satellite planetary mechanism is the main part of an experimental stand that has been designed for coupling 3 power sources. In this way, the behavior of the transmission of a mixed hybrid car with the division of the power flow is simulated, a solution that simulates a continuously variable transmission with electric control. Starting from a dynamic model of the stand, the differential equations of motion of the mechanical system with two degrees of freedom that couple the three machines are determined. The presented mathematical model, by modeling the external characteristics of the engines, can be used to simulate all the operating regimes of a hybrid car. The experimental results are compared with the theoretical ones obtained by integrating the equations of motion, in order to determine the torque in case of regenerative braking.

24.1 Introduction

Car manufacturers have sustained for several decades concerns regarding energy conservation and environmental protection. One of the solutions consists in the use of a hybrid propulsion system [1] that improves energy efficiency, reduces fuel consumption and engine pollutant emissions [2].

The construction of hybrid cars has seen a strong development in recent years [3]. One solution to reduce fuel consumption and pollutant emissions is the use of transmissions with a greater number of gears [4, 5], but this has a direct impact on dimensions and weight. Although hybrid propulsion systems do not converge towards a single solution, mixed systems (series/parallel) whose transmission consists of

D. Popa (✉) · S. Parlac · C.-M. Popa · D.-M. Popa
University of Pitesti, 110040 Pitesti, Romania
e-mail: dinel.popa@upb.ro

N. Herisanu and V. Marinca (eds.), *Acoustics and Vibration of Mechanical Structures—AVMS-2023*, Springer Proceedings in Physics 302,
https://doi.org/10.1007/978-3-031-48087-4_24

one or more planetary mechanisms, are proving to be the most promising hybrid technology [6] and are attracting the attention of car manufacturers [7].

The engines are coupled by planetary mechanisms whose role is to distribute the developed power, to obtain high performance and efficiency but also to contribute to the reduction of fuel consumption [7]. In addition, it offers improved dynamic performance, ensures better efficiency, an extended range of the transmission ratio, has small dimensions and a compact structure [8]. They also manage to ensure a fuel economy by performing functions such as regenerative braking, stop and start, or efficient power flow management [1, 9].

In recent years, numerous studies have been carried out that present various types of mechanical transmissions with power division equipping hybrid electric vehicles [10, 12].

In a previous paper, [11], a planetary mechanism with 4 elements with double satellite was presented that allows the coupling of three power sources. In this paper, for the analysis and evaluation of the kinematic and dynamic behavior of the planetary mechanism as well as the performance obtained, a test bench was designed and built. The behavior of the planetary mechanism in the bench actually describes the behavior of a transmission of a hybrid car in all operating modes.

Particular attention in the experimental determinations was paid to recuperative braking.

24.2 Description of the Experimental Stand

For the study of the dynamic behavior of a mechanical system that couples 3 power sources, the stand with the representation in Fig. 24.1.

The main component of the stand, the mechanical system *SM*, allows the coupling of a thermal engine ME, an electric engine and two alternators GE_1 and GE_2.

In paper [3] it was shown that the mechanical system SM can be used in the transmission of a hybrid car. In order to study the dynamic behavior of the system, there is no need to use engines with the power of a real vehicle. It would result in a voluminous and expensive stand, and some parts would be difficult to procure. The construction companies integrate the electric machines and the power summing mechanism in a single housing. Electric cars are alternating current, high voltage, controlled by an electronic module.

The designed stand allows studying the dynamic behavior of the mechanical system using engines with lower powers.

The main part of the stand, the mechanical system (SM), contains a planetary mechanism with a double satellite (Fig. 24.1). The thermal engine MT drives the sun gear 2, the electric engine drives the port-satellite arm 1 and the electric generator GE drives the sun gear 3. The numbers of teeth of the gear wheels 2, 3 and 4 of the planetary mechanism were noted with z_2, z_3,z_4', z_4''.

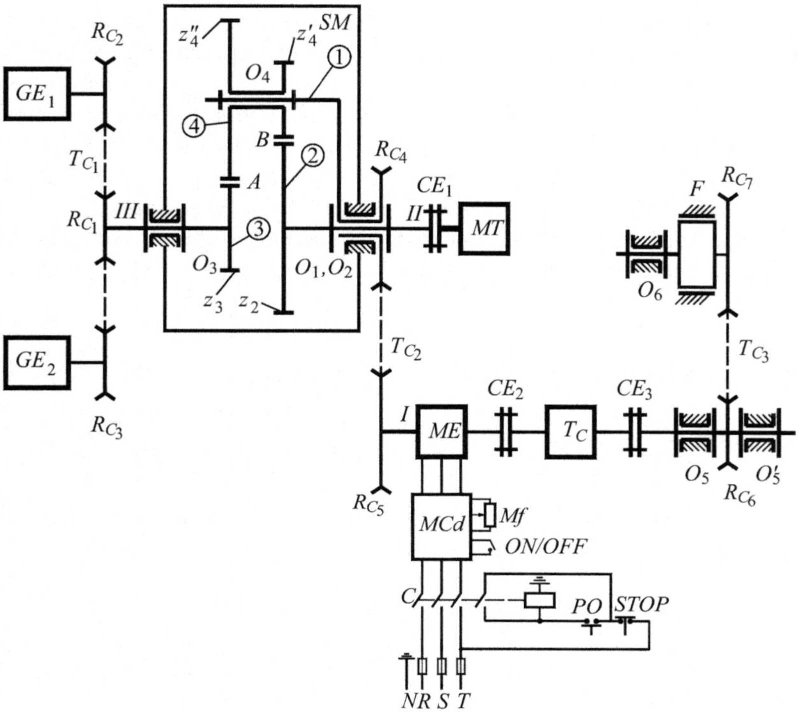

Fig. 24.1 Kinematic diagram of the stand

The thermal engine is an engine with two air-cooled V-cylinders at 90° with a maximum power of 15.3 kW at 3600 rpm and a maximum torque of 46.0 Nm at 2500 rpm, coupled to the mechanical system through the elastic coupling CE_1.

The output shaft *III* from the mechanical system also drives the alternators GE_1 and GE_2 through a belt transmission. The alternators have a rectified voltage of 14.5–15.1 V and a maximum current of 110 A. The three belt wheels R_{C_1}, R_{C_2}, R_{C_3} are driven by a Poly V belt.

The voltages generated by the alternators go to 6 columns of 5 × 12 V car bulbs each, each bulb having a power of 100 W. This is how the alternators are charged, similar to the charging of a traction battery.

The electric motor ME is a three-phase electric motor, type: 132S 38-2 with a power of 7.5 kW, with outputs on both sides. It is coupled through the belt transmission T_{C_2} through the belt wheels R_{C_4} and R_{C_5}. The speed of the electric motor is varied with a control module MC_d composed of a power inverter (SINAMICS) and a control module (CU 240S).

The SINAMICS power module allows the regulation of the speed of the asynchronous electric motor both by changing the supply voltage and by varying the frequency. A particularity of the control module is that of energy recovery during the braking period of the electric motor. The speed of the electric motor ME can be changed by means of a potentiometer M_f or by means of a computer, programming the operation of the control mode.

To measure the torque, the T_rC T20WN type torque transducer is used, mounted between two elastic couplings CE_2 and CE_3, specially designed for this transducer.

Bearings with radial bearings O_5, O_5', support a shaft and pulley R_{C_6}. The motion is further transmitted through the belt drive T_{C_3} to the braking system.

The belt wheel R_{C_7} is solid in rotation with the brake drum F. Actuation of the brake drum shoes is done with a brake cable actuated by a handbrake lever.

24.3 The Differential Equations of Motion

The angular velocities of elements 1, 2, 3 and 4 of the planetary mechanism in Fig. 24.1 are noted with $\omega_1, \omega_2, \omega_3, \omega_4$.

From kinematic considerations, there are the connection relations between the angular velocities:

$$\omega_3 = \omega_1(1 - i_2) + \omega_2 i_2, \quad \omega_4 = \omega_1(1 - i_1) + \omega_2 i_1 \tag{24.1}$$

expressions obtained by applying Willis' relations. It was noted with i_1, i_2, ratios between the numbers of teeth:

$$i_1 = -\frac{z_2}{z_4'}, \quad i_2 = \frac{z_2 z_4''}{z_3 z_4'} \tag{24.2}$$

Noting the torques developed by the electric engine, the thermal engine and respectively the two alternators with M_1, M_2 respectively M_3, and with P_f the power consumed in the brake, the expression of the mechanical power of the system is given by the sum of the powers developed by the thermal engine MT and the electric motor ME from which the powers consumed by the alternators GE1, GE2, and the brake F are subtracted:

$$P = M_1\omega_1 + M_2\omega_2 - M_3\omega_3 - P_F \tag{24.3}$$

Noting with ω_F the angular velocity of the brake F and with ω_F the transmission ratio of the belt transmission T_{C_3}, the expression of the angular velocity ω_F as a function of the angular velocity ω_1 of the electric motor shaft ME is given by the relation:

$$\omega_F = \frac{\omega_1}{i_{TC_3}} \tag{24.4}$$

If we denote by M_f the moment developed in the brake F, the expression of the mechanical power (24.3) becomes:

$$P = M_1\omega_1 + M_2\omega_2 - M_3\omega_3 - \frac{M_f}{i_{TC_3}}\omega_1 \tag{24.5}$$

Taking into account relation (24.1) we obtain:

$$P = \left[M_1 - M_3(1 - i_2) - \frac{M_f}{i_{CT_3}}\right]\omega_1 + [M_2 - M_3 i_2]\omega_2 \tag{24.6}$$

The expression of kinetic energy is:

$$\begin{aligned} T =& \frac{1}{2}\Big[J_1\omega_1^2 + J_2\omega_2^2 + J_3\omega_3^2 + m_4 R_4^2\omega_4^2 \\ & + J_4\omega_4^2 + J_{RC_1}\omega_3^2 + \left(J_{RC_2} + J_{RC_3}\right)\left(\frac{\omega_3}{i_{TC_1}}\right)^2\Big] \\ & + \frac{1}{2}\Big[J_{ME}\omega_1^2 + \left(J_{RC_4} + J_{RC_5}\right)\omega_1^2 + 2J_{CE}\omega_1^2 \\ & + J_{RC_6}\omega_1^2 + \left(J_{RC_7} + J_T\right)\left(\frac{\omega_1}{i_{TC_3}}\right)^2\Big] \end{aligned} \tag{24.7}$$

In relation (24.7) the notations were used: m_4—the mass of element 4 of the planetary mechanism, J_1, J_2, J_3, J_4—the moments of inertia of the elements of the planetary mechanism, R_4—the radius of the satellite carrier arm, J_{RC_1}, J_{RC_2}, J_{RC_3}, J_{RC_4}, J_{RC_5}, J_{RC_6}, J_{RC_7}—the moments of inertia of the wheels of the belt, J_{CE}—the moment of inertia of the elastic coupling, J_T—the moment of inertia of the brake drum, i_{TC_1}, i_{TC_3}—the transmission ratio of the belt transmissions T_{C_1}, T_{C_3}. The belt transmission T_{C_2} has the transmission ratio $i_{TC_2} = 1$ (the belt wheels R_{C_4} and R_{C_5} have the same dimensions). With the notations:

$$\begin{aligned} A_{11} =& J_1 + m_4R_4^2 + J_{ME} + J_{RC_4} + J_{RC_5} + 2J_{CE} + J_{RC_6} + \frac{J_{RC_7} + J_T}{\left(i_{TC_3}\right)^2} \\ & + \left(J_3 + J_{RC_1} + \frac{J_{RC_2} + J_{RC_3}}{\left(i_{TC_1}\right)^2}\right)(1 - i_2)^2 + J_4(1 - i_1)^2, \\ A_{22} =& J_2 + \left(J_3 + J_{RC_1} + \frac{J_{RC_2} + J_{RC_3}}{\left(i_{TC_1}\right)^2}\right)i_2^2 + J_4 i_1^2, \\ A_{12} =& \left(J_3 + J_{RC_1} + \frac{J_{RC_2} + J_{RC_3}}{\left(i_{TC_1}\right)^2}\right)(1 - i_2)i_2 + J_4(1 - i_1)i_1 \end{aligned} \tag{24.8}$$

we obtain the kinetic energy expression:

$$T = \frac{1}{2}\left(A_{11}\dot{\theta}_1^2 + A_{22}\dot{\theta}_2^2\right) + A_{12}\dot{\theta}_1\dot{\theta}_2 \tag{24.9}$$

Applying the Lagrange equations, we obtain the system of differential equations:

$$\begin{cases} A_{11}\ddot{\theta}_1 + A_{12}\ddot{\theta}_2 = M_1 - M_3(1 - i_2) - \frac{M_f}{i_{CT_3}} \\ A_{12}\ddot{\theta}_1 + A_{22}\ddot{\theta}_2 = M_2 - M_3 i_2 \end{cases} \tag{24.10}$$

with the solution:

$$\ddot{\theta}_1 = \frac{\left[M_1 - M_3(1 - i_2) - \frac{M_F}{i_{TC_3}}\right]A_{22} - [M_2 - M_3 i_2]A_{12}}{A_{11}A_{22} - A_{12}^2} \tag{24.11}$$

$$\ddot{\theta}_2 = \frac{-\left[M_1 - M_3(1 - i_2) - \frac{M_F}{i_{TC_3}}\right]A_{12} + [M_2 - M_3 i_2]A_{11}}{A_{11}A_{22} - A_{12}^2} \tag{24.12}$$

The system of differential equations of motion is integrated by knowing the external characteristics, as a function of the revolutions, of the moments developed by the M1, M2 and M3 engines, given by the manufacturer.

By modeling the elements of the stand in AutoCAD to determine the mechanical properties, and substituting the values obtained in relations (24.8), the values of the constants are obtained:

$$\begin{aligned} A_{11} &= 0.340931\,\text{kg}\,\text{m}^2,\ A_{12} = -0.137424\,\text{kg}\,\text{m}^2, \\ A_{22} &= 0.099697\,\text{kg}\,\text{m}^2 \end{aligned} \tag{24.13}$$

24.4 Theoretical and Experimental Results Regarding Regenerative Braking

The stand is operated by the electric motor, while the thermal motor is off. It was aimed that in 20 s the speed of the electric motor would increase to 2000 rpm, remain constant for a few seconds and then decrease to zero.

The speed is varied with the help of the speed variation, in a time interval of 20 s. The speed at the output shaft of the generator is measured with a speed transducer (CA27), the speed of the thermal engine being zero ($n_2 = 0$).

In the time interval [9.5–11.1] s, the speed of the electric motor measured with the speed module of the torque transducer is in the range [1940–2013] rpm, and the speed of the generator drive shaft is 3250 rpm.

Substituting in relation (24.1) $\omega_2 = 0$ and $i_2 = 2625$, we obtain the connection relation:

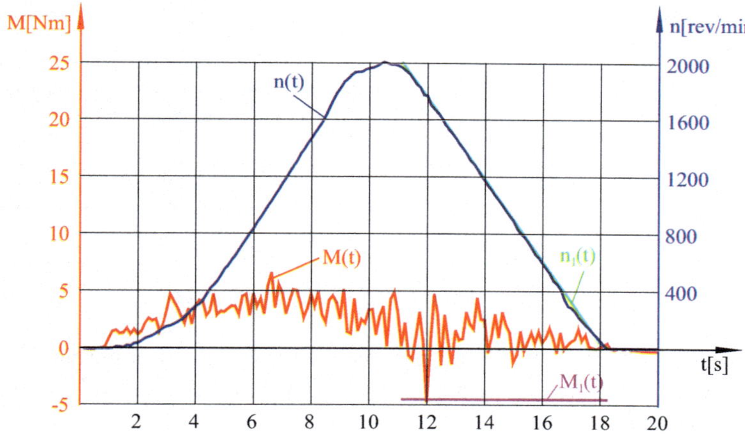

Fig. 24.2 The diagrams $n(t)$, $M(t)$, $n_1(t)$, $M_1(t)$

$$\omega_3 = \omega_1(1 - i_2) = -1625\omega_1 \tag{24.14}$$

For $n_3 = 3250$ rpm, a revolution of the electric motor is obtained of $n_1 = 2000$ rpm.

Recording the moment and speed as a function of time, data provided by the T20WN torque transducer, after processing the data with an AutoLisp function, in AutoCAD the graphs $n(t)$ and $M(t)$ are drawn and from Fig. 24.2.

It was noted with $n(t)$ the speed obtained experimentally, with $n_1(t)$ the theoretical speed after solving the equation of motion and with $M(t)$ the torque measured by the torque transducer.

We follow the decreasing curve in which the speed decreases from 2000 rpm to 0, in the time interval $t \in [11.1 \ldots 18.25]$ s. In this time period the average engine torque calculated is $M_{med} = 1011$ Nm. A linear decrease of the speed in the time interval $[11.1 \ldots 18.25]$ s is considered.

We replace the values of the constants A_{11}, A_{12}, A_{22}, in the differential Eq. (24.11) and obtain:

$$\ddot{\theta}_1 = \frac{M_1 A_{22}}{A_{11}A_{22} - A_{12}^2} = 6.6M_1 \tag{24.15}$$

with the solution:

$$\omega_1 = \dot{\theta}_1 = 6.6M_1(t - t_0) + C \tag{24.16}$$

where $t_0 = 11.1$ s and C is a constant of integration. From the initial conditions, the values of the constants are obtained: $M_1 = -4.45$ and $C = 210$.

The variation of the angular velocity ω_1 is given by the relation:

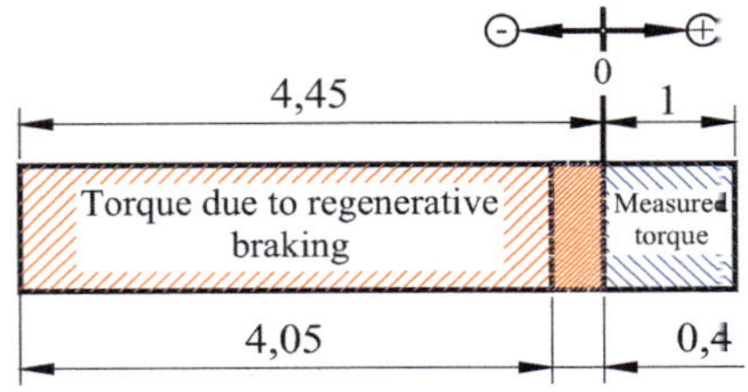

Fig. 24.3 The balance sheet of the couple

$$\omega_1 = \dot{\theta}_1 = -6.6 \cdot 4.45 \cdot (t - 11.1) + 210 \tag{24.17}$$

In Fig. 24.3 was symbolically represented the balance sheet of the couple.

The torque transducer recorded an average torque of $M_{med} = 1011$ Nm. Of this, 1.4 Nm is due to friction ($M_r = 1.4$ Nm).

As can be seen from Fig. 24.3, the negative torque of 4.05Nm is due to the regenerative braking performed by the electric motor control module.

The 4.45Nm at time $t_0 = 11.1$ s are also obtained if the torque of the electric motor $M(t)$ in Fig. 24.2 is averaged in the time interval [9.5–11.1] s. In this time interval, it was considered that the speed remained constant.

Integrating the differential Eqs. (24.11) by the Runge Kutta method with the initial conditions:

$M_1 = -4.45$, $t_0 = 11.1$, $\omega_{10} = 209$, $M_3 = 0$, $M_f = 0$,

is obtained after numerical integration, the variation graphs of torque $M_1(t)$ and speed $n_1(t)$ in the interval [11.1 ... 18.25] s. In Fig. 24.2, the 4 graphs $n_1(t)$ and $M_1(t)$ were displayed superimposed, and being the theoretical results. A very good correlation is observed between the values obtained theoretically and those obtained experimentally.

Practically, the revolution speed $n_1(t)$ is overlapped over the graph of the experimental determinations, and the revolution speed at the end of the 18.25 s is very close to 0 ($\omega_1|_{t=18.25} = 0.02$ rad/s).

24.5 Conclusions

A double-satellite planetary mechanism is the main part of an experimental stand that has been designed for coupling 3 power sources.

The differential equations of motion of the stand allow the study of the dynamic behavior of the mechanism with 2 degrees of freedom that joins a thermal engine and two electric machines. In this way, the behavior of the transmission of a mixed hybrid car with the division of the power flow is simulated, a solution that simulates a continuously variable transmission with electric control.

The motors in the stand have low power, the electric motor being a three-phase motor fed from the laboratory energy grid. It is equipped with a power inverter that allows the speed variation and the recovery of electrical energy during the braking period.

The measuring chain used allows the measurement of torque and revolutions as a function of time for the electric motor and the revolutions of the thermal engine and alternators.

To measure the torque during recuperative braking, only the electric motor was operated, the thermal motor being turned off.

In this case, the system of 2 differential equations is reduced to a single differential equation whose solution is also obtained analytically.

After integration, taking into account the initial conditions, the value of the braking torque is obtained, 4.45 Nm, from which 1.4 Nm is subtracted, the value of the torque due to the friction of the stand components.

The same value was also obtained experimentally considering an average of the torque values measured during the period of decrease in the electric motor speed.

The integration of the differential equations of motion by the Runge–Kutta method of IV-th order with the initial conditions at the beginning of braking, lead to the same result.

These things confirm, first of all, the accuracy of the mathematical apparatus that modeled the entire mechanical system, and, what was important, the fact that the mechanical system with double satellite planetary mechanism can be used in the transmission of a hybrid car.

The presented mathematical model, by modeling the external characteristics of the engines, can be used to simulate all the operating regimes of a hybrid car. This will be done in a future paper.

References

1. A. Alessandrini, F. Orecchini, F. Ortenzi, F. Villatico Campbell, Drive-style emissions testing on the latest two Honda hybrid technologies. Eur. Conf. Transp. Res. Inst. (2009)
2. K.J. Qin, E.H. Wang, H. Zhao, G.K. Shi, R.G. Wang, Z.G. Kong, W. Wang, Development and experimental validation of a novel hybrid powertrain with dual planetary gear sets for transit bus applications—Science China. Technol. Sci. **58**(12), 2085–2096 (2015)
3. D. Popa, N.D. Stanescu, I. Tudor, Dynamic behavior of the mechanical systems from the structure of a hybrid automobile. Int. Cong. Autom. Transp. Eng. Adv. Eng. Piteşti (2017)
4. M. Park, T.W. Park, S.H. Lee, S.W. Han, S.K. Kwon, Analytical study to estimate the performance of the power shift drive (PSD) axle for a forklift. Int. J. Automot. Technol. **11**(1), 49–56 (2010)
5. R. Schuebel, M. Gegner, F. Beeck, in *Solving the Powertrain Puzzle* ed by Schaeffler Technologies GmbH & Co. KG. Turning New Directions: Surprising Potential in Planetary Transmissions. Part 1: Planetary Gear Set Schaeffler Technologies GmbH (2014)
6. S. Wang, S. Zhang, D. Shi, X. Sun, J. He, Research on instantaneous optimal control of the hybrid electric vehicle with planetary gear sets. J. Br. Soc. Mech. Sci. Eng. (2019)
7. C. Xiang, K. Huang, R. Langari, Power-split electromechanical transmission design and validation with three planetary gears for heavy-duty vehicle. Iran. J. Sci. Technol. Trans. Mech. Eng. (2018)

8. C. Huang, J.-X. Wang, K. Xiao, M. Li, J.Y. Li, Dynamic characteristics analysis and experimental research on a new type planetary gear apparatus with small tooth number difference. J. Mech. Sci. Technol. **27**(5), 1233–1244 (2013). www.springerlink.com/content/1738-494x
9. S.J. Kim, Y.-S. Yoon, S. Kim, K.-S. Kim, Fuel economy assessment of novel multi-mode parallel hybrid electric vehicle. Int. J. Automot. Technol. **16**(3), 501–512 (2015)
10. H. Yang, B. Kim, Y. Park, W. Lim, S. Cha, Analysis of planetary gear hybrid powertrain system part 2: output split system. Int. J. Automot. Technol. **10**(3), 381–390 (2009)
11. I. Tudor, D. Popa, N.D. Stanescu, A comparative study of planetary mechanisms usable for power sources coupling. Int. Cong. Autom. Transp. Eng. Adv. Eng. Piteşti (2017)
12. C. Mansour, D. Clodic, Dynamic modeling of the electro-mechanical configuration the Toyota Hybrid system series/parallel power train. Int. J. Automot. Technol. **13**(1), 143–166 (2012)

Chapter 25
Effect of Build Orientation on the Resonant Frequencies of 3D Printed Nylon12 CF35 Specimens

Tiberiu Baldogi and Mircea-Cristian Dudescu

Abstract Vibration is a phenomenon constantly encountered in everyday life. To ensure that the components of a vehicle can withstand these mechanical demands over time, they are tested in specialized laboratories with the help of equipment called shakers. The connecting part between the tested components and the test equipment is called the fixture. Until now, they were made of an aluminum alloy, but with the increasing complexity of the shape of the components, they are increasingly difficult to manufacture. With the emergence of new manufacturing technologies (e.g. additive manufacturing), new solutions appear. Using FDM, the purpose of this work is to compare the influence given by the printing direction (build orientation) of the Nylon12CF35 material, which has the potential to be the right material to keep up with the evolution of technology in this industry. For this scope 20 specimens in 4 different build orientations were printed and tested using sine sweep method, to determine the resonance frequencies. All printed specimens had the same infill rate and infill pattern. X-ray analysis of the specimens reveals the internal structure generated by fused deposition technology, layer disposal with respect to longitudinal axis of the specimens. A comparison of the results in terms of resonant frequencies allows us to determine which direction of printing is more suitable for vibration modes and frequencies to replace the classical massive fixtures. The results also show how the printing parameters, in this case build orientation change the inner structure and which is the connection between these parameters and the frequency response of the structure.

T. Baldogi (✉) · M.-C. Dudescu
Department of Mechanical Engineering, Technical University of Cluj-Napoca, 103-105 Muncii Boulevard, 400641 Cluj-Napoca, Romania
e-mail: baldogi.tiberiu@gmail.com

T. Baldogi
Qualification Laboratory Sibiu, Continental Sibiu, Str. Salzburg Nr. 8, 550018 Sibiu, Romania

N. Herisanu and V. Marinca (eds.), *Acoustics and Vibration of Mechanical Structures—AVMS-2023*, Springer Proceedings in Physics 302,
https://doi.org/10.1007/978-3-031-48087-4_25

25.1 Introduction

Since additive manufacturing (AM) has shown its potential to produce parts with highly complex geometries and is increasingly used in wear-bearing applications in industry, it is important to determine if parts can withstand mechanical stress under both static and cyclic loading. Despite its many advantages, parts created by Fused Deposition Molding (FDM) have inferior mechanical properties due to additional porosity and anisotropy due to the manufacturing process. In this regard, several studies have shown the effect of printing parameters such as infill rate, filling pattern, layer thickness, layer height and other printer settings on the mechanical behavior of 3D printed parts [1–5]. Most of the studies mainly focus on the mechanical behavior of manufactured parts under static loading [6], the characterization of the structural dynamic behavior of additively manufactured parts is relatively less investigated [7–11], so it is useful to understand the effect of the mentioned parameters on the vibration behavior of 3D printed parts. The interest in the characterization of the vibration behavior of additively manufactured materials like polylactic acid (PLA) is evaluated by Medel et al. [11] by Laser Doppler Vibrometry (LDV) characterization of 3D printed rectangular prisms. Thus, natural frequencies of 1st and 2nd bending modes allowed computation of the corresponding equivalent elastic moduli and identification of significant influences of process parameters (raster angle, nozzle temperature, etc.) on the mechanical behavior. Damping ratio data confirmed the correlation between high damping and poor interlaminar bonding, identification of optimal values of 3D printing process parameters for low damping ratios was possible. Modal response of hybrid raster orientation on material extrusion printed acrylonitrile butadiene styrene and polyethylene terephthalate glycol under thermo-mechanical loads is investigated by Almutari et al. in [12]. Effect of infill pattern and ratio on the flexural and vibration damping characteristics of FDM printed PLA specimens is presented in [13]. The panels' flexural behaviors and vibration-damping capabilities produced in different infill patterns (Grid, Cross, and Tri-hexagon) and different infill ratios (0, 20, 40, 60, 80, and 100%) were investigated. The results show that the infill pattern and ratio significantly affect flexural behavior and vibration-damping properties. The best damping properties were obtained on a 40% infill ratio with all the different tested patterns.

The natural frequency, damping ratio, or stiffness of a specific part can be adjusted with the design parameters of additively manufactured parts allowing achievement of both structural and vibrational requirements by selecting the right materials and printing configuration. Most 3D-printed polymers have inconsistent material properties, particularly those related to damping. It is therefore necessary to investigate the effectiveness of 3D-printed structures further.

The purpose of this work is to compare the influence given by the printing direction (build orientation) of the Nylon12CF35 material. For this scope 20 specimens in 4 different build orientations were printed and tested using sine sweep method, to determine the resonance frequencies. After experimental testing a comparison of the results in terms of resonant frequencies allows us to determine which direction of printing is more suitable for vibration modes and frequencies.

25.2 Materials and Methods

25.2.1 3D Printing Material and Specimen's Dimensions

Nylon12 CF35 is a material with high strength properties, which is perfect for applications with long time usage such as pedals, handles or supports. These are designed for high resistance to applied forces and resistance over time. It is light weight and rigid and due to FDM technology, it can be shaped in all kinds of complex shapes (Table 25.1).

The fixture is the component that makes the connection between the testing device and the tested part. This component must meet certain conditions to be considered suitable for vibration testing. The first condition is that the transmissibility between the shaker and the tested component should be 1:1. The second principle refers to the characteristics of the material from which it is made (rigidity, weight, strength, etc.). The third principle refers to the shape of the fixture, which must be easy to fabricate, preferably from a single piece and which corresponds to the installation of the component in the vehicle. Until now, they were made of an aluminum alloy, but with the increasing complexity of the shape of the components, they are increasingly difficult to manufacture.

With the emergence of new manufacturing technologies (e.g. additive manufacturing), new solutions appear. Using FDM, the purpose of this work is to compare the influences given by the printing direction (build orientation) of the Nylon12CF35 material, which has the potential to be the right material to keep up with the evolution of technology in this industry.

For fixture applications complexity is a must, because it needs to be design for many different parts, from small ones (like valves, radars, lidars, etc.) to large parts (like hydronic control units, clusters, washer fluid tanks, etc.). When designing fixtures, we must make sure that the vibration generated by the shaker is the same

Table 25.1 Tensile properties [14]

Mechanical properties	Units	XZ orientation	ZX orientation
Yield strength	MPa	No yield	No yield
Elongation yield	%	No yield	No yield
Strength break	MPa	83.5 (1.7)	32.7 (3.5)
Elongation break	%	2.4	1.2 (0.27)
Modulus (elastic)	GPa	9.46 (0.46)	3.00 (0.43)

with the vibration that the component receives without any absorption or amplification. To achieve this, simulation is a crucial part. If the properties of the material are known, we can test the mechanical resistance and use it in simulations. Then the simulations will be much more accurate. After the fixture creation, a series of testing take part into the laboratory to verify the transmissibility at the fixation points.

After this study, we should be able to understand the behavior of the material under such mechanical loads. For this scope 20 specimens in 4 different build orientations (Fig. 25.1) were printed and tested using sine sweep method, to determine the resonance frequencies (Fig. 25.7). All printed specimens had the same infill rate and infill pattern. X-ray analysis of the specimens (Fig. 25.3) reveals the internal structure generated by fused deposition technology, layer disposal with respect to longitudinal axis of the specimens.

In Table 25.2 are presented the mean values of the dimension of the printed specimens and their corresponding mass. A closer look of the outer layer of the specimens reveals the filaments direction with respect to the tensile direction of the specimens for the investigated build orientations (Fig. 25.4).

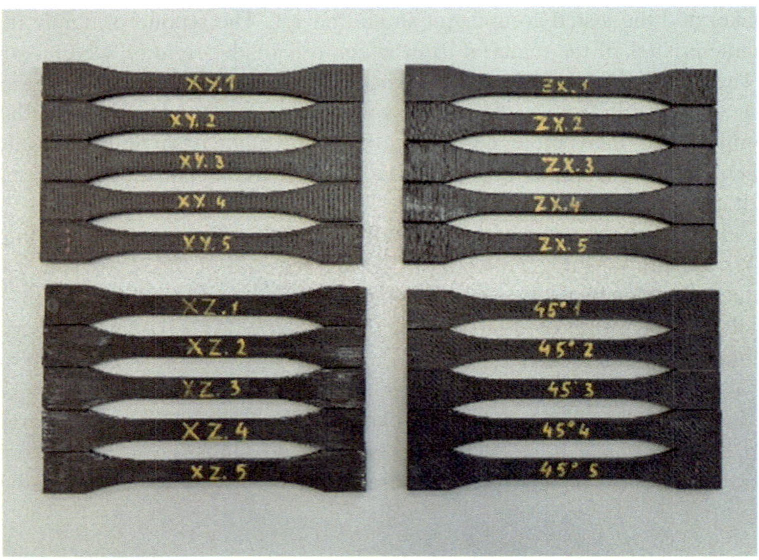

Fig. 25.1 Printed specimens

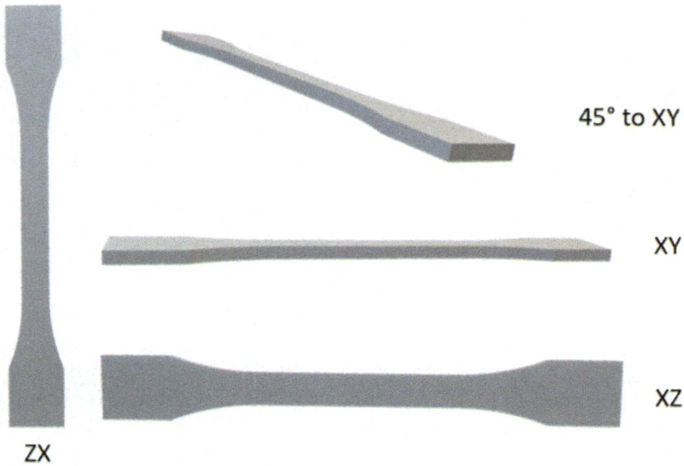

Fig. 25.2 Build orientation planes

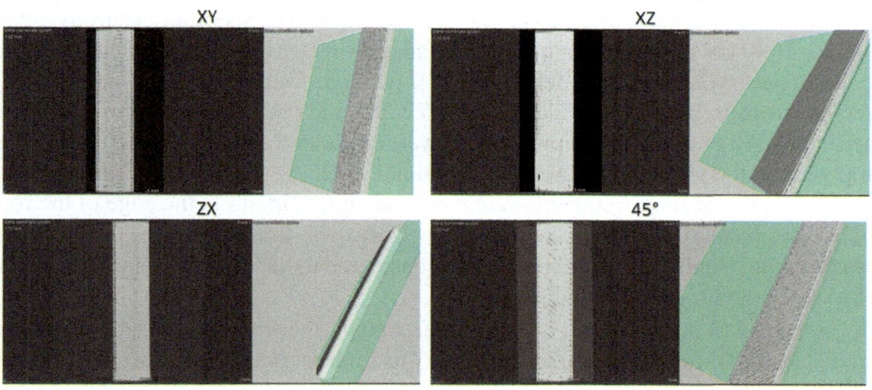

Fig. 25.3 X-ray view of specimen's build orientation

Table 25.2 Dimensional measurements (mean values) of the printed specimens

Orientation	Length (mm)	Width (mm)	Height (mm)	Mass (g)
XY	155.03	9.93	4.26	9.95
XZ	155.01	10.36	3.99	9.19
ZX	155.46	10.08	4.01	9.07
45°	155.1	10.01	4.29	10.05

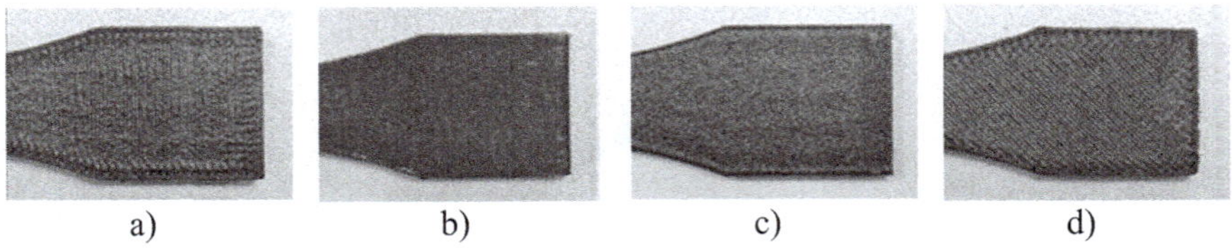

Fig. 25.4 Build orientation of samples: **a** XY; **b** XZ; **c** ZX and **d** 45° to XY

25.3 Vibration Testing of Specimens

For the testing part, a specimen holder was created as shown in the image below. This fixture is designed to hold 2 specimens at the same time. The scope of the test is to identify the first 3 resonant frequencies of each of the specimen. The method for determining the resonant frequencies is sine sweep from 5 Hz (lowest frequency of the shaker capability) to 2400 Hz (highest necessary frequency to determine the first 3 natural frequencies) with a constant (linear) sweep rate of 1 Hz per second.

The testing equipment used for experimental test consists of an electrodynamic shaker of 80 kN force, one control charge accelerometer (type B&K 4384), a signal conditioner (type Nexus, model 2692-0S4), two light weight tear drop accelerometers for measurement (type PCB, model 352C22) and one vibration controller (type VR9500). All the testing equipment used were calibrated and verified before testing, top view of the vibration setup is presented in Fig. 25.5 (vibration on Z direction).

Measurement accelerometers were mounted on specimens at the edge of the free end. For the measurement accelerometers beeswax was used for fixing. The beeswax was weighted and had under 0.001 g (not considered to have any influence) and used the same amount for all tested parts.

The test was performed by sweeping throughout the frequencies with a constant rate starting from lower frequency (5 Hz) to highest frequency (2400 Hz). After

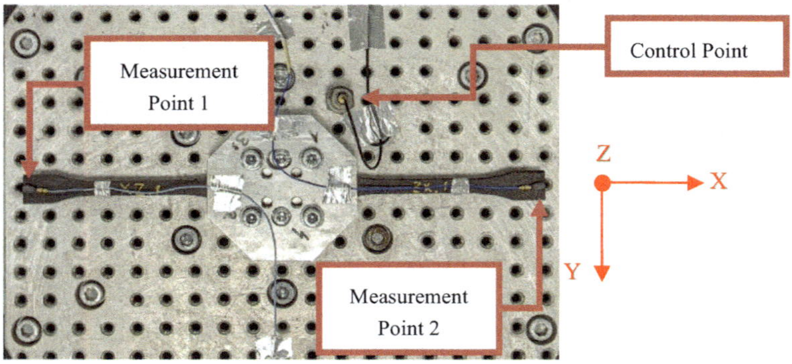

Fig. 25.5 Vibration setup (vibration on Z direction)

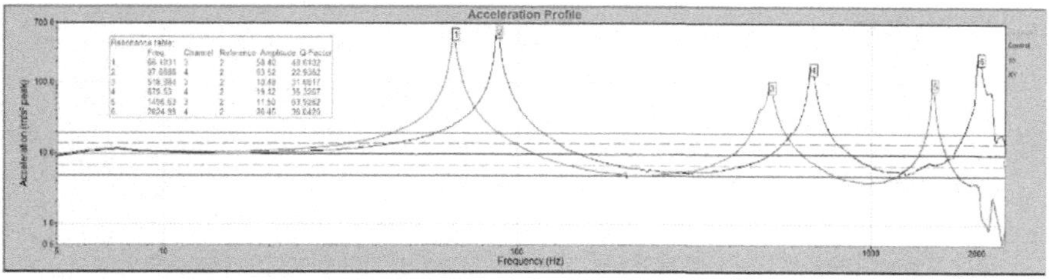

Fig. 25.6 Vibration graph

the sweep ended, specimens were removed and repeated the process with other specimens. In the graph presented in Fig. 25.6 can be observed 3 signals.

The blue line is the control signal which follows the loaded vibration profile. The pink line is from one of the measurement accelerometers that was mounted, in this case, on the specimen with the build orientation of 45 degrees to XY. The brown line is from the other control accelerometer mounted on the specimen with the build orientation of XY. The dotted lines represent the alarm limit of the graph which is $\pm$ 3 dB from the desired amplitude for the control signal, and the red lines represent the abort limit of the profile at $\pm$ 6 dB. If the control signal touches the red line, the electrodynamic shaker is stopped.

On the resonance frequencies results (Tables 25.3, 25.4 and 25.5 and Fig. 25.7) it can be seen, the frequencies where we have the highest amplitude peaks. In the charts we can observe how different build orientation influences the frequency of the resonance points. Knowing this, we can predict the behavior of the fixtures created using this Nylon12CF35. The fixtures can be designed and printed using the build orientation that is most suitable for the vibration direction and mechanical loads of the specific tested part.

Table 25.3 Measured first resonance frequency (Hz)

Specimen/build orientation	XZ	ZX	XY	45°
1	96	42	88	66
2	94	42	86	67
3	88	42	87	66
4	89	42	88	66
5	88	42	88	67

Table 25.4 Measured second resonance frequency (Hz)

Specimen/build orientation	XZ	ZX	XY	45°
1	738	318	686	517
2	735	328	666	519
3	666	329	680	519
4	681	326	681	518
5	653	327	684	513

Table 25.5 Measured third resonance frequency (Hz)

Specimen/build orientation	XZ	ZX	XY	45°
1	2192	961	2027	1479
2	2033	1008	2026	1489
3	1888	1009	2025	1497
4	1964	974	2025	1500
5	1852	969	2027	1494

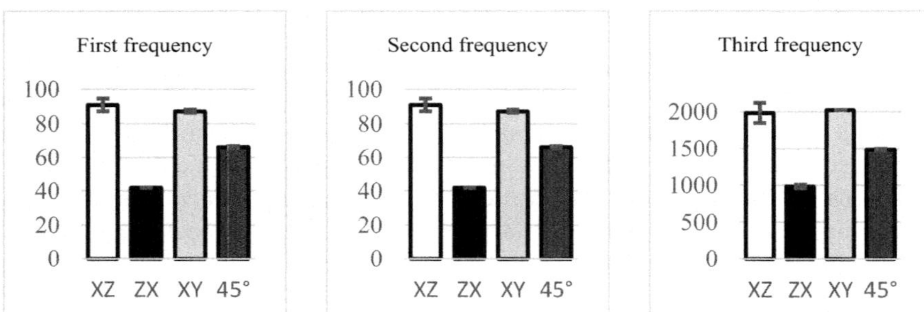

Fig. 25.7 Bar charts of the resonant frequencies

25.4 Conclusions

In this paper 4 groups of specimens were printed using Nylon12CF35 material. Each group was printed on different build orientation (XY, XZ, ZX and 45° to XY), for the purpose of finding their resonance frequency and to determine how different build orientation influence the specimen behavior to vibration. Initially, the specimens were measured and weighted. Then the standard deviation was determined compared to the average value of the measurements for each individual group. Then the samples were rigidly mounted, on one end, on the electro-dynamic shaker and vibrated using sine sweep method. The first 3 resonance frequency points were found for each individual specimen.

From the collected data we can draw the following conclusions:

1. The lowest frequency point detected was at 42 Hz on the specimens with the orientation of ZX and the highest first resonance frequency was at approximately 91 Hz for the build orientation of XZ. The most desired case is for the resonance frequencies to be as high as possible to avoid amplifying the mechanical energy transmitted through the fixture to the tested part. The best behavior in this case had the build orientations XZ and XY. The same pattern can be noticed for the other two frequencies detected.
2. Another observation made for all the 3 identified frequencies is that for the build orientation of XZ the measured points variated more than the others. This led to a higher error of measurement.
3. For the XY orientation, the third resonance frequency measured was above 2000 Hz which is outside the regular testing frequency range in automotive industry.
4. In conclusion, given these results, the most desired build orientation for fixture creation is XY and the worst is ZX.

Acknowledgements The authors gratefully acknowledge the support of Continental Automotive (Qualification Laboratory) for the involvement in this study, by providing test equipment and manufacturing samples for testing.

References

1. A.A. Bakır, R. Atik, S. Özerinç, Mechanical properties of thermoplastic parts produced by fused deposition modeling: a review. Rapid Prototyp. J. **27**(3), 537–561 (2021)
2. S. Teraiya, S. Vyavahare, S. Kumar, in *Advances in Manufacturing Processes*. Experimental Investigation on Influence of Process Parameters on Mechanical Properties of PETG Parts Made by Fused Deposition Modelling (Springer, Singapore, 2021), pp. 283–293
3. I. Khan, N. Kumar, Fused deposition modelling process parameters influence on the mechanical properties of ABS: a review. Mater. Today Proc. **44**, 4004–4008 (2021)
4. D. Popescu, A. Zapciu, C. Amza, F. Baciu, R. Marinescu, FDM process parameters influence over the mechanical properties of polymer specimens: a review. Polym. Testing **69**, 157–166 (2018)
5. S. Khan, K. Joshi, S. Deshmukh, A comprehensive review on effect of printing parameters on mechanical properties of FDM printed parts. Mater. Today Proc. **50**, 2119–2127 (2021)
6. M.Q. Tanveer, G. Mishra, S. Mishra, R. Sharma, Effect of infill pattern and infill density on mechanical behaviour of FDM 3D printed Parts-a current review. Mater. Today Proc. (2022)
7. V. Shanmugam, O. Das, K. Babu, U. Marimuthu, A. Veerasimman, D.J. Johnson et al., Fatigue behaviour of FDM-3D printed polymers, polymeric composites and architected cellular materials. Int. J. Fatigue **143**, 106007 (2021)
8. L. Safai, J.S. Cuellar, G. Smit, A.A. Zadpoor, A review of the fatigue behavior of 3D printed polymers. Addit. Manuf. **28**, 87–97 (2019)
9. S. Wang, Y. Ma, Z. Deng, S. Zhang, J. Cai, Effects of fused deposition modeling process parameters on tensile, dynamic mechanical properties of 3D printed polylactic acid materials. Polym. Testing **86**, 106483 (2020)
10. M. Li, S. Du, F. Li, X. Jing, Vibration characteristics of novel multilayer sandwich beams: Modelling, analysis and experimental validations. Mech. Syst. Signal Process. **142**, 106799 (2020)

11. F. Medel, J. Abad, V. Esteban, Stiffness and damping behavior of 3D printed specimens. Polym. Testing **109**, 107529 (2022)
12. M.D. Almutairi, T.A. Mascarenhas, S.S. Alnahdi, F. He, M.A. Khan, Modal response of hybrid raster orientation on material extrusion printed acrylonitrile butadiene styrene and polyethylene terephthalate glycol under thermo-mechanical loads. Polym. Testing **120**, 107953 (2023)
13. M.Ö. Öteyaka, F.H. Çakir, M.A. Sofuoğlu, Effect of infill pattern and ratio on the flexural and vibration damping characteristics of FDM printed PLA specimens. Mater. Today Commun. **33**, 104912 (2022)
14. Data Sheet—EN A4 FDM Nylon 12CF (2021). https://www.stratasys.com/siteassets/materials/materials-catalog/fdm-materials/nylon-12cf/mds_fdm_nylon-12cf_0921a.pdf?v=48e2e4

Chapter 26
Response of the Vibrating System to Discrete Monotonic Change in Voigt-Kelvin Rheological Stiffness

Polidor Bratu, Cornelia Dobrescu, Ovidiu Vasile, and Oana Tonciu

Abstract The results of the research on the monotonically increasing change in the stiffness of the Voigt-Kelvin viscoelastic materials correspond to real dynamic regimes in the field of vibration compaction technologies. Thus, the models identified by the Voigt-Kelvin schematization both in the laboratory and in situ regarding foundation soils, concretes or asphalt mixtures highlight the discrete modification of the staged stiffness through appropriate compaction technological processes. In this context, the way to modify the dynamic response to the action of the harmonic excitation force is presented.

26.1 Introduction

The one-degree-of-freedom dynamic model is frequently used in technologies for compaction of in-place materials for foundations, precast concrete and asphalt pavements for roads where the forced vibration regime is the determining factor [1–5]. For this, within the road and highway works in Romania, the numerical and experimental results highlighted the important role of changing the stiffness of the material processed by vibration, for technological sequences defined by the degree of settlement or compaction [6–8].

P. Bratu
Romanian Academy of Technical Sciences, 030167 Bucharest, Romania

C. Dobrescu
"Dunărea de Jos" University of Galati, 800008 Galati, Romania

O. Vasile (✉)
National University of Sciences and Technology POLITEHNICA Bucharest, 060042 Bucharest, Romania
e-mail: ovidiu_vasile2002@yahoo.co.uk

O. Tonciu
Technical University of Civil Engineering of Bucharest, 020396 Bucharest, Romania

N. Herisanu and V. Marinca (eds.), *Acoustics and Vibration of Mechanical Structures—AVMS-2023*, Springer Proceedings in Physics 302,
https://doi.org/10.1007/978-3-031-48087-4_26

The dynamic excitation is given by the time-domain varying perturbing force of the form $P(t) = P_0 \sin \omega t$, where the amplitude of the force is proportional to the square of the excitation pulsation ω [9–12]. Thus, $P_0 = m_0 r \omega^2$ where $m_0 r$ is the static moment of the eccentric masses of dynamic imbalance equipping the vibrator also called vibrodyne [13–16].

The aim of the paper is to highlight the method of quantifying the specific parameters of the compaction process according to the Voigt-Kelvin model when, sequentially, the stiffness of the land changes after each pass on the same layer of the structure to be compacted.

26.2 Dynamic Response in Forced Vibration Regime

The dynamic calculation model represented in Fig. 26.1 consists of a body of mass m with a single vertical degree of freedom with the instantaneous displacement $x = x(t)$, with linear viscoelastic connection modeled Voigt-Kelvin having the discretely variable stiffness $k = \lambda k_0$, where $\lambda = 1,2,3,4,5$ and the c linear viscous damping coefficient.

26.2.1 Amplitude in Forced Dynamic Mode

Differential equation of motion is of the form

$$m\ddot{x} + c\dot{x} + kx = P_0 \sin \omega t \tag{26.1}$$

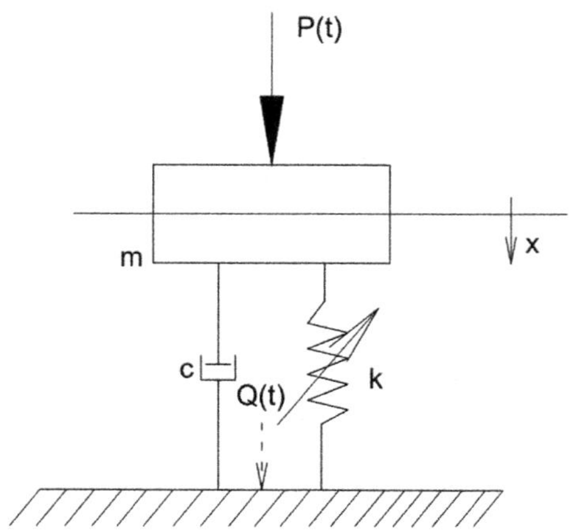

Fig. 26.1 Dynamic model

with the forced dynamic regime solution given by the relation

$$x = A_P^{V-K}\sin(\omega t - \phi) \tag{26.2}$$

in which the vibration amplitude A_P^{V-K} is given by the formula

$$A_P^{V-K} = \frac{P_0}{\sqrt{\left(k - m\omega^2\right)^2 + c^2\omega^2}} \tag{26.3}$$

and the phase shifting φ between displacement and force is of the form:

$$tg\phi = \frac{c\omega}{k - m\omega^2} \tag{26.4}$$

where the pulsation of the excitation force ω can be varied monotonically increasing as the current variable [1, 2].

26.2.2 Transmitted Force

Instantaneous transmitted force $Q(t) = Q_0^{V-K}\sin(\omega t - \alpha)$ represents the ability of the dynamic system to achieve compaction of the Voigt-Kelvin modeled material. Thus, we can write the force $Q(t)$ as [1, 5, 10]

$$Q(t) = kx + c\dot{x} = Q_0^{V-K}\sin(\omega t - \alpha) \tag{26.5}$$

where

Q_0^{V-K} is the amplitude of the transmitted force or the maximum transmitted force Q_{max}.

α the phase difference between the instantaneous displacement $x = x(t)$ and force $Q(t)$.

From relations (26.2) and (26.5) result in the two quantities Q_0^{V-K} and α, as

$$Q_0^{V-K} = A_P^{V-K}\sqrt{k^2 + c^2\omega^2} \tag{26.6}$$

$$tg\alpha = \frac{ktg\phi - c\omega}{k + c\omega tg\phi} \tag{26.7}$$

26.2.3 Energy Dissipated in the Dynamic Process

The dissipated energy is given by the viscous component of the Voigt-Kelvin modeled material having the analytical expression of the form

$$W_d^{V-K} = \pi c\omega\left(A_P^{V-K}\right)^2 \tag{26.8}$$

In the situation of the sequential technological process, where the processing steps are delimited by representative settlements and resumptions of the work cycle, the discrete change of the stiffness k, makes the dissipated energy distinctly representable. For values of k_j, j = 1...5 and the continuous variation of the pulsation ω as a factor of the forced dynamic regime results in the family of dissipated energy curves [1, 2, 10, 14].

26.3 Variation of the Parameters of the Dynamic Process in the Regime of Forced Vibrations

Graphical representation of families of curves for technological vibration amplitude $A_P^{V-K} = A_P^{V-K}(\omega, k)$, the maximum force transmitted to the compacted material $Q_0^{V-K} = Q_0^{V-K}(\omega, k)$ and dissipated energy $W_P^{V-K} = W_P^{V-K}(\omega, k)$ is achieved by the continuous variation of the excitation pulse $\omega = \Omega\omega_n$ and discrete stiffness $k = \lambda k_0$, where $\omega_n = \sqrt{\frac{k}{m}}$ is its own pulsation and $\lambda = 1, 2, 3, 4, 5$ the multiplier of rigidity on processing sequences.

In this case, the continuous variable Ω the essential discrete variable $k = \lambda k_0$ will be the sizes that will individualize the families of curves.

For this, the relations (26.3), (26.6) and (26.8) can be reformulated as follows:

$$A_P^{V-K} = \frac{m_0 r}{m}\Omega^2\sqrt{\frac{mk_0\lambda}{mk_0\lambda(1-\Omega^2)^2 + c^2\Omega^2}} \tag{26.9}$$

$$Q_0^{V-K} = \frac{m_0 r}{m}k_0\lambda\Omega^2\sqrt{\frac{mk_0\lambda + c^2\Omega^2}{mk_0\lambda(1-\Omega^2)^2 + c^2\Omega^2}} \tag{26.10}$$

$$W_P^{V-K} = \pi c\left(\frac{m_0 r}{m}\right)^2\Omega^5\frac{\sqrt{mk_0^3\lambda^3}}{mk_0\lambda(1-\Omega^2)^2 + c^2\Omega^2} \tag{26.11}$$

where we have:

$$\Omega = \Omega(k_0\lambda) = \omega\sqrt{\frac{m}{k_0\lambda}} \tag{26.12}$$

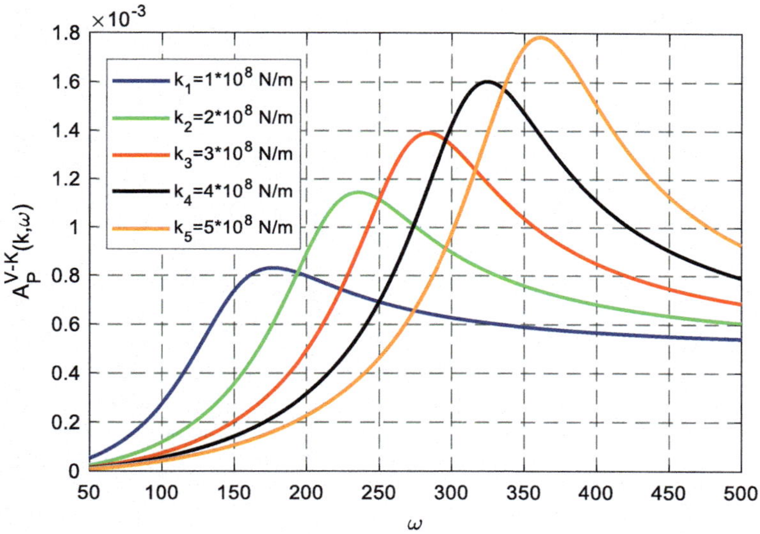

Fig. 26.2 Amplitude variation as function of the continuous variable ω and discrete variable k

- relative pulsation depending on ω and λ

$$A_0 = \frac{m_0 r}{m} \tag{26.13}$$

- amplitude in post-resonance for the condition $\lim\limits_{\Omega \to \infty} A_P^{V-K} = A_0$

The case study was carried out for a vibratory equipment and a viscoelastic material with the following physical–mechanical characteristics [1, 2]:

$$m = 4000\,\text{kg};\, c = 4 \times 10^5 \frac{Ns}{m};\; m_0 r = 2\,\text{kg m};\, k_0 = 10^8 \frac{N}{m};$$
$$\lambda = 1, 2, 3, 4, 5;\; \omega = 0 \ldots 500 \frac{rad}{s}$$

In Figs. 26.2, 26.3 and 26.4 are represented the sizes $A_P^{V-K}(\omega, k)$, $Q_0^{V-K}(\omega, k)$ and $W_P^{V-K}(\omega, k)$ which are measured in m, N and J respectively.

26.4 Conclusions

For the vibratory compaction system, numerical and experimental tests were carried out for various categories of land that can be modeled rheologically according to the Voigt-Kelvin schematization.

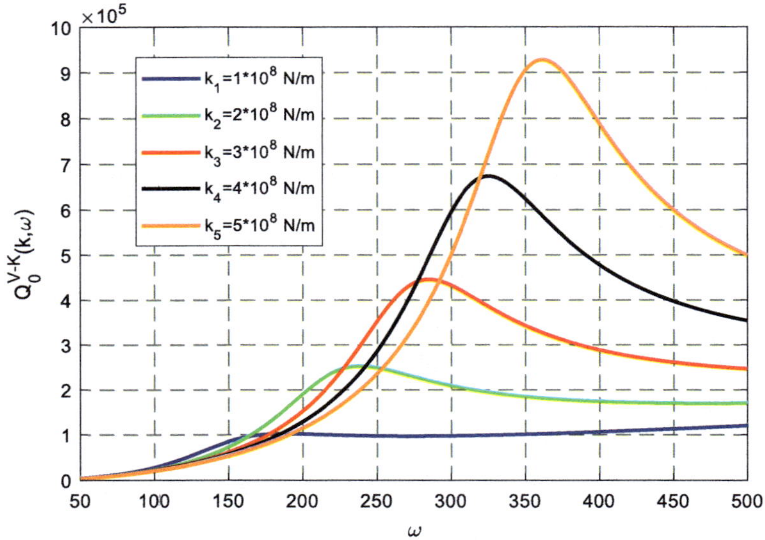

Fig. 26.3 The variation of the transmitted force as a function of the continuous variable ω

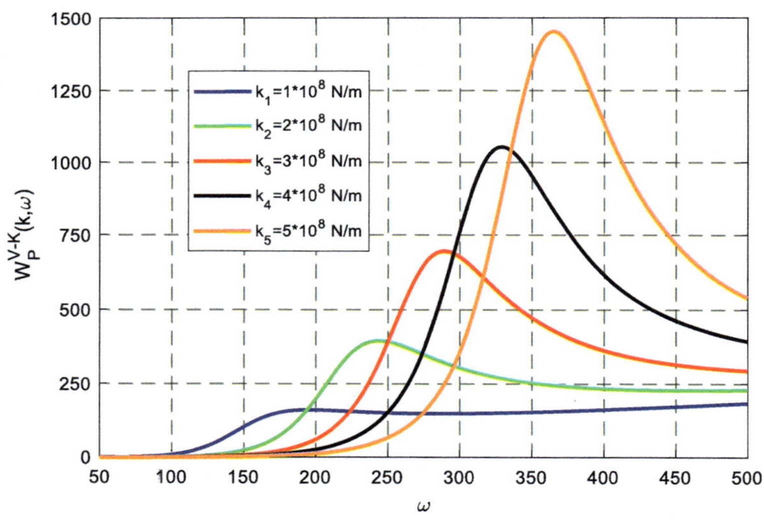

Fig. 26.4 Variation of dissipated energy as a function of pulsation ω and stiffness k

Within the Research Institute for Construction Equipment and Technologies ICECON Bucharest in Romania, the mechanical characteristics of the compactor were measured such as mass m, the static moment m_0r and the excitation pulsation

ω simultaneously with the vibration amplitude, the ground reaction force as a transmitted force as well as the physico-mechanical characteristics of the ground k and c after each processing sequence. Viscous damping c remained at relatively constant values, instead the terrain stiffness changed after each processing sequence, being measured 5 values of $k = \lambda k_0$, where $\lambda = 1, 2, 3, 4, 5$.

Based on the construction of the families of curves, the following conclusions can be synthesized:

1. The amplitude curves are characterized by resonance pulsations that move in an increasing direction with values of the resonance peaks that are inscribed on a straight line (Fig. 26.2);
2. The resonance peaks of the transmitted force (Fig. 26.3) are recorded on an ascending curve with the highest value at 350 rad/s.
3. The dissipated energy reaches maximum values at resonance for the highest natural pulsation corresponding to the last sequence of the compaction process (Fig. 26.4).

Compared to the above, the Voigt-Kelvin rheological behavior of viscoelastic materials can be evaluated based on numerical analysis and/or experimental results.

Thus, at discretely decreasing values of the damping rate ζ, the resonance peaks of the amplitude are increasing reaching the highest value at the lowest damping when the excitation pulse ω is monotonically increasing.

References

1. P. Bratu, *Dynamic Isolation. Vibrational and Seismic Actions* (Impuls Publishing House, Bucharest, 2021)
2. C. Dobrescu, in *22th International Congress on Sound and Vibration.* The Rheological Behaviour of Stabilized Bioactive Soils During the Vibration Compaction Process for Road Structures (Florence, Italy, 2015)
3. J.A. Forrest, in *Annual Conference of the Australian Acoustical Society.* Free-Free Dynamics of Some Vibration Isolators (Adelaide, Australia, 2002).
4. E.A. Johnson, J.C. Ramallo, B.F. Spencer, M.K. Sain, in *2nd World Conference on Structural Control.* Intelligent Base Isolation Systems (Kyoto, Japan, 1998)
5. Y.-S. Kim, Study on the effective stiffness of base isolation system for reducing acceleration and displacement responses. J. Korean Nuclear Soc. **31**(6), 586–594 (1999)
6. K.-S. Park, S.-W. Cho, I.-W. Lee, in *14th KKNN Symposium on Civil Engineering.* A Comparative Study on Aseismic Performances of Base Isolation Systems for Multi-Span Continuos Bridge (Kyoto, Japan, 2001)
7. A.M. Mitu, T. Sireteanu, G. Ghita, Passive and semi-active bracing systems for seismic protection: a comparative study. Rom. J. Acoust. Vib. **12**(1), 49–56 (2015)
8. S.K. Park, K.B. Han, Effects of seismic isolation bearing with sliding mechanism on the response of brige. Mater. Struct. **37**, 412–421 (2004)
9. S. Pinarbasi, K. Dimitrios, J.M. Kelly, in *10th World Conference on Seismic Isolation, Energy Dissipation and Active Vibrations Control of Structures.* Seismic isoLation for soft-Story Buildings (Istanbul, Turkey, 2007)
10. A. Potirniche, Assessments regarding effective configurations of vibration isolators based on displacements analysis. Rom. J. Acoust. Vib. **12**(2), 126–131 (2015)

11. T. Sireteanu, Smart suspension systems. Rom. J. Acoust. Vib. **13**(1), 2–2 (2016)
12. T. Sireteanu, in *9th WSEAS International Conference on Acoustics & Music Theory & Applications (AMTA '08)*. The Effect of Structural Degradation on the Dynamic Behaviour of Buildings (Bucharest, Romania, 2008)
13. N.D. Stanescu, Vibrations of a shell with clearances, neo-Hookean stiffness, and harmonic excitations. Rom. J. Acoust. Vib. **13**(2), 104–111 (2016)
14. O. Vasile, Active vibration control for viscoelastic damping systems under the action of inertial forces. Romanian J. Acoust. Vib. **14**(1), 54–58 (2017)
15. Y.-P. Wang, Fundamentals of seismic base isolation, International training programs for seismic design of building structures hosted by National Center of Research on Earthquake Engineering, Taiwan
16. X-Y Zhou, M. Han, L. Yang,: Study on protection measures for seismic isolation rubber bearings. ISET J. Earthq. Technol. **40**(2–4), 137–160 (2003) (Paper no. 436)

Chapter 27
The Dynamic Response of the Vibrating System in the Case of the Maxwell Rheological Model

Polidor Bratu, Oana Tonciu, Cornelia Dobrescu, and Ovidiu Vasile

Abstract The Maxwell rheological behavior of materials with a high content of fluid fraction with very fine particles in the mixture leads to the predominant character of the viscous component in series connection with the elastic component. For this, laboratory tests were carried out on small-scale samples subjected to a variable-frequency vibrating regime with monotonically increasing command. Thus, from the amplitude-frequency characteristic, results that from the construction of a family of curves, the largest amplitude corresponds to the material with the largest viscous damping, in contrast to the Voigt-Kelvin rheological model. Materials in this category are used in constructions with commissioning through appropriate technological vibrations. In this case, the materials can be listed: lightweight concretes with additives and setting accelerators, bitumens, sandy soils and dusty clays, paints and mastics for road works etc. Research carried out within the Research Institute for Construction Equipment and Technologies ICECON Bucharest, Romania, have highlighted the fact that certain categories of materials have Maxwell rheological behavior such as: soils in filling for road works, asphalt finishing mixtures, composite mastics for filling expansion joints, foundation soils with a high composition of sand and fine clay for bridge piles etc.

P. Bratu
Romanian Academy of Technical Sciences, 030167 Bucharest, Romania

O. Tonciu
Technical University of Civil Engineering of Bucharest, 020396 Bucharest, Romania

C. Dobrescu
"Dunărea de Jos" University of Galati, 800008 Galati, Romania

O. Vasile (✉)
National University of Sciences and Technology POLITEHNICA Bucharest, 060042 Bucharest, Romania
e-mail: ovidiu_vasile2002@yahoo.co.uk

N. Herisanu and V. Marinca (eds.), *Acoustics and Vibration of Mechanical Structures—AVMS-2023*, Springer Proceedings in Physics 302,
https://doi.org/10.1007/978-3-031-48087-4_27

27.1 Introduction

The essential parametric changes in amplitude, transmitted force and energy dissipation to continuous variation of excitation pulsation and discrete variation of rigidity are highlighted in this paper. Thus, the compaction variation curves were raised after each technological step characterized by the change in rigidity. In this way it can be ascertained to what extent the influence of rigidity can change both the dynamic response until the resonance state disappears, as well as the cases of linearity for the transmitted force or dissipated energy.

The novelty of the research consists in the fact that the asymptotic behavior of the amplitude for a certain rigidity with a "threshold" value of k $\geq 2 \times 10^8$ N/m is highlighted, as well as the disappearance of the resonance. It is also noted that the transmitted force and dissipated energy have a linear variation with respect to excitation.

For the application of the Maxwell modeled materials commissioning technology, a device is used that acts on the material in the regime of controlled forced vibrations. In this case, the specific processing of viscoelastic materials can be: homogenization, surface or deep compaction, as well as joint filling with the help of vibrating devices [1].

The vibrations are generated by a dynamic system that produces a force that disturbs the form $P(t) = m_0 r\omega^2 \sin \omega t$ whose amplitude $P_0 = m_0 r\omega^2$ is of the inertial type with an eccentric mass of static moment $m_0 r$ and angular velocity ω [2–5].

Tests carried out in the laboratory and in situ with soil compaction equipment highlighted the Maxwell rheological character and allowed the establishment of stiffness parameters, linear viscous damping at the interaction of vibratory equipment—ground.

As a result of the analysis of the amplitude-pulsation (angular frequency) response curves and the physical measurements for representative points at discrete pulsations of the excitation, a good approximation of the Maxwell rheological behavior could be found [6–8].

27.2 Dynamic Response

The dynamic model (Fig. 27.1) consists of a system with a mass degree of freedom m excited with a perturbing force $P(t) = P_0 \sin \omega t$, where $P_0 = m_0 r\omega^2$ and a "viscoelastic Maxwell medium with monotonically increasing stiffness $k = \lambda k_0$, with $\lambda = 1, 2, 3, 4, 5$ and constant viscous damping coefficient c".

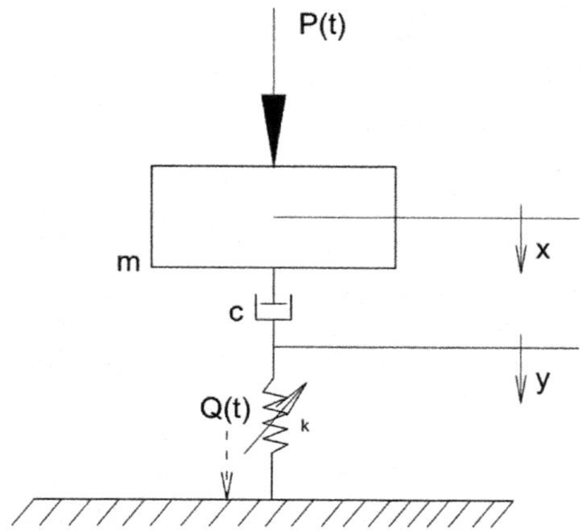

Fig. 27.1 Maxwell rheological model

27.2.1 Forced Mode Vibration Amplitude

Linear differential equations, formulated in complex quantities, can be written as:

$$\begin{cases} m\ddot{\tilde{x}} + c\left(\dot{\tilde{x}} - \dot{\tilde{y}}\right) = P_0 e^{j\omega t} \\ c\left(\dot{\tilde{x}} - \dot{\tilde{y}}\right) = k\tilde{y} \end{cases}, \tag{27.1}$$

where the coordinate $\tilde{x} = \tilde{x}(t)$ is the complex instantaneous displacement of the vibrations of mass m, and $\tilde{y} = \tilde{y}(t)$ is the complex size of the "hidden" coordinate of the displacement connection point (*c*-*k*).

$$\begin{cases} \tilde{x} = \tilde{\bar{x}} e^{j\omega t}, \text{ where } \tilde{\bar{x}} = A_P^M e^{-j\varphi} \\ \tilde{y} = \tilde{\bar{y}} e^{j\omega t}, \text{ where } \tilde{\bar{y}} = B e^{-j\theta} \end{cases}, \tag{27.2}$$

the imaginary unit of complex numbers; φ—the phase difference between the instantaneous displacement $x(t)$ and the force $P(t)$; θ—the phase difference between the instantaneous displacement $y(t)$ and the force $P(t)$.

The depreciation rate can be written in the form

$$\zeta = \zeta(k) = \frac{c}{2\sqrt{km}}, \tag{27.3}$$

For the technological processing of Maxwell modeled materials, only the A_P^M amplitude resulting from the solution of system (27.1) is of interest [2]. Thus, based on the studies carried out, we have:

$$A_P^M(\omega, \zeta) = \frac{m_0 r}{m}\Omega^2\sqrt{\frac{1 + 4\zeta^2\Omega^2}{\Omega^4 + 4\zeta^2\Omega^2(1 - \Omega^2)^2}}. \tag{27.4}$$

27.2.2 Transmitted Force

Dynamic transmitted force $Q^M(t) = Q_0^M \sin(\omega t - \beta)$ is determined using the relation [2]:

$$\tilde{Q}^M(t) = k\tilde{y}(t) \text{ or } \tilde{Q}^M(t) = kBe^{-j\theta}e^{j\omega t}. \tag{27.5}$$

Taking into account the studies carried out, we have for B the following formula:

$$B = \frac{m_0 r}{m}\frac{2\zeta\Omega^3}{\sqrt{\Omega^4 + 4\zeta^2\Omega^2(1-\Omega^2)^2}}, \tag{27.6}$$

in which case for Q_0^M we have the formula

$$Q_0^M(\omega,\zeta) = \frac{m_0 r}{m}\frac{2\zeta\Omega^3 k}{\sqrt{\Omega^4 + 4\zeta^2\Omega^2(1-\Omega^2)^2}}. \tag{27.7}$$

27.2.3 Dissipated Energy

For the energy dissipated on the element of viscous constant c, we have the following relation:

$$W_P^M(\omega,\zeta) = \pi c\Omega\sqrt{\frac{k}{m}}\left[A_P^M(\omega,\zeta)\right]^2, \tag{27.8}$$

where $A_P^M(\omega,\zeta)$ is given by the relation (27.4), and $\Omega\sqrt{\frac{k}{m}} = \Omega\omega_n = \omega$.

27.3 Parametrized Curve Families

The graphical representation of the families of curves is based on the continuous variation of the variable ω and discrete stiffness variation $k = k_0\lambda$; $\lambda = 1, 2, 3, 4, 5$. Taking into account that relative pulsation $\Omega = \frac{\omega}{\omega_n} = \omega\sqrt{\frac{m}{k}}$, and $\zeta = \zeta(k) = \frac{c}{2\sqrt{km}}$ are functions of ω and k results that the families of curves parametrized by ζ can be obtained at the continuous variation of the excitation pulsation ω.

The case study is characterized by a vibrating Maxwell model system with the following mechanical characteristics [9]:

$$m = 4000\,\text{kg};\ c = 4\times 10^5\,\frac{\text{Ns}}{\text{m}};\ m_0 r = 2\,\text{kg m};$$

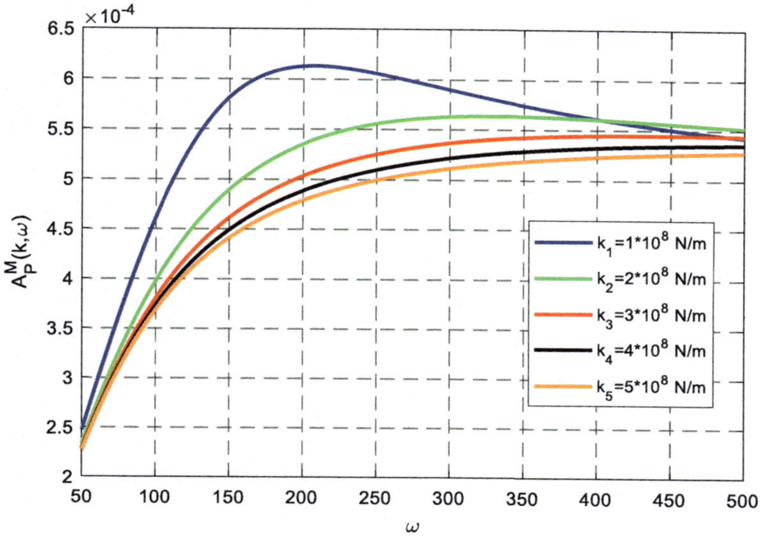

Fig. 27.2 Amplitude variation as a function of ω and k

$$k_0 = 10^8\,\frac{\mathrm{N}}{\mathrm{m}};\ \lambda = 1, 2, 3, 4, 5;$$
$$k = k_0\lambda;\ \omega = 0\ldots500\,\frac{\mathrm{rad}}{\mathrm{s}}.$$

The families of curves are given in Figs. 27.2, 27.3 and 27.4, where $A_P^M(k,\omega)$ in m, $Q_0^M(k,\omega)$ in N and $W_P^M(k,\omega)$ in J.

27.4 Conclusions

The Maxwell rheological system that models the behavior of viscoelastic materials with particularities regarding the proportion of viscous and elastic components. Thus, based on the elevation of the amplitude-frequency characteristic for five stiffness values, it is highlighted that at dynamic response in harmonic regime the amplitude reaches the highest values for stiffness below 10^8 N/m, and at higher values the resonance peak attenuates until it disappears.

As a result of the discrete change of the stiffness k and at a constant damping rate ζ, after each technological sequence, the characteristic curves were raised as follows:

- The amplitude $A_P^M(\omega,\zeta)$, for each stiffness value $k = k_0\lambda$ with $\lambda = 1, 2, 3, 4, 5$, shows decreasing resonance maxima (peaks) according to the family of curves in Fig. 27.2;

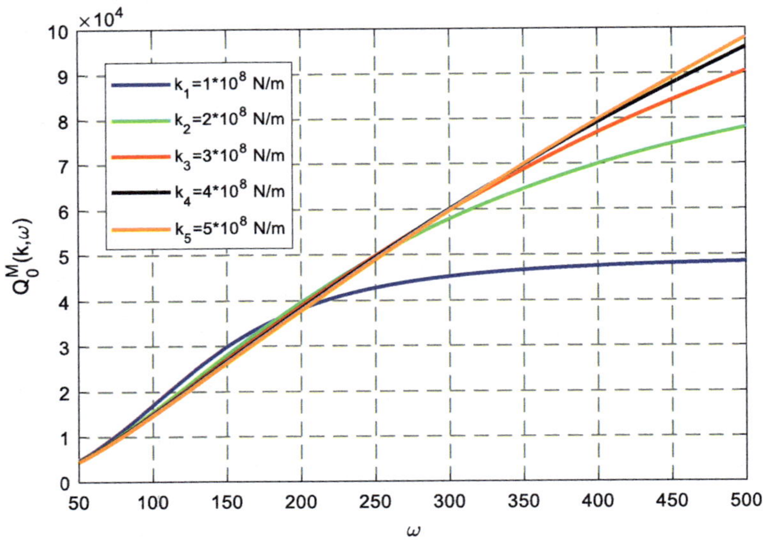

Fig. 27.3 Variation of transmitted force as a function of ω and k

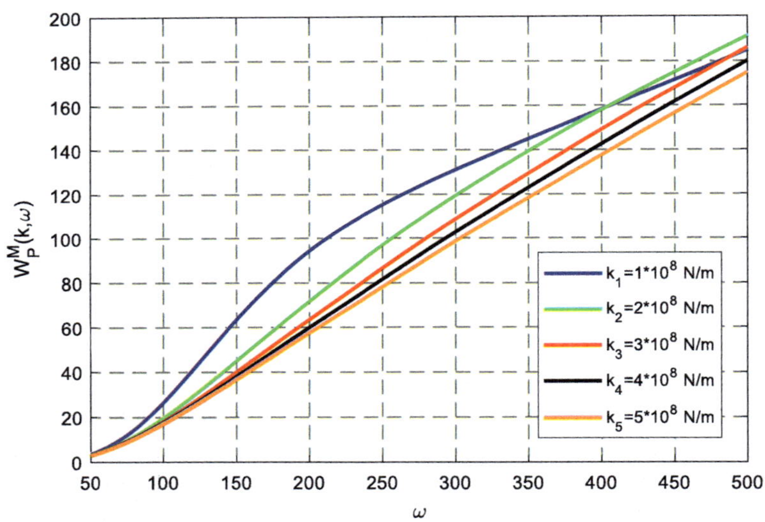

Fig. 27.4 Variation of dissipated energy as a function of ω and k

- The maximum transmitted force for values of $k > 2 \times 10^8$ N/m has a linear behavior in relation to the continuous variation of the excitation pulsation ω. The force variation tends asymptotically from the value 4×10^5 N when the pulsation $\omega > 200\,\mathrm{rad}/\mathrm{s}$ (Fig. 27.3).

Compared to the above, it is found that both analytically and experimentally, depending on the distribution of the parametrized curves according to the decreasing values of stiffness k and for the monotonically increasing variation of the excitation pulsation ω, the Maxwell rheological behavior of viscoelastic materials can be evaluated.

The numerical and experimental evaluation method can be used to identify the rheological behavior of viscoelastic materials under the action of technological vibration.

References

1. P. Bratu, C. Dobrescu, Dynamic response of Zener-modelled linearly viscoelastic systems under harmonic excitation. Symmetry **11**(8), 1050 (2019). https://doi.org/10.3390/sym11081050
2. P. Bratu, *Dynamic Isolation. Vibrational and Seismic Actions* (Impuls Publishing House, Bucharest, 2021)
3. R. Morariu-Gligor, A. Crișan, F. Șerdean, Optimal design of an one-way plate compactor. ACTA Technica Napocensis Appl. Math. Mech. Eng. **60**(4), 557–564 (2017)
4. S. Nastac, C. Debeleac, P. Nechita, Assessments on shock absorption properties of foam-formed low density cellulose composites. ACTA Technica Napocensis Appl. Math. Mech. Eng. **60**(4), 565–572 (2017)
5. R.M. Morariu-Gligor, Factors influencing the degree of soil compaction. ACTA Technica Napocensis Ser. Appl. Math. Mech. **60**(1), 125–130 (2017)
6. A. Potîrniche, S. Năstac, Functional correlations regarding passive isolation of symmetrical systems. ACTA Technica Napocensis Ser. Appl. Math. Mech. Eng. **56**(4), 777–784 (2013)
7. O. Vasile, S. Vlase, C. Năstac, M.L. Scutaru, Experimental analysis of a mechanical system composed by two identical parts. ACTA Technica Napocensis Ser. Appl. Math. Mech. Eng. **61**(3), 333–338 (2018)
8. S. Bejan, Rheological models of the materials for the road system in the compaction process. Romanian J. Acoust. Vibr. **11**(2), 167–171 (2014)
9. "In situ" tests of the Bucharest-Constanța highway, Romania. Test report, ICECON, Bucharest (2005)

Chapter 28
Dynamic Study of Bucket Lifting Mechanism on Heavy Excavators for Pits

Gina Diana Muscă

Abstract On mechanical excavators for pits the digging is usually performed by lifting the bucket in the same time with its pushing (pressing) in the ground. Depending on category of the ground to be dig in, the pushing mechanism of bucket handle provides suitable thickness of the furrow. After the bucket is filled in with dug ground, the entire work equipment together with upper platform slew towards unloading point. Besides static forces, dynamic stresses may also frequently come up in mechanisms of excavators with one bucket. These stresses can have various causes. To dynamically study a mechanism, it is necessary to make the elastic system scheme of the mechanism, based on which the behaviour of each element of the system is studied. When making the calculation schemes, for dynamic systems chosen, we considered the constructive scheme and forces group acting on the mechanism. We got complex general schemes for which it is quite difficult to establish dynamic forces and oscillation frequencies. Studying the certain way of mechanism operation, as well as behaviour of the component elements, simplified calculation schemes can be admitted, generally with one or two degrees of freedom, which theoretically can easily be solved.

28.1 Introduction

An excavator is a machinery for earth digging and extraction of ballast, sand, crushed stone or other materials from pits or hardly accessible places. The excavator also secondarily or implicitly performs loading operation for excavated or extracted material. Such excavators are manufactured in many constructive versions, namely: excavators with boom consisting of a single beam and handle made of two beams, excavators with single beam handle and boom made of two jointed sections and excavators with single beam handle and oscillating auxiliary boom for handle pushing. The main

G. D. Muscă (✉)
Faculty of Engineering and Agronomy in Braila, Research Centre for Machines and Technological Equipment's Mechanics, "Dunarea de Jos" University of Galati, Galati, Romania
e-mail: diana.anghelache@ugal.ro

N. Herisanu and V. Marinca (eds.), *Acoustics and Vibration of Mechanical Structures—AVMS-2023*, Springer Proceedings in Physics 302,
https://doi.org/10.1007/978-3-031-48087-4_28

Fig. 28.1 Mechanical excavator with handle made of two beams for pit works

work mechanisms of such excavators are: bucket lifting mechanism, handle pushing mechanism and slewing mechanism [3].

Within this paperwork, the research is performed on an excavator with boom consisting of a single beam and handle made of two beams (Fig. 28.1). The handle consists of two parallel beams, supported on boom, at their rear side, by means of a jointed guide. The boom is only intended to support the bucket lifting hoist. Mechanical excavators for pits usually perform digging by lifting the bucket while pushing (pressing) it into earth [2].

In Fig. 28.1 the following notations are made: 1—rotating upper platform; 2—crawlers; 3—lower frame; 4—bucket; 5—bucket lifting hoist; 6—boom; 7—handle; 8—jointed guide; 9—boom supporting hoist.

Pushing mechanism of the bucket handle provides suitable thickness of the furrow, depending on category of dug earth. After filling the bucket with dug earth, the entire work equipment together with upper platform are slewing towards discharge point. By help of a cable the lock of bucket lower wall is pulled, and it discharges into the vehicle for transport or onto the heap. The work equipment is brought back to start position of digging and the cycle is repeated.

28.2 Dynamics of Bucket Loading Mechanism

Besides static forces, dynamic stress may also frequently come up in the mechanisms of excavators with one bucket. Such stress may be determined by various causes. Therefor when digging cohesive earth with big and hard stone-like inclusions sudden increases of resistance to digging may come up, leading to bucket blockage. Also, when loading rocks which are not enough deployed by explosion, very high dynamic stress may occur on bucket and driving mechanisms. High dynamic stresses usually come up during acceleration-braking processes of various mechanisms. To dynamically study a mechanism, it is necessary to issue the elastic system scheme of the mechanism, based on which the behavior of each element of the system is studied.

When issue the calculation schemes, for dynamic systems chosen, the constructive scheme as well as forces assembly stressing the mechanism are taken into account.

General complex schemes are obtained this way, for which determination of dynamic loads and oscillations frequency is difficult enough. But, studying the certain way of operation of the mechanism, as well as behavior of components, simplified calculation schemes can be admitted, generally with one or two degrees of freedom, which theoretically can be easily solved. By dynamic study of the elements of excavators with periodic action, we notice that is possible to reduce the general schemes to simplified ones and the results obtained are not too much different from the real ones [8]. It is known that it is necessary for the heavy pit excavators to be equipped with one electric motor for each drive.

28.2.1 Dynamic System Diagram of the Bucket Lifting Mechanism

The bucket lifting mechanism can be made in various versions, including as elements one or two motors, gearboxes and lifting trolleys, or linear or rotating hydraulic motors. Dynamically speaking, the schemes of the various versions have many common elements [7].

Figure 28.2a, shows the complete scheme of dynamic system for bucket lifting mechanism in case of one electric motor drive [4]. In Fig. 28.2, the following notations are made: m_1—is the sum of reduced masses of motor rotor and semi-clutch for motor shaft drive; m_2—the sum of reduced masses of the motor shaft driven semi-clutch, gear wheels of the first transmission and semi-clutch for intermediate shaft drive; m_3—reduced mass of intermediate shaft driven semi-clutch; m_4—reduced mass of the boom; m_5—reduced mass of the handle and bucket; m_6—reduced mass of excavator without work equipment; C_1—reduced stiffness of the motor shaft, motor shaft clutch and wheel shaft; C_2—reduced stiffness of the coupling and intermediate shaft; C_3—stiffness of lifting mechanisms cables; C_4—stiffness of boom supporting elements; P—reduced effort for motor driving; S_r—earth digging strength reduced to lifting hoist and Q—reduced effort given by excavator weight to ensure its stability. Figure 28.2b, shows the complete scheme of the same mechanism driven by two motors installed in parallel. In this case, the drive being symmetrical, it can be considered that: $m_1 = m_{1'} + m_2''$; $m_3 = m_{3'} + m_3''$; $C_1 = C_{1'} + C_1''$; $C_2 = C_{2'} + C_2''$ and $P = P' + P''$.

The scheme in Fig. 28.2b. can be reduced to the one in Fig. 28.2a. Simplified schemes of dynamic system for the bucket lifting mechanism driven by individual electric motors are given in Fig. 28.2c.

In Fig. 28.2 we noted with: m_1—reduced mass of the entire lifting system; m_2—reduced mass of the excavator in case of its oscillation against stability axis.

Scheme 2c, shows excavator operation in stable position, and scheme 2d, the case when machinery starts oscillating against stability axis.

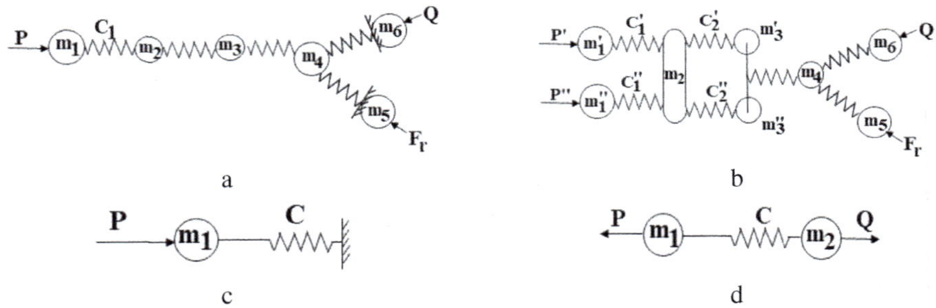

Fig. 28.2 Schemes of dynamic system for bucket lifting mechanism of one bucket excavators: **a** complete scheme for driving with one electric motor; **b** complete scheme for driving with two electric motors; **c**, **d** simplified calculation schemes

When making such schemes the issue is that for simplifying and generalizing the schemes, both the mass moments of inertia and stiffness of connection elements are considered as being reported to the element with the greatest influence within the mechanism, respectively dynamic system [5].

The excavators for pits are driven by individual motors for each mechanism [1]. We study dynamic behavior of the mechanism for bucket loading for a fixed angle α between equipment main boom and soil as shown in Fig. 28.3.

The following notations are made in above figure: l_b—main boom length; α—rotation angle of the main boom with a parallel plan with soil; β—jib (handle) rotation

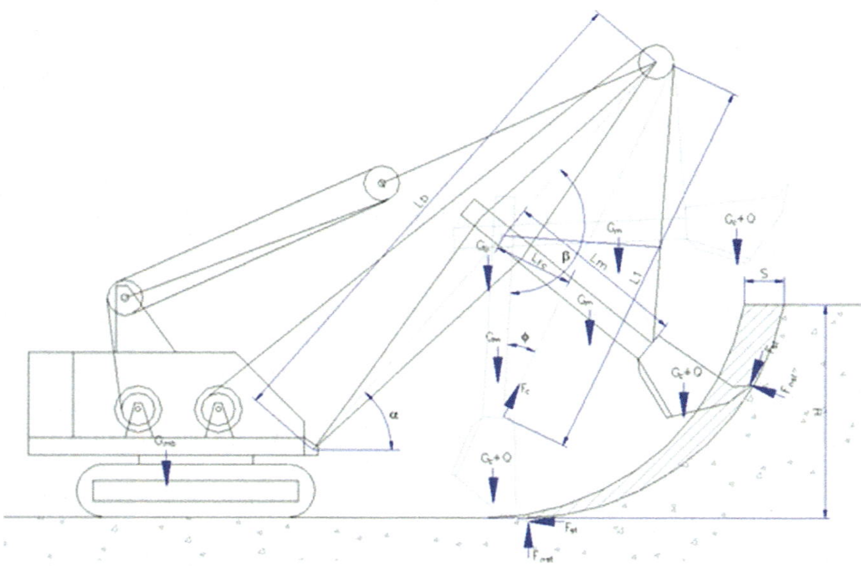

Fig. 28.3 Bucket loading mechanism for a fixed angle

angle with median axis of main boom; l_m—handle length from jointing point of the boom until anchoring point of the driving cable.

Jointing point of the handle on main boom is considered to be at ½ l_b.

The force in cable depends on: handle rotation angle $\beta \cong 90^\circ + \alpha$; digging force R_{st} from bucket ($R_{st} = c_1 \cdot b \cdot H$ si $R_{nst} = \Psi \cdot R_{st}$); bucket capacity, implicitly the load Q ($Q = \gamma_{earth} \cdot V_{bucket} \cdot \cos(\beta' - \alpha)$).

To determine the force in driving cable of the handle and bucket, F_c we write the moments equilibrium equation from O_1 point of handle joint:

$$\sum M_{o1} = 0 : R_{st} \cdot l_m + \frac{l_m}{2} \cdot G_m \cdot \cos(\beta - \alpha) + (G_c + Q) \cdot l_c \cos(\beta - \alpha) + F_c \cdot l_{FC}$$

$$F_c = \frac{F_{st} \cdot l_m + \frac{l_m}{2} \cdot G_m \cdot \cos(\beta - \alpha) + (G_c + Q) \cdot l_c \cdot \cos(\beta - \alpha)}{l_{FC}} \tag{28.1}$$

where: l_{FC} is variable depending on β

$$l_1^2 = l_m^2 + \frac{l_b^2}{4} - 2l_m \cdot \frac{l_b}{2} \cos \beta \tag{28.2}$$

$$\Psi = \arccos\left(\frac{l_m^2 + l_1^2 - \frac{l_b^2}{4}}{2 \cdot l_m \cdot l_1}\right) \tag{28.3}$$

$$l_{Fc} = l_m \cdot \beta \tag{28.4}$$

The angle β starts from β_{0max} until β_{0min} .

Movement equations describing the physical process are:

$$\begin{cases} J_m \cdot \ddot{\phi}_b + R_{st} \cdot L_b + Q \cdot l_Q = k \cdot \Delta l;\ \Delta l = \frac{F_c}{k_c} \quad k = \frac{EA}{l} \\ J_{red} \cdot \ddot{\phi}_{tamb} + M_{rez} \cdot (sgn\dot{\phi}) + k \cdot \Delta l = M_{mot}; \end{cases} \tag{28.5}$$

The main boom (for handle support) was considered stiff.

Within above relations we used the following notations: J_m—boom moment of inertia; $\ddot{\phi}$—boom rotation angle; R_{st}—tangent digging resistance; L_b—boom length; Q—load in the earth bucket; l_Q—length from joint to excavator mass centre (bucket centre); k—elastic coefficient in the cable; Δl—cable elongation; J_{red}—moment of inertia reduced to motor shaft; $\ddot{\phi}_{tam}$—drum rotation angle; M_{rez}—friction moment of the drum-motor transmission; sgn($\dot{\phi}$)—mathematical function giving the sign and M_{mot}—motor moment.

After writing the equations describing the movement, these were numerically solved, by using the Runge Kutta of order 4 method and a specially performed software [6]—as solving method.

28.3 Results Obtained

This numerical program displays dynamic behavior of some parameters (drum shaft speed, load travel, load speed, tension on drum, tension in bucket and cable force), for the first two seconds of the transitory mode.

In Fig. 28.4—the drum speed has a strongly irregular behavior and a time increasing amplitude, which is undesirable during operation. Load movement shows a slight irregular increase.

In conclusion, the 14 kW motor power is not indicated for driving mechanism of the pit excavator bucket.

In Fig. 28.5—motor power is 15 kW, drum speed has a pulsatory evolution with slightly increasing amplitude.

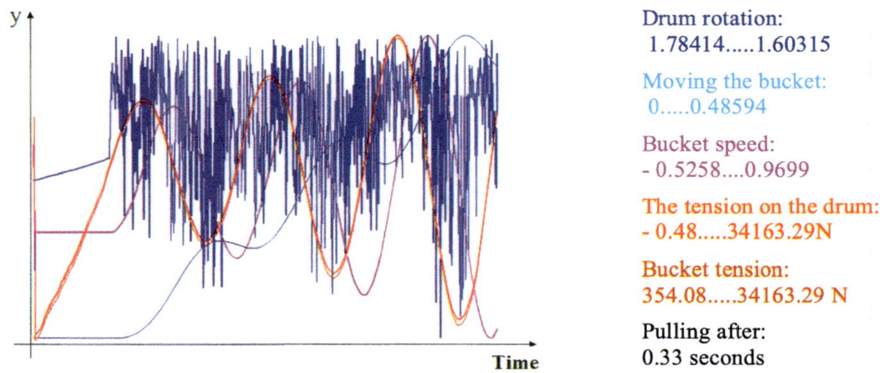

Fig. 28.4 Graphic representation for 14 kW motor

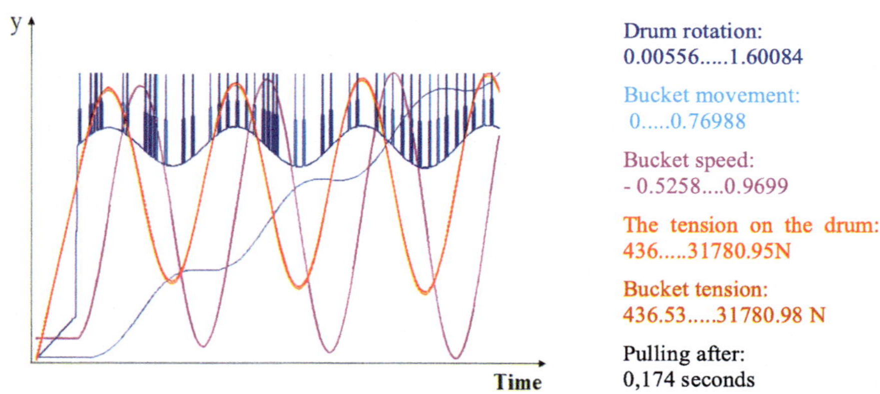

Fig. 28.5 Graphic representation for 15 kW motor

The other shapes represent travel, bucket speed and tension at drum, for the bucket being similar to previous case except for the pullout which was faster namely 0.174 s comparing to 0.33 s in the previous case.

In Fig. 28.6—motor power of 26.5 kW—drum speed shows the same pulsatory behavior, just the amplitude is kept or shows a slight decrease. Due to pullout—0.0978 s the tension in cable reaches zero value too.

In Fig. 28.7—for a 30 kW motor, the drum speed has a strongly increasing evolution afterwards having an approximately linear evolution.

Duet o pulling out that takes place within 0.07485 s, the tension in the cable as well as on drum and bucket reaches also zero value, but comparing to previous case, both the time and tension in the cable are zero, during consecutive periods these decrease.

In the last representation, Fig. 28.8, the motor power is 45 kW, speed on drum has a strongly increasing behavior as in previous case, after which it has a pulsatory

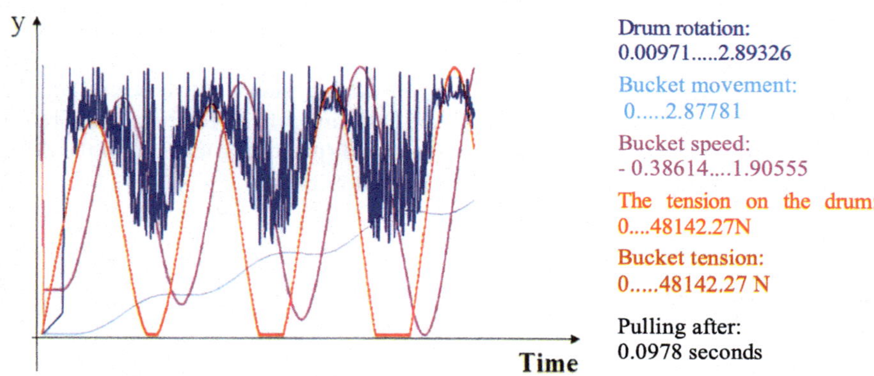

Fig. 28.6 Graphic representation for 26.5 kW motor

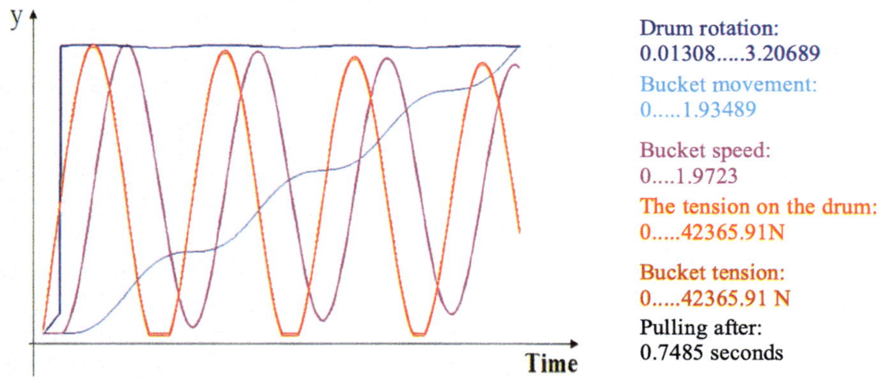

Fig. 28.7 Graphic representation for motor of 30 kW

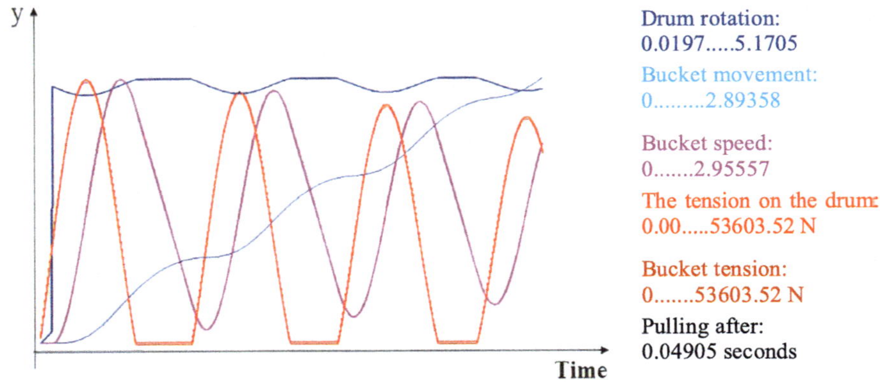

Fig. 28.8 Graphic representation for 45 kW motor

evolution. It is noticed that for two consecutive pulsations the period of time decreases while the tension in drum cable is zero.

Y-axis is the vertical axis on wich the following variables have been scaled in particular cases illustrated in 4, 5, 6, 7 and 8 figures: drum rotation (dark blue), movement of the bucket (light blue), bucket speed (magenta), tension on the drum (red), bucket tension (light brown) and pulling after (black).

28.4 Conclusions

This paperwork proposes a new method to determine and choose the drive motor for the bucket slewing mechanism, reported to the classic method for establishing the motor power to drive such mechanism.

In Fig. 28.4—drum speed has a strongly irregular behavior and a time increasing amplitude, which is undesirable during operation. Load movement shows a slightly irregular increase.

In conclusion, the 14 kW motor power is not indicated for bucket drive mechanism of the pit excavator.

In Fig. 28.5—the motor power is 15 kW, drum speed has a pulsatory evolution with slightly increasing amplitude. The other shapes represent the movement, bucket speed and tension on drum, on bucket being similar to previous case except for the pulling out was faster, meaning 0.174 s comparing to 0.33 s—in previous case.

In Fig. 28.6—the motor power of 26.5 kW—drum speed shows the same pulsatory behavior, just the amplitude is kept or shows a slight decrease.

Due to pulling out of 0.0978 s the tension in the cable reaches also zero value.

In Fig. 28.7—for a 30 kW motor, the drum speed has a strongly increasing evolution after which it has an approximately linear evolution.

Due to pulling out which takes place within 0.07485 s, the tension in cable as well as on drum and bucket reaches also zero value, but comparing to previous case both the time and tension in cable are zero, on consecutive periods these decrease.

In the last representation, Fig. 28.8, the motor power is 45 kW, speed on drum has a strongly increasing behavior like in the previous case, after which it has a pulsatory evolution.

The time periods while the tension is zero in drum cable and decreases in comparison to the previous pulsation.

The most favourable case is the one represented in Fig. 28.7, and the most unfavorable is the one shown in Fig. 28.6.

Therefor, to equip the excavator as to bucket driving mechanism is concerned, the motor with 30 kW power will be chosen.

References

1. G. Căpățână, A. Potîrniche, Aspects regarding 3D modelling and finite element analysis of a crusher bucket, in *MODTECH International Conference IX, Book SERIES: IOP Conference Series: Materials Science and Engineering*, vol. 1182, Article Number: 012012 (2021)
2. L. Curduman, C. Debeleac, S. Năstac, Analysis of working body track vibrations of a single bucket excavating equipment during the bank-sloping and differential excavation processes, in *Proceedings of the 23rd International Congress on Sound & Vibration, (paper 793), Congress orga nized by the International Institute of Acoustic and Vibration—IIAV* (2016)
3. C.N. Debeleac, *Machinery and Technological Equipment for Building* (Publishing House Galati University Press, GUP, 2019), 231 p.
4. Ş. Mihăilescu, P. Bratu, A. Vlădeanu, Ș. Aramă, *Machinery for Constructions, Construction, Calculation and Testing of Machinery for Earthworks*, vol. II (Technical Publishing House, Bucureşti, 1984)
5. D.G. Muscă (Anghelache), A. Potîrniche, G.F. Căpățână, Analyse of dynamic operation mode for drilling machine equipped with drill for pits digging, in *MODTECH International Conference VIII, Book SERIES: IOP Conference Series: Materials Science and Engineering*, vol. 916(1), Article Number: 012070 (2020)
6. S. Năstac, *Numerical Analysis with Applications in Mechanical Engineering* (Impuls Publishing House, Bucuresti, 2004), 100 p.
7. A. Potîrniche, G.F. Căpățână, Finite element analysis of an excavator bucket embedding ripper teeth. Ann. "Dunărea de Jos" Univ. Galați Fascicle XIV Mech. Eng. **24**(1), 5–10 (2017). ISSN 1224-5615
8. A. Potîrniche, G. Căpățână, Aspects regarding the behaviour of the work equipment of a loader during the work process. Ann. "Dunărea de Jos" Univ. Galați Fascicle XIV Mech. Eng. **26**(1), 33–36 (2019). ISSN 1224-5615

Chapter 29
The Assessment of the Seismic Answer of Structures with Viscous Fluid Dissipators Included in the Construction and Functionally

Andrei Tudor and Polidor Bratu

Abstract The global tendency is to increase the safety level for new constructions, as well as to consolidate the existing buildings. Once the demand grows, it is also necessary to tackle solutions able to comply with this desire. The dissipation of energy induced by earthquakes to frame structures can be obtained through various methods. The herewith article analyses the ability of viscous fluid devices to fight against the effects of seismic actions, by increasing the fraction of critical damping. Adopting dampers can be an alternative to taking over structural damage, and can be seen as an alternate way to absorb the energy induced by the earthquake. The assessment of the seismic answer to the structures equipped with viscous dampers was made through reference to the seismic hazard and the specificity of location conditions in Romania.

29.1 Classification into Regulations, Local Specific Requirements

29.1.1 Design Regulations and Codes

Implementation of viscoelastic and viscous dampers in structures is well spread in countries like Japan, USA, U.K., Canada, actually a full list is revealed in Appendices A, B, C and D of Christopoulus book [1] (Fig. 29.1).

For the situation in which the equipment of buildings' structural frames with dampers is analyzed, it is necessary to comply with the requirements provided by the Romanian code, named P100-1 [2], in the sense that the structural design principles, as well as the specific provisions regarding resistance and stability, must be respected.

A. Tudor (✉) · P. Bratu
The Institute of Solid Mechanics of the Romanian Academy, Bucharest, Romania
e-mail: andrei.tudor@imsar.ro

N. Herisanu and V. Marinca (eds.), *Acoustics and Vibration of Mechanical Structures—AVMS-2023*, Springer Proceedings in Physics 302,
https://doi.org/10.1007/978-3-031-48087-4_29

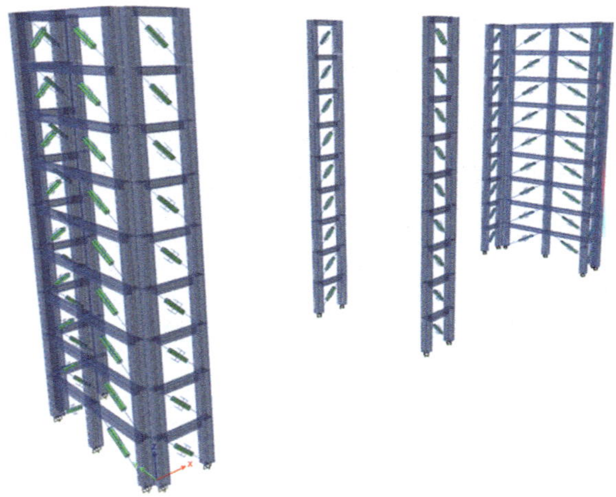

Fig. 29.1 A typical installation of viscous dampers, case study

Classic mechanisms for energy dissipation through plastic deformation under seismic action are mentioned among the design strategies directly related to ductility and strength.

The subject of reducing the seismic response through energy dissipation outside the main structural system is detailed in a single chapter dedicated to base isolation. Chapter 11 does not bring any additional provisions compared to Eurocode, as can be highlighted by its own text: "*Note:* Chap. 11 *reproduces practically unchanged the text of the chapter with the same title from SR EN 1998-1:2004. Only the general principles of seismic isolator design are presented here. Where appropriate, the provisions appropriate to seismic conditions in our country, especially in the capital area, have been adopted*".

To control compliance with the code requirements, two associated limit states are defined, ULS (ultimate limit state) and SLS (serviceability limit state). In this version of the code, a higher level of seismic hazard has been established compared to the 2006 version, namely by adopting values corresponding to an earthquake with a mean recurrence interval of IMR = 225 years for LS (life safety) and IMR = 40 years for LD (limitation of damage). The code also mentions the need for future revision of this recurrence value, up to IMR = 475 years, values provided by European or American standards.

29.1.2 Seismic Hazard Level, Evolution, and Perspectives

The values of the average recurrence intervals can be expressed through the probabilities of exceeding and categorized as follows:

- IMR = 225 years, 20% probability of exceeding in 50 years (P100-1)

- IMR = 475 years, 10% probability of exceeding in 50 years (Eurocode)
- IMR = 2475 years, 2% probability of exceeding in 50 years (International Building Code)

The evolution of design codes from 1963 to the present has led to an increase in the intensity of seismic design forces. Currently, according to the P100-1 code [2], the seismic hazard is defined by the peak value of horizontal ground acceleration associated with an average recurrence interval of IMR = 225 years for Bucharest, with a value of $a_g = 0.30$ g, which leads to PGA = 2.94 m/s^2 (as Peak Ground Acceleration). Aligning with European requirements will likely lead to the development of hazard maps for higher parameters, namely IMR of 475 or 2475 years, as can be seen in certain unofficial maps [3] (Figs. 29.2 and 29.3).

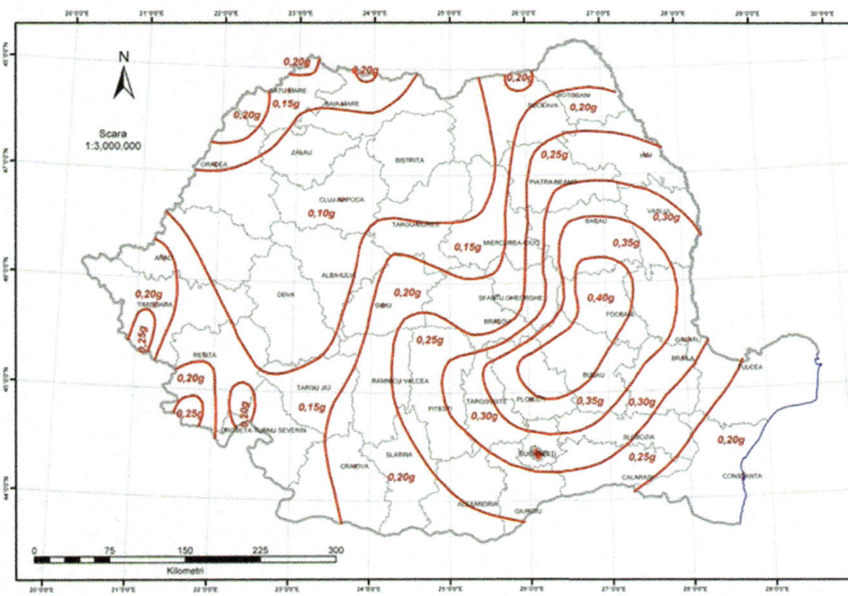

Fig. 29.2 Seismic hazard map, IMR 225 years, ag = 0.30 g, Bucharest

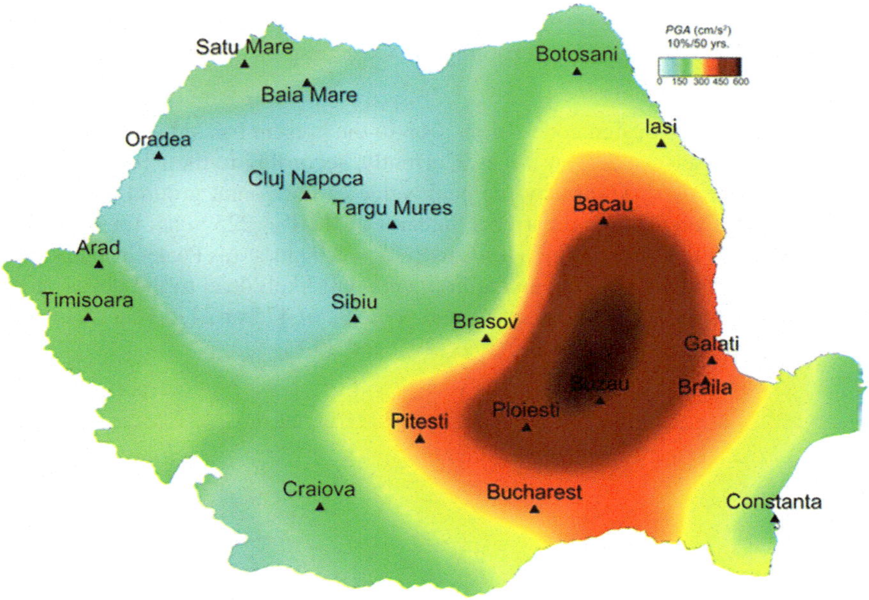

Fig. 29.3 Seismic hazard map, IMR 475 years, $a_g \sim 0.35 \ldots 0.40$ g, Bucharest

29.2 The Principle of Energy Dissipation

29.2.1 The Influence of Energy Dissipation on the Induced Energy Balance

The concept of energy dissipation affects the balance of externally induced energy, such as seismic energy, by redirecting it towards dissipaters, thus reducing the demand on the structural system. From a mathematical point of view, the balance in a structure can be formulated [1], without including the static energy induced in the structure:

$$E_k^a(t) + E_{vd}(t) + E_a(t) + E_{sd}(t) = E_{in}^a(t) \tag{29.1}$$

Thus, during seismic action, the absorbed energy E_a is dissipated both elastically (E_{es}) and inelastically (E_h), through structural deformations:

$$E_a(t) = E_{es}(t) + E_h(t) \tag{29.2}$$

the equation details the following terms:

- E_k^a—kinetic energy
- E_{vd}—damping energy

- E_a—absorbed energy
- E_{es}—energy dissipated elastically
- E_h—energy dissipated hysteretically by the structure
- E_{sd}—energy dissipated by dampers
- E_{in}—energy induced by external (seismic) action.

In the case of structures equipped with viscous dampers, E_{vb}, vibrational energy can be defined, which corresponds to that part of the induced energy that has not been dissipated by the viscous damping of the structure or by the additional damping of the dampers. The goal is to maintain the vibrational energy as low as possible.

$$E_{vb}(t) = E_k^a(t) + E_a(t) \tag{29.3}$$

Thus, from the equation above, the vibration energy E_{vb} is defined as the difference between the induced energy E_{in} and the sum of the energy dissipated by the inherent damping of the structural system E_{vd} combined with that dissipated by the dampers E_{sd}.

$$E_{vb}(t) = E_{in}^a(t) - E_{sd}(t) - E_{vd}(t) \tag{29.4}$$

It can be inferred that maximizing the energy dissipated by dampers does not necessarily lead to obtaining minimal vibrational energy, as it can also increase induced energy. Therefore, the strategy for sizing dampers consists of minimizing the difference between induced energy and the energy dissipated by the dampers.

29.2.2 Damping in Single-Degree-Of-Freedom Systems

Single degree of freedom systems, without external excitation source, i.e. for free vibrations, are governed by the following equation of motion [4]:

$$m\ddot{u} + c\dot{u} + ku = 0 \tag{29.5}$$

and for an armonic motion, the right side term is replaced by armonic force:

$$m\ddot{u} + c\dot{u} + ku = p_o \sin \omega t \tag{29.6}$$

and dividing by mass, we can rewrite the equation:

$$\ddot{u} + 2\zeta\omega_n\dot{u} + \omega_n^2 u = \frac{\mathrm{p_o}}{m} \sin \omega t \tag{29.7}$$

with the following solution:

$$u_p(t) = C \sin \omega t + D \cos \omega t \tag{29.8}$$

and the particular solutions:

$$C = \frac{p_0}{k} \frac{1 - (\omega/\omega_n)^2}{\left[1-(\omega/\omega_n)^2\right]^2 + 2\zeta(\omega/\omega_n)^2};$$
$$D = \frac{p_0}{k} \frac{-2\zeta(\omega/\omega_n)}{\left[1-(\omega/\omega_n)^2\right]^2 + 2\zeta(\omega/\omega_n)^2} \tag{29.9}$$

and complementary solution (the response of free vibrations):

$$u_c(t) = e^{-\zeta \omega n t}(A \cos \omega_D t + B \sin \omega_D t); \omega_D = \omega_n \sqrt{1 - \zeta^2} \tag{29.10}$$

complete solution, with the two vibration components, the excitation frequency "ω" and the natural frequency "ω_n":

$$u(t) = e^{-\zeta \omega n t}(A \cos \omega_D t + B \sin \omega_D t) + C \sin \omega t + D \cos \omega t \tag{29.11}$$

The simplest definition of viscous damping is based on measuring the response of a system subjected to a harmonic excitation of frequency ω equal to the natural frequency of the system ω_n, experimentally using the relationship:

$$\zeta = \frac{1}{2} \frac{(u_{st})_0}{(u_0)_{\omega=\omega n}}; \; where \; u_o = \frac{(u_{st})_0}{2\zeta}, \; for \; \omega = \omega_n \tag{29.12}$$

We can define equivalent damping "ζ_{eq}" by experimentally measuring the values of "$(u_{st})_0$" and "u_0" in a loading cycle, which takes into account all energy dissipation mechanisms applicable to the measured system, and equating it with an equivalent damped system. In a real system, the measured energy, E_D, is represented by the area contained in the hysteresis loop in a force–displacement graph. By equating this obtained energy, E_D, with the energy dissipated by damping:

$$E_D = 2\pi \zeta \frac{\omega}{\omega_n} k u_0^2. \tag{29.13}$$

and the relation for elastic energy:

$$E_{S0} = \frac{k u_0^2}{2} \tag{29.14}$$

we get the equivalent damping:

$$4\pi \zeta_{eq} \frac{\omega}{\omega_n} E_{S0} = E_D; \zeta_{eq} = \frac{1}{4\pi} \frac{1}{\omega/\omega_n} \frac{E_D}{E_{S0}} \tag{29.15}$$

thus, for values $\omega = \omega_n$:

$$\zeta_{eq} = \frac{1}{4\pi}\frac{E_D}{E_{S0}} \tag{29.16}$$

29.2.3 Structural Response to Seismic Action

We rewrite the equation of motion for single-degree-of-freedom systems, by applying the ground acceleration excitation term on the right-hand side, over time:

$$m\ddot{u} + c\dot{u} + ku = -m\ddot{u}_g \tag{29.17}$$

and dividing by mass:

$$\ddot{u} + 2\zeta\omega_n\dot{u} + \omega_n^2 u = -\ddot{u}_g(t) \tag{29.18}$$

thus, the displacement response for a given ground acceleration is strictly related to the natural frequency of the system, ω_n, that is, the natural vibration period (T_n) and the damping ζ, without any direct dependence on the mass of the system.

Therefore, two systems with the same natural period of vibration and the same damping, regardless of the difference in mass or stiffness, will have the same structural response. Starting from a given acceleration, the response of each single-degree-of-freedom system can be plotted for a range of natural vibration periods, from 0 to, for example, 5 s, the interval in which most resistant structures are found, either using a constant damping or varying it. This approach provides real responses, or spectra.

The seismic recording from the Engineering Strong Motion Database [5] of the 1977 earthquake event recorded in a single location in Bucharest, with a maximum acceleration of ≈ 0.20 g, was used, and the real spectral response in terms of acceleration and displacement was plotted for two dampings, $\zeta_1 = 0.05$ and $\zeta_2 = 0.25$. The recording was made for two directions, named in the database as "HNN" and "HNE" (Figs. 29.4, 29.5 and 29.6).

After obtaining the spectral displacement response, the other two responses can also be obtained, namely the spectral pseudo-velocity response and the spectral pseudo-acceleration response:

$$V = \omega_n D = (2\pi / T_n)D \tag{29.19}$$

$$A = \omega_n^2 D = (2\pi / T_n)^2 D \tag{29.20}$$

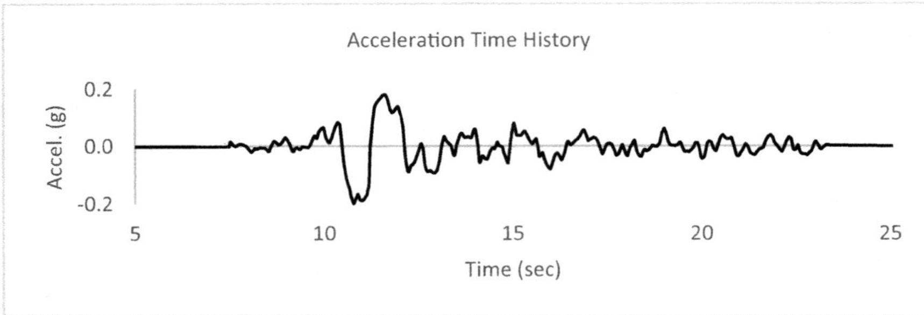

Fig. 29.4 Recorded accelerogram of the 1977 event HNE

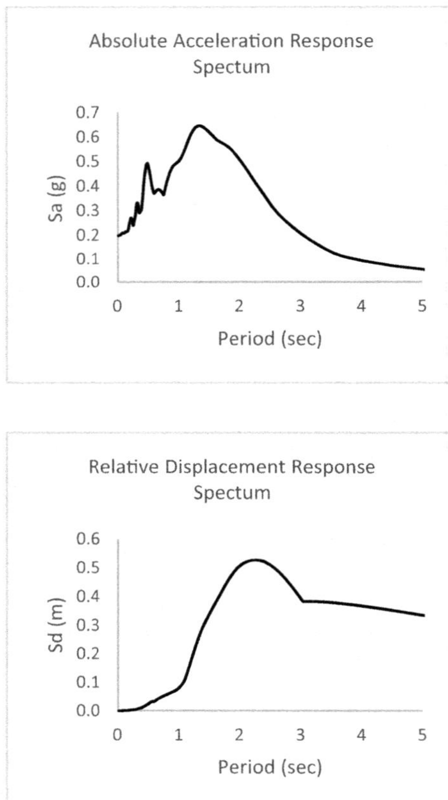

Fig. 29.5 Real spectral response $\zeta_1 = 0.05$ (acceleration and displacement)

Fig. 29.6 Real spectral response $\zeta_1 = 0.25$ (acceleration and displacement)

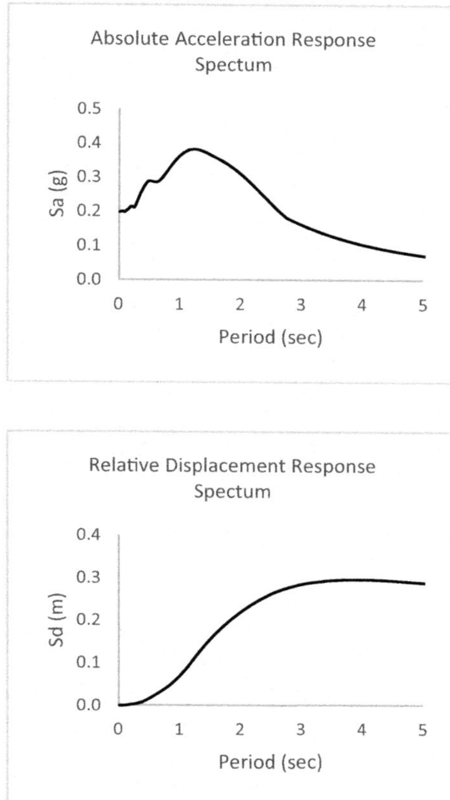

The elastic response spectrum is mainly used in the evaluation of both new and existing resistance structures. The above response spectrum is derived from a single recorded event, namely an accelerogram, but in design codes, a smoothed elastic response spectrum is often used. Its derivation is based on previous events and should also be representative of the analyzed location. The establishment of an elastic response spectrum for a site is made on a statistical basis, having as indicative parameters in the analysis the previous seismic events, the distance of the site from the seismic focus, the fault mechanism, and the geological conditions between the source and the site for which the smoothed response spectrum is constructed. Below is the smoothed elastic response spectrum for the Bucharest site, as well as the comparison with the previously defined spectral acceleration responses for the 1977 event, with both recordings, HNN and HNE (Fig. 29.7).

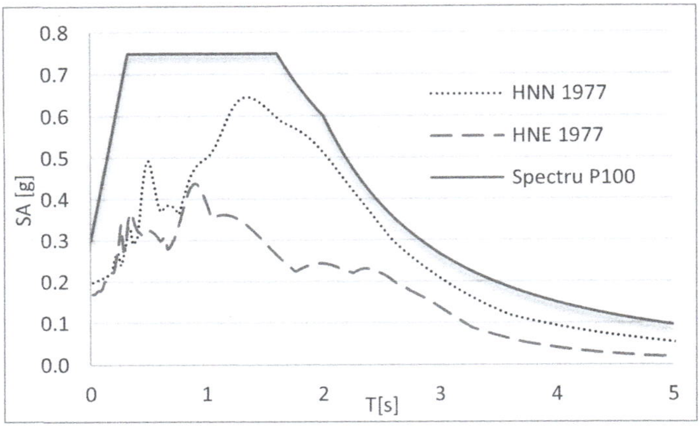

Fig. 29.7 Response elastic code spectrum, $\zeta = 0.05$

29.2.4 *Structural Response to Modified Seismic Action*

The influence of viscous damping on reducing the structural seismic response is recognized in structural dynamics and seismic engineering. Below, the application of various values for critical damping, ζ, on the elastic response spectrum is plotted for a site characterized by $a_g = 0.35$ g and $T_c = 1.6$ s, as well as the reduction of the elastic response spectrum with 5% damping by applying random values of the behavior factor "q" defined in the Romanian code [2] (Fig. 29.8).

From the analysis of the above graph, it can be deduced that the elastic response of a structure with 30% damping, for a defined location, is similar to the inelastic response of a structure with a behavior factor $q = 2$, damped with 5%, for the same

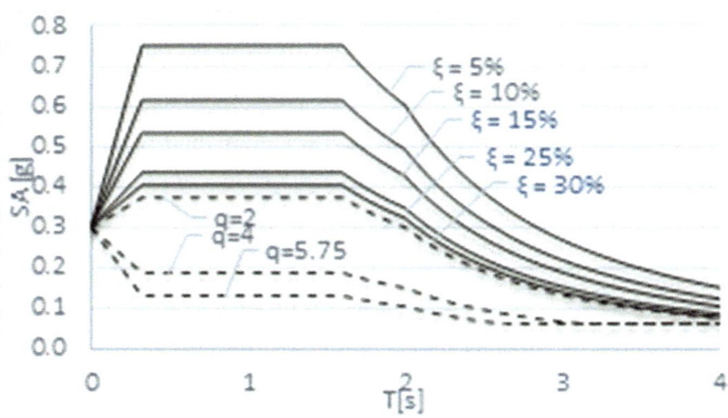

Fig. 29.8 Elastic response spectrum for different damping values, $\zeta = 0.05 \ldots 0.30$, and reduced design spectrum $q = 5.75, 4.00, 2.00$

location. Additional damping of the structural system can lead to a reduction in the seismic response of the structure expressed in base forces and displacements. The application of fluid viscous dampers, as supplementary damping devices, represents a system that is frequently used lately and has very good dissipation capabilities as well as multiple utilization possibilities. The idea that started it all was a common one in the automobile industry, namely the shock absorbers of cars. Starting from this idea and through detailed studies on the behavior of structures equipped with these devices, their usefulness for the protection of civil structures has been demonstrated.

The vast majority of these dampers use fluids with viscous characteristics in their composition. These devices offer very little resistance when subjected to forces applied at a very low speed but have very good damping capabilities when these forces are applied at high speeds. The equation governing their behavior, which is commonly found in several specialized documents, is formulated as follows:

$$F = sign(v)c \times /v/^{\alpha} \tag{29.21}$$

where it was noted that:

F—the force developed by the damper,
v—the relative velocity between the ends of the damper,
c—the damping coefficient,
α—the damping exponent.

The American standard mentions that the validity of the calculation relationship is limited for those types of dampers for which there is no stiffness development in the frequency range of $0.5f$ to $2.0f$, where "f" is defined as the fundamental natural frequency of the structure.

The Maxwell modeling type is also valid for the same principle. Outside the predefined interval, specific models of viscoelastic dampers can be used. Viscous dampers are used because (as also observed from the above equation) they are directly dependent on the relative velocity developed between the attachment points of the damper. This fact also explains the damping system used, namely: the damping force is not coupled with the inertial forces of the structure. From the graphical representation of the velocity and displacement amplitude, as well as the fact that the velocity is equal to the first-order derivative of displacement, it follows that these two components are out of phase by $\pi/2$ (in other words, when the displacement is maximum, the velocity is zero, and when the displacement is zero, the velocity is maximum).

In theory, the forces from the damper do not add to the forces that develop in the structure because they occur when the structure's deformations are very small.

29.3 Conclusions

The sizing strategy of dampers consists of minimizing the difference between induced and dissipated energy by the dampers. The objective is to increase the mechanical work performed by the dampers, thus reducing the degradation in the structure by decreasing this difference. At present, according to the P100-1 code [2], a reduction of seismic requirements is allowed by modifying the damping, i.e., the elastic spectrum by up to 55%, by the correction factor $\eta \geq 0.55$. However, in future revisions, alignment with other design codes from countries where the use of viscous dampers is widespread is desirable, both in the strategy to improve the performance of new buildings and in the retrofitting of existing ones. It has been shown that the elastic response of a structure dampened with a 30% fraction, for a defined location, is similar to the inelastic response of a structure with a behavior factor $q = 2$, dampened with a 5% fraction, for the same location. Thus, the implementation of a viscous damper system can be expected to result in an elastic response for certain structural models, without entering the energy dissipation domain through structural degradation.

References

1. C. Christopoulos, A. Filiatrault, *Principles of Passive Supplemental Damping and Seismic* (IUSS Press, Italy 2006)
2. "Prevederi de proiectare pentru clădiri", indicativ P 100-1/2013, revizuit in 2019 prin Ordinul 928/2019 (Romanian Seismic Code)
3. F. Pavel, R. Vacareanu, J. Douglas et al., An updated probabilistic seismic hazard assessment for Romania and comparison with the approach and outcomes of the SHARE project. Pure Appl. Geophys. **173**, 1881–1905 (2016). https://doi.org/10.1007/s00024-015-1223-6
4. A.K. Chopra, *Dynamics of structures, Theory and Applications to Earthquake Engineering* (Pearson Education Limited, 2020)
5. https://esm-db.eu/ (Engineering Strong Motion DataBase)

Chapter 30
Assessment of Weak Segments in Cantilever Beams Using an Artificial Neural Network

Alexandra-Teodora Aman, Cristian Tufisi, Codruta Oana Hamat, and Gilbert-Rainer Gillich

Abstract Damage detection is a fundamental task in the safety assessment of engineering structures. Various nondestructive damage detection techniques have been proposed to accomplish this task. The research presented herein aims to prove the reliability of an original method for detecting damage in beams using analytically generated data. The method consists of dividing the beam in several segments, among which one or two are considered as damaged. The damage is induced by decreasing the modulus of elasticity of the segment. For each damage scenario we derive the natural frequencies of a few numbers of vibration modes and the resulted frequency shifts (RFS). Taking the RFS values as input data and the damaged segments' number and the modulus of elasticity alteration as output data, we train an Artificial Neural Network (ANN), which can locate the damaged segment and assess the stiffness degradation. The accuracy of the ANN model is validated using analytical data and Finite Element Method (FEM) simulations. Since we succeeded in assessing damage accurately, we demonstrated that the proposed analytical method provides the means for generating large databases that can be used to train ANN's that can effectively detect damage.

30.1 Introduction

Damage assessment is critical to ensure the safety and functionality of engineering structures, and several nondestructive damage detection techniques have been proposed for this task [1]. The modulus of elasticity is a fundamental material property that defines the stiffness of a material and in the context of structural

A.-T. Aman · G.-R. Gillich
Doctoral School of Engineering, Babes-Bolyai University, 400084 Cluj-Napoca, Romania

C. Tufisi (✉) · C. O. Hamat · G.-R. Gillich
Department of Engineering Science, Babes-Bolyai University, 400084 Cluj-Napoca, Romania
e-mail: cristian.tufisi@ubbcluj.ro

N. Herisanu and V. Marinca (eds.), *Acoustics and Vibration of Mechanical Structures—AVMS-2023*, Springer Proceedings in Physics 302,
https://doi.org/10.1007/978-3-031-48087-4_30

engineering, it can be considered as a key factor that influences the natural frequencies of a structure [2]. When the modulus of elasticity of a material is decreased, the stiffness of the material decreases accordingly causing the natural frequencies of the structure to decrease. In the case of damage detection, changes in the natural frequencies can be indicative of damage. Specifically, the natural frequencies of a structure will change when there is a reduction in the stiffness of the structure due to damage, such as cracking or corrosion. Therefore, by measuring changes in the modal parameters, it is possible to detect and locate damage in the structure [3].

In [4] is proposed a method for detection of damages using natural frequency and mode shape data with deep learning techniques. The method involves training a deep neural network using a natural frequency dataset and mode shape measurements obtained from structures both in damaged and undamaged state. The trained network is then used to localize damage in new structures based on their measured natural frequencies and mode shapes. The paper [5] presents a Deep Neural Network (DNN) method for damage identification of steel–concrete composite beams that is also based on changes in natural frequencies caused by damage. The DNN is trained on a dataset of natural frequency measurements obtained from both intact and damaged composite beams and is used to predict the location and to evaluate the damage. The ANN is used also for detecting multiple damages in a beam [6]. Other authors use the ANN in combination with different algorithms to successfully locate damages in plates [7] or in composite structures [8, 9]. An explainable deep learning damage prediction for prismatic cantilever steel beams is proposed in [10]. In the latter article, the prediction is accompanied by explanations in human terms, which enhances the trust in the results and gives the possibility of correction for the future functioning conditions.

In prior studies, we succeeded to train an ANN to identify single damages in beams [11]. This paper presents our latest research results, proposing new features to train the ANN to assess multiple damages. It brings a substantial contribution to the development of reliable and efficient techniques for structural damage assessment, which is essential for ensuring the safety and durability of engineering structures.

30.2 The Proposed Damage Detection Method

In the current section, we describe the proposed damage detection method using AI; the scheme is shown in Fig. 30.1. This method to detect damage offers a novel approach that leverages the ANN's ability to recognize complex patterns in the dynamic behavior if the structure. The method consists in generating a database, by calculating the natural frequencies and the subsequent RFSs of the cantilever beam with one or two damaged segments. The generated data is used to train an ANN that will be able to detect altered beam segments and evaluate their severities.

The INPUT data consists of the Relative Frequency Shifts (RFS), a feature that considers the changes in the natural frequencies due to damage in certain beam segments. The OUTPUT data consists of the stiffness of all beam segments.

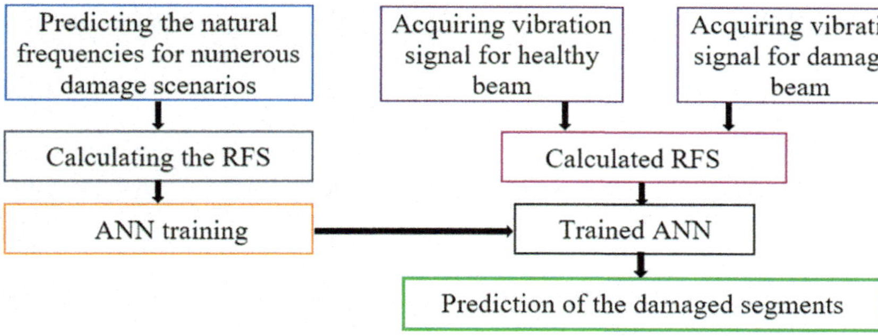

Fig. 30.1 The proposed method

30.2.1 Generation of Training Data

The first step in realizing the database is to develop a beam model, as illustrated in Fig. 30.2. The beam is divided into 20 segments. The scenarios presume reducing the stiffness of one, respectively two segments at a time. The steps encased in blue rectangles described in Fig. 30.1 are applied for the considered scenarios to calculate the natural frequencies of the out-of-plain vibration modes and extract the RFS's.

The beam (Fig. 30.2) considered in the current research is a cantilever beam with following dimensions: the length $L = 1$ m, the width $B = 0.05$ m, and the thickness $H = 0.005$ m. The material is Structural Steel S235 JR with standard Young's modulus of $2{\cdot}10^{11}$ Pa.

We know that the bending moment for the cantilever is given by the relation:

$$\overline{M}_i(x) = 0.5\left\{\frac{\cos\alpha_i L + \cos h\alpha_i L}{\sin\alpha_i L + \sin h\alpha_i L}[\sin(\alpha_i x) - \sin h(\alpha_i x)] - \cos(\alpha_i x) + \cos h(\alpha_i x)\right\}. \tag{30.1}$$

The energy stored in the intact beam can be expressed, in function of the bending moment, as is given by the relation:

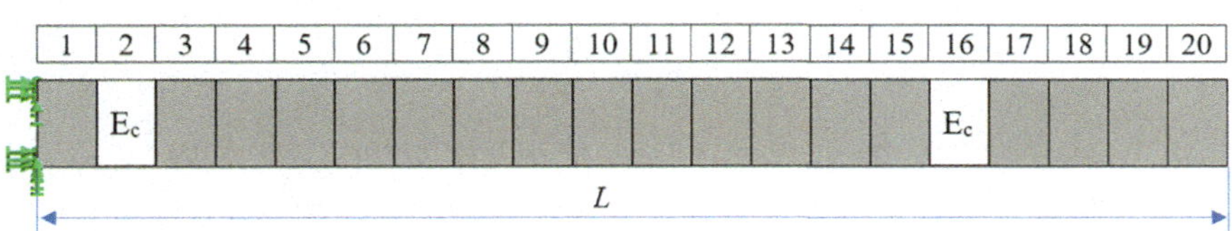

Fig. 30.2 The cantilever beam, with reduced stiffness on segments 2 and 16

$$U_i^{0-L} = \frac{M_{i\max}^2}{2EI}\int_0^{x_1}\left[\overline{M_i}(x)\right]^2 dx + \frac{M_{i\max}^2}{2EI}\int_{x_1}^{x_2}\left[\overline{M_i}(x)\right]^2 dx + \frac{M_{i\max}^2}{2EI}\int_{x_2}^{L}\left[\overline{M_i}(x)\right]^2 dx. \tag{30.2}$$

If the stiffness is variable, the capacity of the beam to store energy is obviously smaller. The bending moment that produces to the damaged beam a deflection similar with that of the healthy beam is:

$$\overline{M}_{ef-i}(x) = \overline{M}_i(x)\frac{E(x)I}{EI} \tag{30.3}$$

If the stiffness is constant except on the segment with the limits $x_1 - x_2$, where it decreases due to the decrease of the Young modulus, which is here constant and equals E_C, the energy stored in the beam is becoming:

$$\begin{aligned} U_{Ci}^{0-L} = {} & \frac{M_{i\max}^2}{2EI}\int_0^{x_1}\left[\overline{M_i}(x)\right]^2 dx + \frac{M_{i\max}^2}{2E_C I}\int_{x_1}^{x_2}\left[\frac{E_C}{E}\overline{M_i}(x)\right]^2 dx \\ & + \frac{M_{i\max}^2}{2EI}\int_{x_2}^{L}\left[\overline{M_i}(x)\right]^2 dx \end{aligned} \tag{30.4}$$

or, after grouping the terms, it is:

$$\begin{aligned} U_{Ci}^{0-L} = {} & \frac{M_{i\max}^2}{2EI}\int_0^{x_1}\left[\overline{M_i}(x)\right]^2 dx \\ & + \frac{M_{i\max}^2}{2EI}\frac{E_C}{E}\int_{x_1}^{x_2}\left[\overline{M_i}(x)\right]^2 dx \\ & + \frac{M_{i\max}^2}{2EI}\int_{x_2}^{L}\left[\overline{M_i}(x)\right]^2 dx \end{aligned} \tag{30.5}$$

It is known that the natural frequency for a vibration mode is proportional to the energy stored in the beam in that mode. Consequently:

$$\frac{f_{Ci}}{f_i} = \left(\frac{U_{Ci}^{0-L}}{U_i^{0-L}}\right)^{0.5}. \tag{30.6}$$

In the general case, we denote the integral with the limits a and b as:

$$\varsigma_i^{a-b} = \int_a^b\left[\overline{M}_i(x)\right]^2 dx. \tag{30.7}$$

After substituting Eqs. (30.2) and (30.5) in Eq. (30.6), and after performing simple mathematical operations, it results the mathematical relation to calculate the frequency of the damaged beam as:

$$f_{Ci} = f_i \left(\frac{\varsigma_i^{0-x_1} + \frac{E_C}{E} \varsigma_i^{x_1-x_2} + \varsigma_i^{x_2-L}}{\varsigma_i^{0-L}} \right)^{0.5}. \tag{30.8}$$

In Eq. (30.8) we exemplified the case of a beam divided into three segments. In this work, from here on, we consider the beam divided into 20 segments. From the Eq. (30.7) we can calculate the coefficients ς_i^{a-b} for all segments, and consequently the natural frequencies of the beams damaged in respect to all possible scenarios. Finally, we calculate the Relative Frequency Shifts (RFS) for the cantilever beams with altered stiffness using a procedure described in [11].

For the current study, the damage scenarios involve segments that have a lower modulus of elasticity, the ratios E_C/E being 0.9, 0.8, 0.7 and 0.6. As mentioned, one or two segments of the cantilever beam have a lower stiffness at one time; the position of the segment/segments is iteratively changed until all possible scenario (1983 cases) are achieved. After performing the calculus, we obtain the INPUT data, which is a matrix with 6 rows (the RFS values for the six bending vibration modes calculated for one damage scenario) and 1983 columns (the number of the considered scenarios). The OUTPUT data is a matrix with 20 rows (the Young's modulus of the twenty segments of the beam for one damage scenario) and 1983 columns (the number of the considered scenarios). It can be observed that each scenario involves one column in the INPUT matrix and one column in the TARGET matrix.

30.2.2 Development of the ANN Model

After generating the training data, a feedforward neural network with backpropagation and Bayesian regularization is developed involving several steps, as described in the orange rectangle in Fig. 30.1. The network architecture consists of an Input with 6 neurons (the six RFS values), two Hidden layers with 40 neurons each, one Output layer with 20 neurons, and the Output consisting of 20 neurons (the predicted value of the Young's modulus of each beam segment). The complete architecture is presented in Fig. 30.3.

This architecture was chosen based on previous studies and empirical evidence that such a structure could provide good performance for the task at hand.

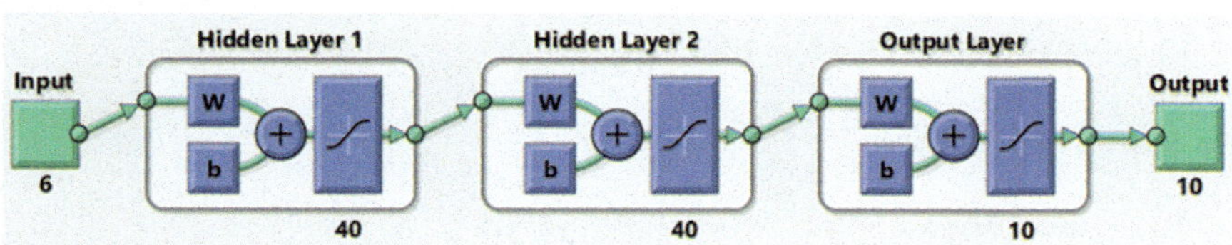

Fig. 30.3 ANN architecture

The training process used the NNTool from MATLAB. The backpropagation algorithm was used to adjust the weights of the network to minimize the difference between the actual and desired output. Bayesian regularization was applied to the network to prevent overfitting and improve generalization. This technique adds a penalty term to the error function, which encourages the weights of the network to be small. This helps to avoid large weight values that may cause overfitting and improve the network's ability to generalize to new data. 70% of the RFS data is used for training, 15% for validation and 15% for testing. After the training phase, the performance plot after training the feedforward artificial neural network is evaluated by displaying the mean squared error (MSE) for each epoch or iteration of the training process, for assessing whether the network is overfitting, underfitting, or has converged to an optimal solution as shown in Fig. 30.4.

The regression plots are generated after training for evaluating the correlation between the predicted and actual outputs, with data points close to the fitted line, as shown in Fig. 30.5.

The ANN performance and regression plots indicates that the neural network model is accurately predicting the target values and furthermore testing with FEM generated data is performed.

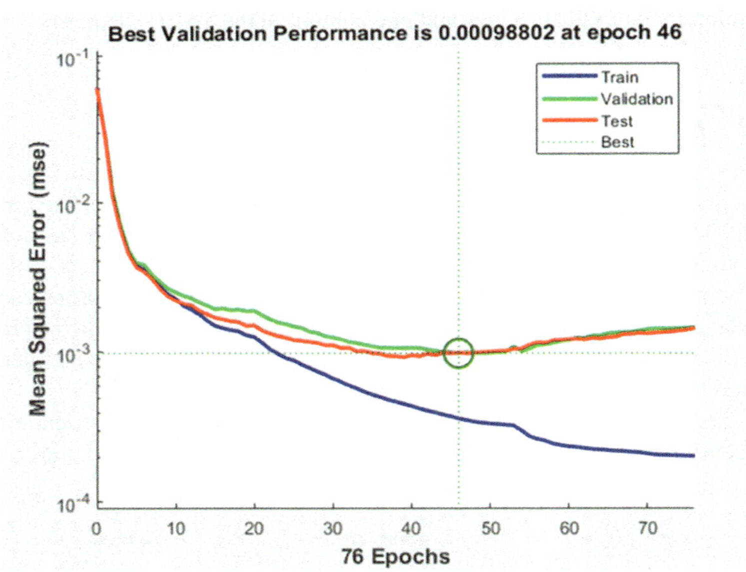

Fig. 30.4 ANN performance

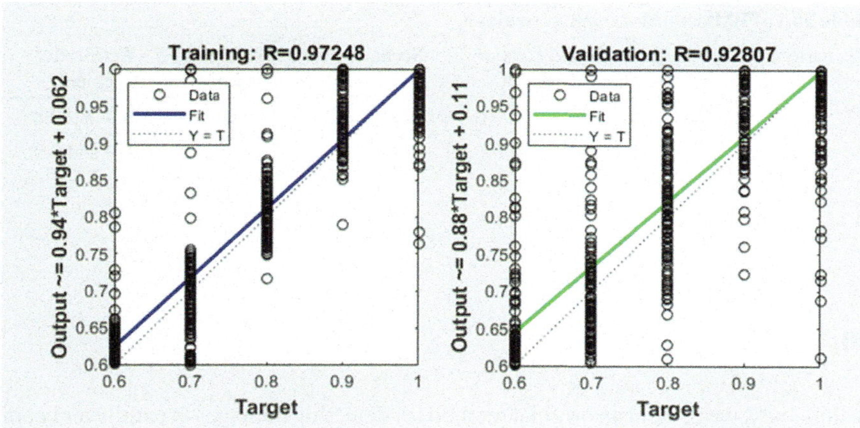

Fig. 30.5 ANN regression plot

30.3 Assessing the Accuracy of the Artificial Neural Network

For assessing the accuracy of the developed ANN, the last step of the method, described with purple rectangles, is applied. FEM damage scenarios are defined in ANSYS, by generating the beam's geometry and dividing it into 20 segments, as shown in Fig. 30.2. Then, damage scenarios are generated by iteratively modifying the elasticity modulus on one or two segments at a time. After each modification, modal simulations are performed to depict the first 6 transverse modes of vibration for the cantilever. The natural frequencies of these modes are then used to calculate the RFS values caused by the damage, that are then introduced into the ANN for assessing its accuracy in detecting the damage.

The boundary conditions are set to simulate the real-world behavior of the cantilever beam by applying a fixed constraint at the face of the beam. The meshing of the beam is set to hexahedral elements of maximum size of 2 mm. The mesh size is chosen to balance the computational cost and accuracy of the simulation.

After the model is parameterized, the damage scenarios are defined by altering the Young's modulus in one or two known segments. The defined damage scenarios are illustrated in Table 30.1.

The results are given in Fig. 30.6. One can observe in this figure that in nine cases the segment or segments with altered modulus of elasticity are correctly identified. It can also be observed that the predicted modulus of elasticity is close to the generated one, which means the method provides accurate predictions.

As an exception, for scenario 7, the segment five is found as damaged, instead of segment 6. However, it is remarkable that the ANN identified a small stiffness decrease at the free end of the cantilever beam (scenario 8).

Table 30.1 FEM defined damage scenarios

Scenario no	Affected segment	*Ec/E* ratio/ segment	Scenario no	Affected segment	*Ec/E* ratio/ segment
1	2 and 16	0.6 and 0.6	6	15 and 19	0.9 and 0.8
2	1 and 11	0.6 and 0.6	7	2 and 6	0.9 and 0.9
3	5 and 6	0.8 and 0.7	8	19	0.9
4	1 and 19	0.8 and 0.8	9	14	0.8
5	10 and 11	0.9 and 0.7	10	15	0.6

30.4 Conclusions

In this study, we present an original method for detecting damages in cantilever beams using an ANN. The proposed approach involves dividing the beam into segments among which one or two are damaged. The damage is induced by decreasing the modulus of elasticity of the segment, and the resulting frequency shifts are used as input data to train the ANN. Using the presented method, large databases with different damage scenarios considering one or more damaged locations along the beam can be generated and used for training ANN's.

In the current paper, the ANN is trained with analytically generated data, using the proposed method, to accurately predict the location and severity of one or two damages in beam-like structures. After the training performance is validated using analytical data, ten damage scenarios have been created by means of FEM analysis with the purpose of testing the developed ANN. The obtained results show that the model can successfully locate the damages for the scenarios with one or even two altered segments, for 9 out of 10 scenarios, as presented in Fig. 30.6. Only one prediction fault was obtained for Scenario 7, where the ANN indicated sector 6 as altered instead of segment 5, which is very near the real position. Therefore, the proposed method can be used as a reliable and efficient alternative to traditional damage detection techniques.

This ANN-based approach can be further extended for efficient and accurate damage localization and stiffness degradation assessment in beams with different boundary conditions, multiple altered segments and by dividing the beam into smaller segments.

Fig. 30.6 Comparison of the predicted results using the developed ANN

References

1. G. Sha, M. Radzieński, M. Cao, W. Ostachowicz, A novel method for single and multiple damage detection in beams using relative natural frequency changes. Mech. Syst. Signal Proc. **132**, 335–352 (2019)
2. H. Chen, F.-K. Chang, Damage detection and identification of structures using changes in flexibility: A review. J. Sound Vib. **419**, 247–269 (2018)
3. C. Tufisi, V.-C. Rusu, G.-R. Gillich, Locating transverse cracks in prismatic beams using random forest method and the frequency drop. Rom. J. Acoust. Vibr. **18**(2), 119–125 (2021)
4. R.A. Alashti, H. Souri, H. Haddadpour, Structural damage detection using natural frequency and mode shape data: a deep learning approach. Mech. Syst. Signal Process. **123**, 640–654 (2019)
5. S. Tian, L. Li, D. Li, J. Huang, Damage identification of steel-concrete composite beams based on changes in natural frequencies using a deep neural network. J. Civ. Struct. Heal. Monit. **11**(2), 307–326 (2021)
6. H. Chen, F.-K. Chang, O. Kwon, Detection of multiple damages in cantilever beams using neural network approach. Struct. Health Monit. **19**(5), 1447–1459 (2020)
7. N.P. Yelve, I.B. Mulla, Locating damage in thin metallic plates using lamb waves and Artificial Neural Network, in 2017 *International Conference on Nascent Technologies in Engineering (ICNTE)* (Vashi, India, 2017), pp. 1–7
8. H. Tran-Ngoc, S. Khatir, H. Ho-Khac, G. De Roeck, T. Bui-Tien, M.A. Wahab, Efficient Artificial neural networks based on a hybrid metaheuristic optimization algorithm for damage detection in laminated composite structures. Compos. Struct. **262**, 113339 (2021)
9. S. Khatir, S. Tiachacht, C.L. Thanh, T.Q. Bui, M.A. Wahab, Damage assessment in composite laminates using ANN-PSO-IGA and Cornwell indicator. Compos. Struct. **230**, 111509 (2019)
10. D. Onchis, G.R. Gillich, Stable and explainable deep learning damage prediction for prismatic cantilever steel beam. Comput. Ind. **125**, 10339 (2021)
11. N. Gillich, C. Tufisi, C. Sacarea, C.V. Rusu, G.R. Gillich, Z.I. Praisach, M. Ardeljan, Beam damage assessment using natural frequency shift and machine learning. Sensors **22**(3), 1118 (2022)

Chapter 31
Case Study on the Dynamic Response Evaluation for a Full-Scale Building with Additional Viscous Damping

Andrei Tudor and Polidor Bratu

Abstract The paper deals with a specific topic regarding the effect of additional viscous damping with linear fluid dampers incorporated in the structure of an existing S + P + 8 floor building, for a county hospital. In this case, the maximum floor displacement values are presented, decreased compared to the situation where the same building with the initial structural composition is unequipped with viscous dampers.

31.1 Compliance with Codes and Specific Local Requirements

31.1.1 Seismic Hazard Level, Evolution and Perspectives

The values of the mean recurrence intervals can be expressed in terms of probabilities of exceeding and classified as follows:

- IMR = 225 years, 20% probability of exceeding in 50 years (P100-1)
- IMR = 475 years, 10% probability of exceeding in 50 years (Eurocode)
- IMR = 2475 years, 2% probability of exceeding in 50 years (Int. Building Code)

At present, according to the P100-1 code [2], the seismic hazard is defined by the peak value of the horizontal ground acceleration associated with a mean recurrence interval, IMR = 225 years, for Bucharest, with the value of $a_g = 0.30$g, which leads to a PGA of 2.94 m/s^2 (as Peak Ground Acceleration). Probably, the alignment with European requirements will lead to the elaboration of hazard maps for higher parameters, namely IMR of 475 or 2475 years, as can be seen in some unofficial maps [5].

A. Tudor (✉) · P. Bratu
The Institute of Solid Mechanics of the Romanian Academy, Bucharest, Romania
e-mail: andrei.tudor@imsar.ro

N. Herisanu and V. Marinca (eds.), *Acoustics and Vibration of Mechanical Structures—AVMS-2023*, Springer Proceedings in Physics 302,
https://doi.org/10.1007/978-3-031-48087-4_31

31.2 Design Code Provisions and Damper Sizing Methods in Current Practice

31.2.1 Linear Damper Characteristics Establishing Method

To calculate the damping coefficient of the damper, a calculation method is used that starts from "tail to head", more precisely a target damping of the structure ξ_{eff} is established, which could lead to a reduced response in the structure, using the modal or spectral calculation method and elastic response spectra modified (according to P100-1 [2]) with the relationship:

$$S_e(T)_{\xi \neq 5\%} = S_e(T)_{\xi = 5\%} \times \eta. \tag{31.1}$$

defining terms:

- $S_e(T)_{\xi=5\%}$ —the elastic response spectrum for ground acceleration components at the site corresponding to the fraction of conventional critical damping $\xi_0 = 5\%$
- $S_e(T)_{\xi\neq 5\%}$ —the elastic response spectrum for ground acceleration components at the site corresponding a different fraction of conventional critical damping $\xi_0 \neq 5\%$
- η—correction factor that takes into account damping, which is determined by the equation

$$\eta = \sqrt{\frac{10}{5 + \xi_d}} \geq 0.55 \tag{2}$$

After determining the effective damping value of the structure, by adding the additional damping provided by the dampers to the inherent damping of the structural system:

$$\xi_{eff} = \xi_s + \xi_d \tag{3}$$

where:

- ξ_d—damping of the energy dissipation system (dampers)
- ξ_s—inherent damping of the structure (has values that vary between 2 and 5% depending on the type of structure, connections between structural elements, etc.)

Given the code limit regarding the correction factor that takes into account damping, of min. 0.55, a maximum damping of $\xi_d = 25\%$, can be deduced, for which $\eta = 0.58 \geq 0.55$.

For linear dampers, after replacing the relationships regarding the energy dissipated by the damper and the energy of elastic deformation in the relation for a linear system, considering:

$$\zeta_{eq} = \frac{1}{4\pi}\frac{E_D}{E_{SO}} \tag{31.4}$$

energy dissipated by a damper in one loading cycle:

$$\mathrm{W_j} = \frac{2\pi^2}{T}C_j\delta_{rj}^2 \tag{31.5}$$

elastic deformation energy:

$$\mathrm{W_k} = \frac{1}{2}\sum_i F_i\delta_i(\text{inertial forces and displacements}) \tag{31.6}$$

the following calculation relationship can be deduced, also provided in FEMA 356 [3]:

$$\xi_d = \frac{T\sum C_j\phi_{rj}^2\cos^2\theta_j}{4\pi\sum m_i s_i^2}. \tag{31.7}$$

defining terms:

- ξ_d—the damping ratio of the energy dissipation system, which is additional and comes from the use of viscous energy dissipation elements
- T—the elastic vibration period of the structure
- C_j—the damping constant of damper "j"
- ϕ_{rj}—the relative displacement in the horizontal direction of damper "j", which can be in the form of modal drift
- θ_j—the angle made by damper "j" with the horizontal
- m_i—mass of level "i"
- s_i—the component of the fundamental vibration mode in the direction of dynamic translation degree of freedom at level "I", which can also be ϕ_i (Fig. 31.1).

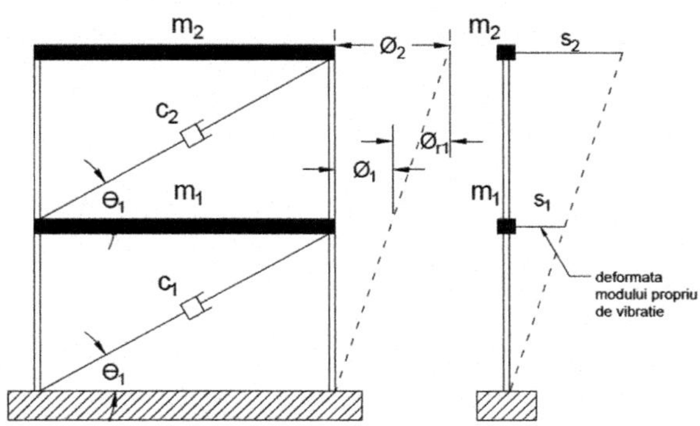

Fig. 31.1 Explanation of terms in the definition of additional damping equation

31.3 Application of the Design Spectrum Modification Principle

31.3.1 Application for an Existing Building

In the case of existing structures that no longer meet current code requirements, it is necessary to establish a consolidation strategy. The concept of seismic structural rehabilitation of an existing building, in this case, which falls into the class II importance category and is located in the seismic zone with $a_g = 0.35$ g, based on the strategy of reducing seismic requirements by increasing the critical damping ratio of the structure, could be considered appropriate.

The concept is materialized by installing energy dissipation devices, such as seismic fluid viscous dampers, speed-dependent devices, within the load-bearing structure to ensure the safety and post-seismic functioning of the building, while complying with the Romanian regulation P 100–3/2013 [1] and attempting to minimize the impact on the building's functioning during consolidation.

Below is a presentation of the increase in critical damping ratio obtained by introducing seismic fluid viscous dampers into the load-bearing structure of the building. As stated [4] it can be deduced that the elastic response of a structure with 30% damping, for a defined location, is similar to the inelastic response of a structure with a behavior factor $q = 2$, damped with 5%, for the same location (Fig. 31.2).

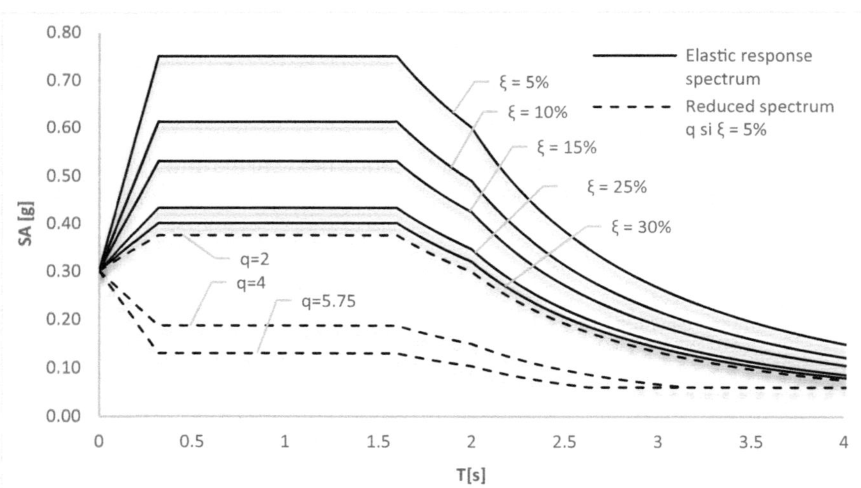

Fig. 31.2 Different strategies for increasing critical damping

31.3.2 Limits and Conditions

The analysis based on response spectra in the Romanian code P100-1 [2], allows the use of modifed spectrum, for critical damping $\xi_0 \neq 5\%$, by applying the factor "η", which can take values starting from minimum 0.55, thus resulting in a first condition that $\xi_d = 0.25$, therefore $\xi_{eff} = \xi_s + \xi_d = 0.30$, starting from an inherent damping of the structure of $\xi_s = 0.05$.

31.3.3 Description of the Analyzed Structure

The analyzed building in this case is an existing hospital located in Onești Municipality, Bacau County. As it is a municipal hospital, the functions are divided into several buildings and the subject of the study refers to the tallest buildings, with a height of 7–8 floors, namely buildings C, D, and E.

The location is within the seismic zone with a ground acceleration for design $a_g = 0.35$ g and the corner period $T_c = 0.7$ s, for earthquakes with a medium recurrence interval of IMR = 225 years. The project on which the hospital was built was prepared by ISCAS Bucuresti around 1961, a period when there were no detailed codes for seismic design in Romania. Considering the appearance of the first seismic design code with the code P13/63, it is possible that the ISCAS project did not benefit from its provisions.

The structural system is of the dual system type, with reinforced concrete walls and preponderant frames. The following materials provided in the original project were considered in the analysis:

- B70 concrete in non-reinforced walls
- B170 concrete in reinforced walls of the infrastructure and entire superstructure
- OL38 reinforcement, unribbed reinforcement, with $f_{yk} = 236$ Mpa
- The concrete class B170 was considered an intermediate class between C8/10 and C12/15, with the following characteristics: $f_{ck} = 10$ MPa, $f_{cd} = 6.5$ MPa, $f_{ctd} = 0.65$ MPa, $E_b = 21.000$ MPa (Fig. 31.3).

For the building was made a structural audit, and as a result, strength deficiencies were identified, resulting in a safety factor according to P100-3 [1] of approximately $R_3 = 0.5$. Therefore, the capacity of the existing structure is estimated to be 50% of the actual code force for SLU (Ultimate Limit State).

Fig. 31.3 Old image after construction, and tridimensional model

31.3.4 Finite Element Method (FEM) Model, Numerical Analysis, Loads, and Modal Analysis Results

The load sets for each floor were evaluated based on the load cases with the values established in Table 31.1, and the stiffness reduction for the elements in the model was done according to Table 31.2.

- "*D*"—dead load
- "*SD*"—superdead load, additional permanent loads from finishes, fixed installations, partition walls
- "*L*"—transient, non-permanent loads.

The load combination sets, based on the load cases are:

Table 31.1 Loads considered in FEM model

Level	D (−)	SD (kN/m^2)	L (kN/m^2)
1–7	**Automat**	**3.5**	**3**
8	**Automat**	**5**	**3**
9	**Automat**	**0.5**	**2**
Stairs	**Automat**	**2.5**	**5**

Bold text was used to highlight the information

Table 31.2 Coefficients of stiffness reduction for modeled elements

Element	Coefficent	Modified parameter
Columns	**0.70 * I_g**	**I_{22}, I_{33}**
Beams	**0.35 * I_g**	**I_{22}, I_{33}**
Slabs (membranes)	**0.25 * I_g**	**f_{11}, f_{22}, f_{12}**
Walls (shell)	**0.70 * I_g** **0.25 * I_g**	**f_{11}, f_{22}** **m_{11}, m_{22}, m_{12}**

Bold text was used to highlight the information

"G_S"—long-term load combination = 1.00 × (D + SD) + 0.4 × L.

"G_F"—fundamental gravitational load combination = 1.35 (D + SD) + 1.5 × L.

"GS_x si GS_y"—combinatie de incarcari seismice = 1.00 (D + SD) + 0.4 × L + S_x/S_y.

G_S, considering $\gamma_G = 1.00$, $\gamma_{Q,1} = 0.4$:

$$\mathrm{E_D} = \sum_{j=1}^{n} \gamma_{G,j} G_{k,j} + \gamma_p P + \gamma_{Q,1} Q_{k,1} + \sum_{i=2}^{m} \gamma_{Q,i} \psi_{0,i} Q_{k,i} \tag{31.8}$$

G_F, Considering $\gamma_G = 1.35$, $\gamma_{Q,1} = 1.5$:

$$E_D = \sum_{j=1}^{n} \gamma_{G,j} G_{k,j} + \gamma_p P + \gamma_{Q,1} Q_{k,1} + \sum_{i=2}^{m} \gamma_{Q,i} \psi_{0,i} Q_{k,i} \tag{31.9}$$

G_{Sx} and G_{Sy}, Considering $G_k = 1.0$ and Coefficient $\psi_2 = 0.4$

$$E_D = \sum_{j=1}^{n} G_{k,j} + P + A_{Ed} + \sum_{i=1}^{m} \psi_{2,i} Q_{k,i} \tag{31.10}$$

The results of the modal analysis, considering 6 vibration modes for which a sufficient modal mass participation can be estimated, are plotted below. Vibration mode 1 is represented by translation in the "U_y" direction with a directional participation of 0.769, while vibration mode 2 is represented by translation in the "U_x" direction with a directional participation of 0.857. Modal deformations are approximately linear throughout the total height of the building. The total mass considered, 7.874 t, was distributed across floors (Fig. 31.4; Tables 31.3 and 31.4).

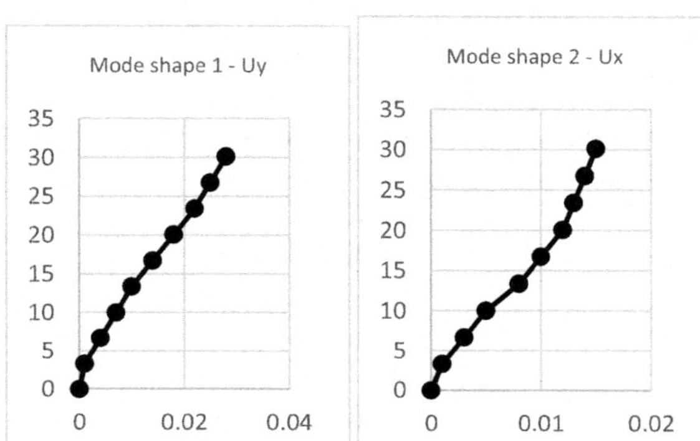

Fig. 31.4 Modal displacements for the first two modes and directions

Table 31.3 Natural vibration periods and floor masses

Case	Mode	Period	UX	UY
		sec		
Modal	**1**	0.452	**0.088**	**0.769**
Modal	**2**	0.404	**0.857**	**0.13**
Modal	**3**	0.356	**0.056**	**0.104**
Modal	**4**	0.132	**0.975**	**0.009**
Modal	**5**	0.121	**0.014**	**0.945**
Modal	**6**	0.107	**0.011**	**0.045**

Bold text was used to highlight the information

Table 31.4 Floor masses

Story	Ux – Uy
kg	kg
8th floor	**229,736**
7th floor	**875,753**
6th floor	**940,932**
5th floor	**947,402**
4th floor	**945,295**
3th floor	**942,796**
2th floor	**947,402**
1th floor	**947,362**
Gth floor	**952,958**
Base	**144,987**

Bold text was used to highlight the information

31.3.5 Design Characteristics for Linear Dampers

We take the conditions mentioned above, namely $\xi_d = 0.25$, so $\xi_{eff} = \xi_s + \xi_d = 0.30$, and from the previously detailed relation, we can establish the characteristics of the dampers based on:

$$\xi_d = \frac{T\sum C_j \phi_{rj}^2 \cos^2\theta_j}{4\pi \sum m_i s_i^2};\ \text{thus} \sum C_j = \frac{\xi_d 4\pi \sum m_i s_i^2}{T\sum \phi_{rj}^2 \cos^2\theta_j} \tag{31.11}$$

The determination of the characteristics is done for both directions, as the natural periods are different, and also the frames where the dampers are placed have different dimensions. The damper system was arranged in an adjacent frame outside the building, considering the poor quality of the concrete in the existing structural system, as well as the desire to reduce the impact of intervention during the implementation

Fig. 31.5 Exterior structural frame, linked to the main building, for installing the dampers

period, while the hospital is functioning. The location of the dampers was established so that there is an equidistance in placement with respect to the center of masses.

The new frame system proposed consists of 70 × 70 cm columns and 40 × 50 cm beams, with superior grade concrete, namely C20/25. These dimensions are taken into account in the mathematical model, and the frame is finally attached by characteristic details to the existing frames, through the interlocking of the existing columns. "Rigid" link-type connections were defined to link the existing system with the newly proposed one, at each floor level (Fig. 31.5).

Taking into account the proposed layout, the initial approximation of damping constants values are summarized as follows:

- 6 braced frames per level with a damping constant of 70.000 kN-s/m in the X direction
- 4 braced frames per level with a damping constant of 35.000 kN-s/m in the Y direction.

31.3.6 The Results of Implementing Linear Dampers

Linear analyses were re-performed to confirm the characteristics of the dampers and their influence by modeling them in the analysis and model. In the first part, the modal analysis did not report significant differences in the stiffness of the consolidated model by decreasing the period. At this stage, the dampers were considered as "exponential damper links" with the damping constants previously mentioned.

Although these elements are speed-dependent, the calculation program required an assessment of stiffness, roughly estimated from the damper extension element and the Maxwell stiffness provided by a damper manufacturer, based on the relationship:

$$\frac{1}{K^*} = \frac{1}{\mathrm{K}_d} + \frac{1}{\mathrm{K}_b} \tag{31.12}$$

where the terms are defined:

- K^*—link element stiffness,
- K_d - damper stiffness, $\mathrm{K_b}$—bracing element stiffness
- E—steel's modulus of elasticity
- A—section area for the extension of the damper, considered as the area of a HEA200 profile
- l—length of the element (damper and damper extension)

In terms of the efficiency of implementing linear dampers, first the performance criteria need to be established. In this case, it was desired to limit the base shear force to an acceptable value for the existing structural system.

Below is defined the design base shear force corresponding to the fundamental mode, considering a behavior factor $q = 2.5$, based on the calculation methodology with equivalent lateral forces:

$$F_b = \gamma_{1,e} \times S_d(T_1)\,\mathrm{m} \times \lambda = 1.2 \times (0.35\,g \times 2.5)/2.5 \times 0.85 \times G/g$$
$$F_b = 0.36 \times G = 0.36 \times 78746\,\mathrm{kN} = 28112\,\mathrm{kN} \tag{31.13}$$

According to the audit report prepared for the analyzed building, an approximate 50% capacity of the existing structural system was highlighted, compared to the code design force:

$$F_{b,capacity} = 50\% \times 28112\,\mathrm{kN} = 14056\,\mathrm{kN}$$

At first, a difference in values for the base shear design forces is observed strictly from the applied analysis method, namely the equivalent static forces method and the response spectrum method. If in the approximate method $\mathrm{F_b}$ = 28,112 kN, through the second method the resulting basic force is around the values $\mathrm{F_{b,X}}$ = 25,400 kN and $\mathrm{F_{b,Y}}$ = 22,100 kN (Fig. 31.6).

31.4 Conclusions

After modelling the linear dampers, in the configuration detailed above, a decrease in the response of the structure can be appreciated, applying the same design spectrum, reduced with $q = 2.5$ and $\xi_0 = 5\%$, of about 28% for the Y direction and 23% for

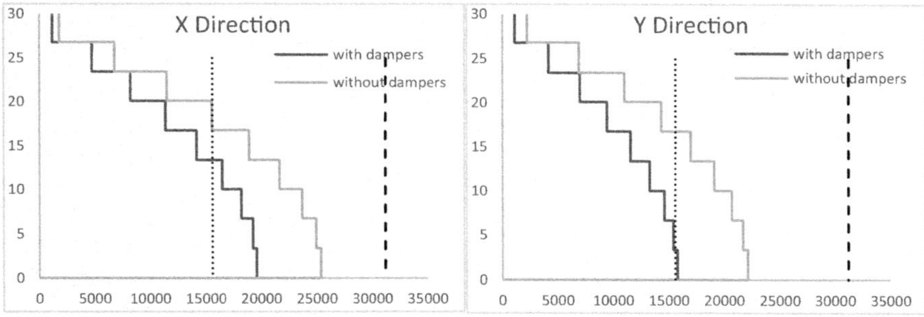

Fig. 31.6 Distribution of story shear, with and without dampers

the *X* direction. This resulted in $F_{b,X,\ FVD} = 19{,}606$ kN and $F_{b,Y,FVD} = 15{,}808$ kN, both values close to the capacity for the undamped (initial structure) estimated by the expert report.

The methodology of placing the bracing frames equipped with viscous fluid dampers, i.e. as far away from the centre of stiffness as possible, and externally, in addition to correcting the response of the structure for torsional effects, also allows the hospital to function during the execution of the strengthening works, with a restricted area being affected for the intervention.

The application of external bracing frames to the building, as a structural subsystem type, equipped with dampers, attached to the main system, with the aim of improving the performance of behaviour to seismic actions, are a pioneer in Romania and can be a viable solution for the possibility of installing fluidic damping devices.

References

1. Cod de proiectare seismică—Partea a III-a—Prevederi pentru evaluarea seismică a clădirilor existente, indicativ P 100–3/2019, aproved by Ordinul 2834/2019 (Romanian Seismic Code for rehabilitation of existing buildings)
2. Prevederi de proiectare pentru clădiri, indicativ P 100-1/2013, revised in 2019 by Ordinul 928/ 2019 (Romanian Seismic Code)
3. FEMA 356 NEHRP Prestandard and commentary for the seismic rehabilitation of buildings (2010)
4. A. Tudor, P. Bratu, The assessment of the seismic answer of structures with viscous fluid dissipators included in the construction and functionally (AVMS, 2023)
5. F Pavel R Vacareanu J Douglas 2016 An updated probabilistic seismic hazard assessment for romania and comparison with the approach and outcomes of the SHARE project Pure Appl. Geophys. 173 1881 1905

Chapter 32
Numerical and Experimental Results Obtained from the Study of the Mechanical Coupling Systems of Power Sources

Dinel Popa, Irina-Virginia Tudor, Dragos-Mihai Popa, and Claudia-Mari Popa

Abstract In the beginning of the paper is presented an experimental stand used in the study of the operation of a mechanical system that connects a heat engine, an electric motor and a generator. This stand simulates the operation of a hybrid car. The mechanical system for coupling the power sources has in its composition a planetary mechanism with double satellite. The mathematical model of the stand consists in a system of 2 differential equations of the second degree, in which the unknowns are the angular displacements and the angular velocities of the electric motor and of the thermal motor. This reduces the entire system to the two inputs of the planetary mechanism. The measuring chain used in the experimental determinations consists of a torque and speed transducer, a measuring amplifier, an integrated data acquisition system and an analysis software. The stand is loaded with a brake. At the end of the paper, following the analysis of the experimental results, the conclusions are presented.

32.1 Introduction

The energy crisis and the measures imposed for the constant reduction of vehicle emissions made the study of hybrid propulsion systems a constant concern of researchers.

The diversity of hybrid propulsion systems has imposed a classification according to criteria such as: coupling mode of power sources, complementary functions, degree of hybridization. A solution adopted by builders is the use of mechanical coupling systems of thermal and electrical power sources.

D. Popa (✉) · I.-V. Tudor · D.-M. Popa · C.-M. Popa
University of Pitesti, 110040 Pitesti, Romania
e-mail: dinel.popa@upb.ro

N. Herisanu and V. Marinca (eds.), *Acoustics and Vibration of Mechanical Structures—AVMS-2023*, Springer Proceedings in Physics 302,
https://doi.org/10.1007/978-3-031-48087-4_32

In some papers [1–5] studies are made on the architecture and performance of mechanical systems in the construction of hybrid transmissions, and in other papers [6–8], their dynamic behavior is also studied. The experimental results are the object of other researches [9–12], highlighting the reduction of fuel consumption, noise, etc. Another advantage of hybrid propulsion systems consists in energy recovery and its storage during the braking phase.

In papers [3] and [6], solutions of planetary mechanisms with 4 elements that can couple 3 power sources are studied from a kinematic and dynamic point of view, comparatively: a thermal engine and 2 electric machines.

In this paper, in order to simulate the operation of a hybrid car, a stand was designed that has as its main component a planetary mechanism that joins a thermal engine and 2 electric machines. In a first phase, the system of two differential equations of motion obtained for the stand is presented. The unknowns are the angular displacements and angular velocities of the electric motor and the thermal motor. The whole system is thus reduced to the two inputs of the planetary mechanism. With the help of a measurement chain, several experimental determinations are obtained. Their main purpose is the dynamic validation of the planetary mechanism with double satellite used in the coupling of thermal and electric power sources.

32.2 Description of the Experimental Stand

In Fig. 32.1 and 32.2, are shown the front and side view of the experimental stand. The mechanical system *SM*, with the kinematic diagram in Fig. 32.3, connects the thermal engine *MT*, the electric engine *ME* and two alternators.

The satellite port arm 1 (Fig. 32.3) is driven by the *ME* electric motor (Fig. 32.1 and 32.2) by means of a *TC2* belt transmission. The electric motor is a three-phase electric motor with a power of 7.5 kW at a load speed of 2890 rev/min and a frequency of 50 Hz.

The solar wheel 2 (Fig. 32.3) is driven by the thermal motor *MT* (Fig. 32.1) through an elastic coupling *CE1* mounted on the output shaft II of the mechanical

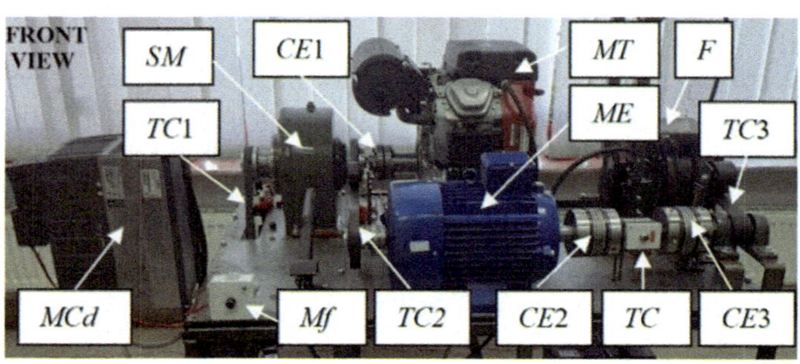

Fig. 32.1 Experimental stand, front view

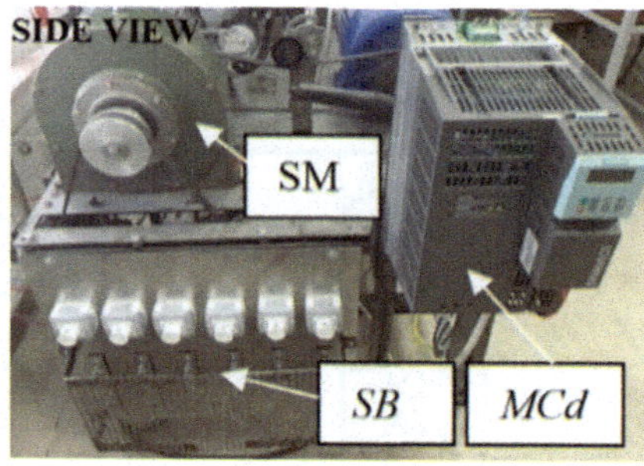

Fig. 32.2 Experimental stand, side view

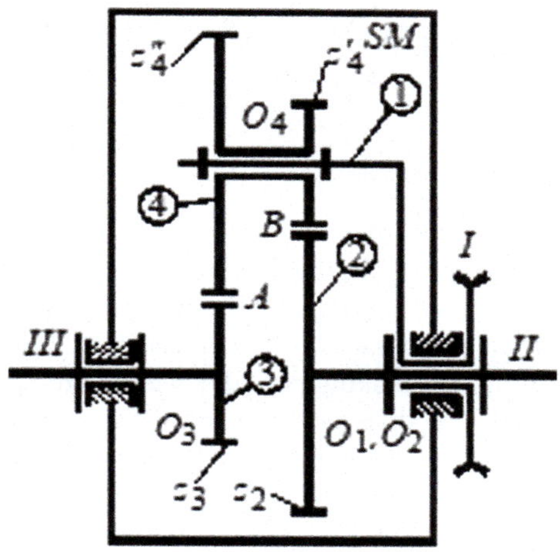

Fig. 32.3 Kinematic diagram

system *SM*. The thermal engine has a power of 15.3 kW at a speed of 3600 rev/min and a torque of 46 Nm at a speed of 2500 rev/min.

The solar wheel 3 (Fig. 32.3) drives, through the *TC1* belt transmission, two alternators: *GE1*, *GE2* positioned in the lower part of the stand (not shown in the figure). The voltages generated by the alternators are distributed to the 6 columns ($C_1 - C_6$, Fig. 32.4) by 5 car bulbs, *BE*, of 12 V/100 W and 2 car batteries of 12 V. At the end of each column of light bulbs, a socket ($S_1 - S_6$, Fig. 32.4) was mounted, which allows the connection of an ammeter connected in series with consumers or the installation of a shunt. The *ME* electric motor, with outputs on both sides, has a torque transducer *TC* coupled between two elastic couplings *CE2*, *CE3*, specially designed for the protection of the transducer.

The *TC* torque transducer is of the T20WN type and allows the simultaneous measurement of torque and speed. The movement is transmitted through the *TC3* belt transmission to the brake *F* of the stand. This is a drum brake; the brake shoes being actuated by a cable through a handbrake lever. Brake *F* is the one that simulates the forward resistance of the vehicle.

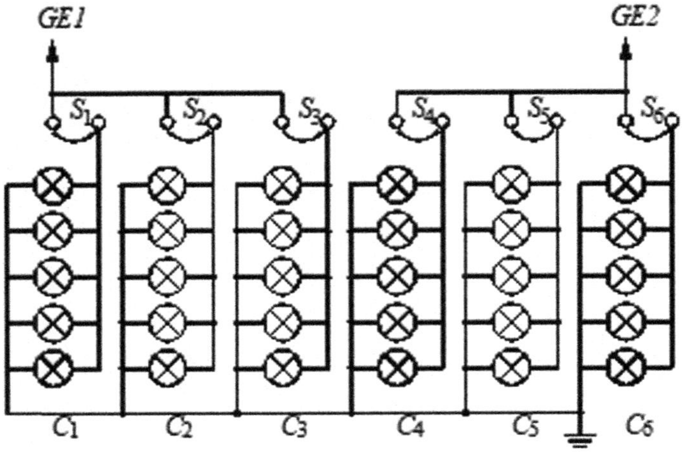

Fig. 32.4 SB bulb system

The *MCd* control module, composed of a SINAMICS power inverter and a CU 240S control module, varies the speed of the three-phase electric motor. It is powered from the laboratory's three-phase grid. The control mode also allows energy recovery during the braking period. Changing the speed is done, either by programming the control module, or by means of a potentiometer *Mf* (Fig. 32.1).

In the lower part of the stand, the 3 × 12 V car batteries were positioned, two for the electrical circuit of the alternators and the third for the electrical circuit of the thermal engine.

32.3 Mathematical and Numerical Model

In Fig. 32.3, the numbers of teeth of the gear wheels were denoted by z_2, z_3, z_4', z_4''.

The reports are denoted by i_1, i_2:

$$i_1 = -\frac{z_2}{z_4'},\, i_2 = \frac{z_2 z_4''}{z_3 z_4'} \tag{32.1}$$

Applying the Willis relations for the relative motion with respect to the satellite port 1, we obtain the connection relations between the angular velocities:

$$\omega_3 = \omega_1(1 - i_2) + \omega_2 i_2,\, \omega_4 = \omega_1(1 - i_1) + \omega_2 i_1 \tag{32.2}$$

We use the notations: i_{TC_1}, i_{TC_3} the transmission ratio of belt transmissions T_{C_1}, T_{C_3}; i_{TC_2} the transmission ratio of the belt drive T_{C_2}; m_4 the mass of element 4 of the planetary mechanism; J_1, J_2, J_3, J_4 the moments of inertia of the elements of the planetary mechanism; R_4 the radius of the satellite carrier arm; J_{RC_1}, J_{RC_2}, J_{RC_3}, J_{RC_4}, J_{RC_5}, J_{RC_6}, J_{RC_7} the moments of inertia of the pulleys; J_{CE} the moment of inertia of the elastic coupling; J_T the moment of inertia of the brake drum.

In order to obtain the equations of motion, Lagrange's equations for holonomous links are used. The expression of the kinetic energy $\underline{T}$ of the stand is:

$$T = \frac{1}{2}\Big[J_1\omega_1^2 + J_2\omega_2^2 + J_3\omega_3^2 + m_4R_4^2\omega_4^2 + J_4\omega_4^2 + J_{RC_1}\omega_3^2 + \left(J_{RC_2} + J_{RC_3}\right)\left(\frac{\omega_3}{i_{TC_1}}\right)^2\Big] + \frac{1}{2}\left[J_{ME}\omega_1^2 + \left(J_{RC_4} + J_{RC_5}\right)\omega_1^2 + 2J_{CE}\omega_1^2 + J_{RC_6}\omega_1^2 + \left(J_{RC_7} + J_T\right)\left(\frac{\omega_1}{i_{TC_3}}\right)^2\right] \tag{32.3}$$

and the expression of generalized forces:

$$Q_1 = M_1 - M_3(1 - i_2) - \frac{M_f}{i_{CT_3}},\ Q_2 = M_2 - M_3 i_2 \tag{32.4}$$

It was noted with M_1 the torque of the ME electric motor, M_2 the thermal engine torque and M_3 the torque of the GE1 and GE2 alternators.

Denoting with A_{11}, A_{12} and A_{13}, and the expressions of the constants:

$$A_{11} = J_1 + m_4R_4^2 + J_{ME} + J_{RC_4} + J_{RC_5} + 2J_{CE} + J_{RC_6} + \frac{J_{RC_7} + J_T}{\left(i_{TC_3}\right)^2} + \left(J_3 + J_{RC_1} + \frac{J_{RC_2} + J_{RC_3}}{\left(i_{TC_1}\right)^2}\right)(1 - i_2)^2 + J_4(1 - i_1)^2,$$

$$A_{22} = J_2 + \left(J_3 + J_{RC_1} + \frac{J_{RC_2} + J_{RC_3}}{\left(i_{TC_1}\right)^2}\right)i_2^2 + J_4 i_1^2,$$

$$A_{12} = \left(J_3 + J_{RC_1} + \frac{J_{RC_2} + J_{RC_3}}{\left(i_{TC_1}\right)^2}\right)(1 - i_2)i_2 + J_4(1 - i_1)i_1 \tag{32.5}$$

with the generalized coordinates θ_1, θ_2 ($\omega_1 = \dot{\theta}_1, \omega_2 = \dot{\theta}_2$) and taking into account relations (32.2), the kinetic energy expression given by relation (32.3) becomes:

$$T = \frac{1}{2}\left(A_{11}\dot{\theta}_1^2 + A_{22}\dot{\theta}_2^2\right) + A_{12}\dot{\theta}_1\dot{\theta}_2 \tag{32.6}$$

In the Lagrange equations, replacing the expressions of the partial derivatives, a system of 2 differential equations is obtained, with the solutions:

$$\ddot{\theta}_1 = \frac{\left[M_1 - M_3(1 - i_2) - \frac{M_F}{i_{TC_3}}\right]A_{22} - [M_2 - M_3 i_2]A_{12}}{A_{11}A_{22} - A_{12}^2},$$
$$\ddot{\theta}_2 = \frac{-\left[M_1 - M_3(1 - i_2) - \frac{M_F}{i_{TC_3}}\right]A_{12} + [M_2 - M_3 i_2]A_{11}}{A_{11}A_{22} - A_{12}^2} \quad (32.7)$$

The values of the constants A_{11}, A_{12},A_{13}, depend on the mechanical properties of the stand elements. For this, the components of the mechanical system and the other elements of the stand are modeled with solids, in AUTOCAD. The values of the findings from relation (32.5) are obtained:

$$A_{11} = 0,340931\,\mathrm{Kg \cdot m^2},\ A_{12} = -0,137424\,\mathrm{Kg \cdot m^2},\ A_{22} = 0,099697\,\mathrm{Kg \cdot m^2} \quad (32.8)$$

The external characteristics of the 3 engines are given by the manufacturer. To model them, polynomials are used, choosing points on the characteristic.

For the thermal engine, a 2nd degree polynomial of the form $M_2(\omega) = a\omega^2 - b\omega + c$ was chosen. To determine the 3 constants, 3 points were chosen on the external characteristic of the engine: $M_{21} = 42$ Nm at $\omega_1 = 209.44$ rad/s, $M_{22} = 45$ Nm at $\omega_2 = 256.5$ rad/s and $M_3 = 40.5$ Nm at $\omega_3 = 377$ rad/s. After solving the system of 3 equations, the values of the constants a, b, c are obtained, and finally the expression of the external characteristic:

$$M_2 = \begin{cases} 0 & If\omega < 100\,\mathrm{rad/s} \\ -0,00063\omega^2 + 0,363 \cdot \omega - 6,25 & If\omega > 100\,\mathrm{rad/s} \end{cases} \quad (32.9)$$

In the case of alternators, the characteristic is indicated: the intensity of the electric current as a function of speed, at a constant voltage. Starting from this characteristic and taking into account that the external characteristic $P_3 = M_3\omega = U \cdot I$ ($U = 14$ V) is similarly obtained:

$$M_3 = \begin{cases} 0 & If\omega < 100\,\mathrm{rad/s} \\ -0,00565\omega + 6,45547 - \frac{284,0378}{\omega} & If\omega > 100\mathrm{rad/s} \end{cases} \quad (32.10)$$

For the electric motor, a fixed value $M_1 = 25$ Nm is considered (by programming the CU240S control module) in the [0..3000] rev/min range.

With these data, the equations of motion (32.7) can be integrated to obtain the variations of angular velocities at the shafts of the planetary mechanism.

The integration is done by the Runge–Kutta numerical method of order IV with the help of a calculation program. It generates script files based on which the graphs of variation of angular velocities are drawn in AutoCAD. The notations are used:

$$\theta_1 = Y_1, \frac{dY_1}{dt} = Y_2, \theta_2 = Y_3, \frac{dY_3}{dt} = Y_4 \tag{32.11}$$

An integration step $\Delta t = 0, 01$ s and a time interval $t \in [0...20]s$ are chosen.

32.4 The Measuring Chain Used

The measurement chain made is composed of: T20WN torque transducer ⇒ VK20 amplifier ⇒ Sound Book integrated acquisition system ⇒ Samurai analysis software. The Samurai software allows obtaining values simultaneously for torque and speed, memorizing them and writing the values to text files.

A data acquisition time period of 20 s is chosen. The stored data are: time, voltage generated by the torque transducer, number of pulses generated by the speed transducer.

The Samurai software allows, in four quadrants, the display of diagrams for the moment and information about the speed. For the accuracy of the interpretations, it was preferred to process the data provided by the analysis software, with a calculation program made in AutoLisp, a calculation program integrated in AutoCAD.

The calculation program reads the values from the text files, interprets them and processes them so that they can be displayed in AutoCAD. In this way, the double calculation precision of a programming language is combined with the precision of displaying the acquired data in the form of graphics.

The graphs: speed as a function of time $n(t)$ and moment as a function of time $M(t)$, are obtained by drawing in AutoCAD some polylines between the points read from a script file.

32.5 Interpretation of Numerical and Experimental Results

Several experimental determinations will be made to validate from a dynamic point of view the planetary mechanism with double satellite used in the coupling of thermal and electric power sources.

(a) Determination of the moment due to friction.

In the construction of the stand, transmissions with belts, transmissions with toothed wheels, bearings with bearings, elastic couplings, etc. were used. All this introduces friction and therefore power loss. In the specialized literature, yields are indicated for each category of torques, but to be more precise in the determinations, it is preferable to experimentally measure the torque due to friction. For this we will train the stand with the electric motor ME at various constant speeds, without applying the brake and without charging the alternators. With the help of the speed variator, the electric motor speeds of 500, 950, 1500 and 2000 rev/min are obtained, speeds that are in the speed range at which the experimental determinations will be made.

These values will be averaged to study the influence of the friction torque as a function of speed. The period in which the torque was determined is 20 s.

Table 32.1 presents the obtained results.

We will consider the losses due to friction as constant, amounting to $M_r = 1,4\,\text{Nm}$.

Since in the theoretical model we did not take into account the frictions from torques and yields, these now being determined will be taken into account.

(b) Stand operated by the electric motor, thermal motor off, brake progressively applied.

Before the experimental determination, the speed of the electric motor is adjusted to 1100 rev/min with the speed variator. The thermal engine is switched off. Maintain this constant speed for approximately 5 s, apply the brake in two stages. In a first stage, brake progressively for about 3 s with a medium force and in the second stage, in approximately 2.5 s, brake with a greater force to try to stop the electric motor. At the end, the brake is released and the recordings are stopped when the 20 s have elapsed (time interval initially selected). After processing the data and plotting the diagrams, the representation from Fig. 32.5 is obtained. It was noted with $M(t)$. the torque measured by the torque transducer and with $n(t)$. the measured revolutions.

The value of the torque (22 Nm) is noted in the time interval $[8 \div 10, 5]$. s, a value that approaches the value of the maximum torque of the electric motor (25 Nm). The torque variation is mainly due to the low operating speed of the electric motor (120 rev/min) and the pulsed operation mode of the speed variator.

In the time interval $[12 \div 20]$. s, the measured torque is higher than in the time interval $[0 \div 5]$ s, the speed being the same, the shoes of the braking system staying longer in contact with the brake drum.

During the measurements, the thermal engine is not operated ($n_2 = 0$) and the shaft of the mechanical system that drives the alternators has a variation from -1780 rev/min to -200 rev/min (Fig. 32.6).

Figure 32.6 presented the graphs as a function of time of the revs: the electric motor (n_1), the thermal engine (n_2) and the alternator drive shaft (n_3), obtained with the calculation program. A very good correlation between the theoretical and experimental values is observed.

(c) Stand driven by the thermal engine, electric motor not driven, brake locked, consumers.

Before starting the determinations, the thermal engine is brought to a constant speed $n_2 = 2200$ rev/min. Apply the brake until it locks ($n_1 = 0$ rev/min). The

Table 32.1 Dependence of the friction torque on the speed function

Speed [rev./min]	500	950	1500	2000
Torque [Nm]	1,42	1,41	1,4	1,4

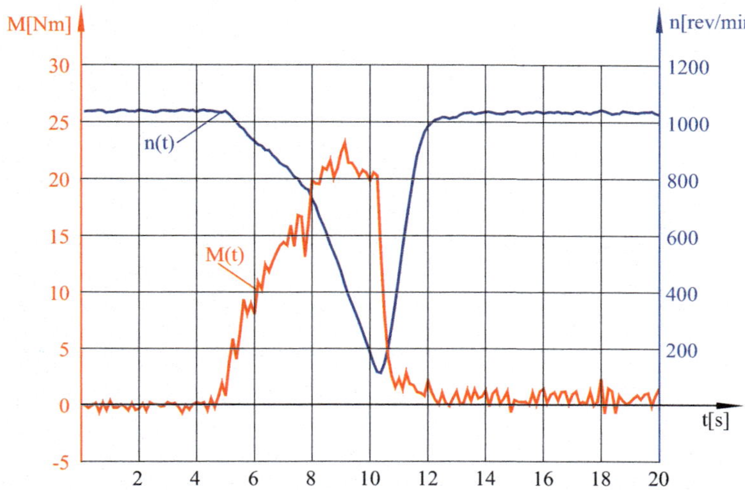

Fig. 32.5. $n(t)$, $M(t)$ diagrams

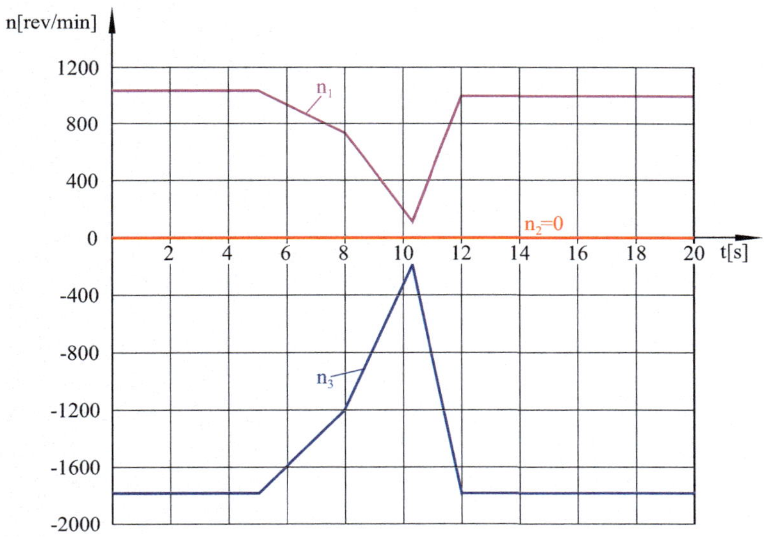

Fig. 32.6 Speed diagrams

electric motor control module is switched off and the alternators are connected to the accumulator batteries. In this case, the speed of the shaft of the mechanical system that drives the alternators (measured with the speed transducer CA27) is $n_3 = 5775$ rev/min and the alternators have a speed of $n_{alt} = 10462$ rev/min. Alternators are charged with the 6 columns of 12 V car bulbs.

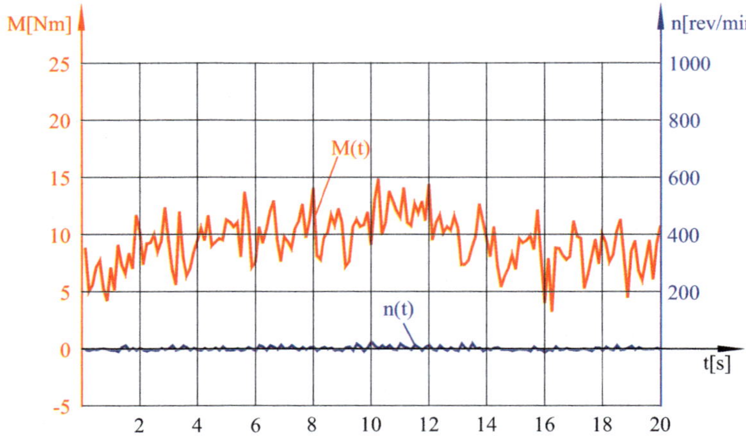

Fig. 32.7 Thermal engine at constant speed with locked brake and consumers

The T20WN transducer records torque and speed which, in this case, is 0. During the experiment, consumers are introduced in turn for the two alternators. After processing the experimental data, the diagrams in Fig. 32.7.

Analyzing the torque variation, it is observed that its values are influenced by the load on the alternators. The brake being locked, the engine torque was transferred only to the alternators.

The torque $M(t)$, in the 20 s time interval has an average of $M_{med} = 9.46\,\text{Nm}$ value calculated with the AutoLisp function. This torque was produced by the thermal engine and transferred to the alternators.

As mentioned, the speed $n_2 = 2200$ rev/min of the thermal engine was constant ($\omega_2 = 230.38\,\text{rad/s} = ct.$). The recorded moment 1.4 Nm is due to friction ($M_r = 1.4\,\text{Nm}$). With these observations, the average value of the power transferred to the alternators can be calculated: $P = (M_{med} - M_r)\omega_2 = 1856.8\,\text{W}$. Part of the energy is transferred to the columns of light bulbs and another part of the energy is transferred to the storage batteries for charging. In the case of a hybrid vehicle, this is the case of parking and charging the traction battery.

(d) Stand operated by the thermal engine, electric motor operated, progressively operated brake.

Before starting the determinations, the thermal engine is brought to a constant speed $n_2 = 2200$ rev/min, the brake is released and the control module of the electric engine is turned on. With the help of the control mode, the electric motor is brought up to speed $n_1 = 1150$ rev/min. The speed of the shaft n_3 of the mechanical system to which the alternators are connected, measured with the speed transducer CA27, is $n_3 = 3906$ rev/min. During the experiment, the brake is applied in several stages. After the acquisition and processing of the data, the representation from Fig. 32.8.

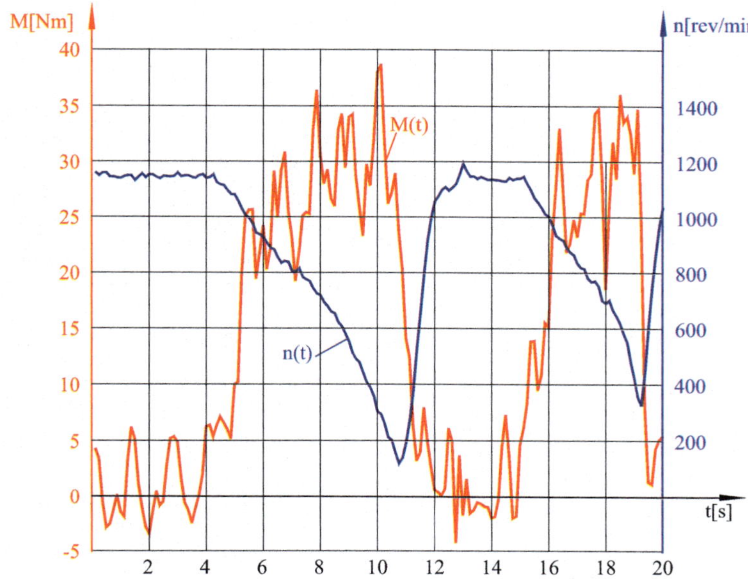

Fig. 32.8 Thermal engine at constant speed with brake operated without consumers

The main conclusion of this experiment is that the power of the electric motor is supplemented by the heat engine by means of the planetary mechanism. The electric motor has a maximum torque of 25 Nm and the peak values of the torque in some cases exceed the value of 35 Nm. When the stand was operated by the electric motor, the thermal motor being turned off, when the brake was actuated, a torque of 22 Nm was recorded at a speed of 120 rev/min. In the present case, the torque measured at 150 rev/min was 35 Nm. It can be concluded that the 35 Nm were provided by the thermal engine and the electric motor.

If a torque transducer had been mounted between the thermal engine and the planetary mechanism, the torque transferred to the stand by the thermal engine could have been measured. In this way, the weight of the two couples in the transferred couple would have been known exactly.

(e) Stand operated by the thermal engine, electric motor, consumers, progressively operated brake.

Compared to the previous experiment, consumers also connected. The thermal engine has a constant speed $n_2 = 2200$ rev/min, the brake is released, the electric motor has speed $n_1 = 1150$ rev/min and two columns of light bulbs (1000 W) are powered. The speed of the shaft n_3 of the mechanical system to which the alternators are connected is $n_3 = 3906$ rev/min.

During the experiment, the brake is applied in several stages.

After the acquisition and processing of the data, the representation from Fig. 32.9. Compared to the previous experiment, the additional torque can be observed to

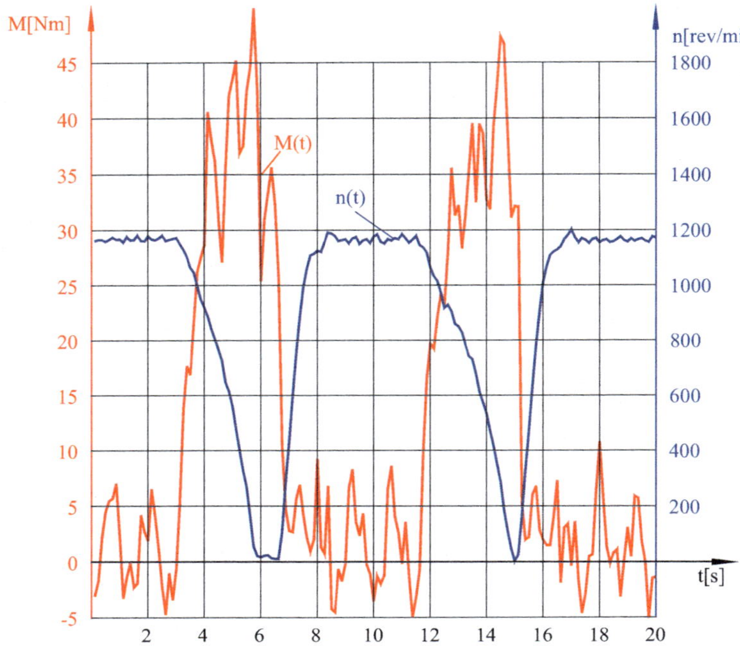

Fig. 32.9 Thermal engine and electric engine at constant speeds, progressively actuated brake and consumers

compensate for the load on the alternators. The measured torque has peak values exceeding 40 Nm.

Considering the maximum torque provided by the thermal engine at 2200 rev/min of 44 Nm and the torque of 25 Nm provided by the electric motor, it follows that in this experiment the stand could be loaded with a maximum torque of 69 Nm with the help of consumers and by operating the brake. The recorded values approach the maximum possible values in proportion of more than 70%.

32.6 Conclusions

The purpose of the experimental determinations was to validate from a dynamic point of view the planetary mechanism with double satellite used in the coupling of thermal and electric power sources. In this sense, a stand was designed and made that allows the coupling of an electric motor, a heat engine and two alternators charging electrical consumers.

To study the dynamic behavior of the mechanism, the following were used: a thermal engine with a power of 15.3 kW, an electric motor with a power of 7.5 kW, and instead of the second electric machine, two alternators were used that have load 6 columns of car bulbs of 500W each.

The electric motor is a three-phase electric motor fed from the three-phase network of the laboratory. Its speed control is carried out with an electronic module that also allows energy recovery during braking.

5 experiments were chosen as representative.

In the first experiment, losses due to friction in the stand couplings were determined. After studying the dependence of the friction torque as a function of speed, it turned out that it can be considered constant, in the rpm range $500 \div 2000$. The experimental measurement of the torque due to friction and its subtraction from the experimentally measured value was preferred.

To study the movement of the hybrid vehicle in the urban cycle with sudden braking, experiment *b* was carried out. The thermal engine is turned off, the stand is operated only by the electric motor. It is noticeable in case of violent braking, close to stopping, the torque of the electric motor is close to the maximum (22 Nm out of 25 Nm available).

In experiment *c*, the case of parking a hybrid car and charging the traction battery by the thermal engine was studied. The experimental results confirm that the energy generated by the thermal engine is transmitted to the alternators. They consume the energy feeding the 6 columns of light bulbs, thus simulating the transfer of the energy of the traction battery.

In experiment *d*, both engines are used, thermally and electrically, with the stand brake being operated progressively. In this way, it simulates the movement of a hybrid vehicle at high speed, when the forward resistance is high. It also simulates an emergency braking with a significant reduction in speed. The conclusion of the experiment is that the power of the electric motor is supplemented by the thermal engine through the planetary mechanism.

In the last experiment, the stand is operated by the thermal engine and the electric motor, there are consumers and the brake is operated. In this way, the electric motor's torque is simulated by the thermal engine, the latter also charging the traction battery. The experiment demonstrates the ability of the mechanical system to couple the thermal and electrical power sources close to the maximum possible values.

The final conclusion is that the presented mechanical system can be successfully used in the structure of a hybrid vehicle, providing high torque at low revs without the need for torque amplification with a gearbox.

References

1. M. Park, T.W. Park, S.H. Lee, S.W. Han, S.K. Kwon, Analytical study to estimate the performance of the power shift drive (PSD) axle for a forklift. Int. J. Automot. Technol. **11**(1), 49–56 (2010)
2. R. Schuebel, M. Gegner, F. Beeck, Turning new directions: surprising potential in planetary transmissions. Part 1: Planetary gear set, in *Schaeffler Technologies, Solving the Powertrain Puzzle* eds by GmbH & Co. KG. (Schaeffler Technologies GmbH, 2014)

3. I. Tudor, D. Popa, N.D. Stanescu, A comparative study of planetary mechanisms usable for power sources coupling, in *International Congress of Automotive and Transport Engineering-Advanced Engineering* (Piteşti, 2017)
4. C. Huang, J.-X. Wang, K. Xiao, M. Li, J.-Y. Li, Dynamic characteristics analysis and experimental research on a new type planetary gear apparatus with small tooth number difference. J. Mech. Sci. Technol. **27**(5), 1233–1244 (2013). www.springerlink.com/content/1738-494x
5. S. Wang, S. Zhang, D. Shi, X. Sun, J. He, Research on instantaneous optimal control of the hybrid electric vehicle with planetary gear sets. J. Braz. Soc. Mech. Sci. Eng.
6. D. Popa, N.D. Stanescu, I. Tudor, Dynamic behavior of the mechanical systems from the structure of a hybrid automobile, in *International Congress of Automotive and Transport Engineering—Advanced Engineering* (Piteşti, 2017).
7. C. Mansour, D. Clodic, Dynamic modeling of the electro-mechanical configuration the Toyota hybrid system series/parallel power train. Int. J. Automot. Technol. **13**(1), 143–166 (2012)
8. H. Yang, B. Kim, Y. Park, W. Lim, S. Cha, Analysis of planetary gear hybrid powertrain system part 2: output split system. Int. J. Automot. Technol. **10**(3), 381–390 (2009)
9. A. Alessandrini, F. Orecchini, F. Ortenzi, F. Villatico Campbell, *Drive-style Emissions Testing on the Latest Two Honda Hybrid Technologies* (European Conference of Transport Research Institutes, (ECTRI) 2009)
10. S.J. Kim, Y.-S. Yoon, S. Kim, K.-S. Kim, Fuel economy assessment of novel multi-mode parallel hybrid electric vehicle. Int. J. Automot. Technol. **16**(3), 501–512 (2015)
11. K.J. Qin, E.H. Wang, H. Zhao, G.K. Shi, R.G. Wang, Z.G. Kong, W. Wang, Development and experimental validation of a novel hybrid powertrain with dual planetary gear sets for transit bus applications—Science China. Technol. Sci. **58**(12), 2085–2096 (2015)
12. C. Xiang, K. Huang, R. Langari, Power-split electromechanical transmission design and validation with three planetary gears for heavy-duty vehicle. Iran J. Sci. Technol. Trans. Mech. Eng.

Chapter 33
The Influence of the Modulus of Elasticity in the Numerical Modal Analysis of Truss Elements Which Are Defined as Beam, Truss or Solids

Dan Alexandru Pîrşan, Zeno-Iosif Praisach, and Cornel Haţiegan

Abstract In the numerical modal analysis, the elements of a truss can be defined as beam, truss or solids. The natural frequencies, as well as the modal shapes obtained from these analyzes differ for each case considered separately. The paper presents the natural frequency changes obtained through numerical modal analysis using SolidWorks software for cases where the structural elements are assimilated as beams, trusses, and solids. For each individual case, the changes in natural frequencies are analyzed for the case where an element of the structure becomes much more elastic. In other words, for this element, the value of the modulus of elasticity was intentionally reduced. The study demonstrates how changes in the modulus of elasticity can significantly impact the dynamic response of these structures. The findings of this study can be used to better understand the behavior of these structures under varying conditions and to optimize their design for better performance.

33.1 Introduction

The diagnosis and location of structural deterioration using FEM has seen significant advancements. Young's modulus, commonly referred to as the modulus of elasticity, is a measurement of a material's stiffness or resistance to deformation. The modulus of elasticity of a material can have a major impact on how that material responds to dynamic situations like vibrations or impacts [1–3].

Materials with higher moduli of elasticity are typically stiffer and more resistant to deformation, which can lead to higher natural frequencies and reduced vibrations. These materials may also transmit more force and generate higher stresses during

D. A. Pîrşan (✉) · Z.-I. Praisach
Doctoral School of Engineering, Babes-Bolyai University, 400084 Cluj-Napoca, Romania
e-mail: dan.pirsan@ubbcluj.ro

Z.-I. Praisach · C. Haţiegan
Department of Engineering Science, Babes-Bolyai University, 400084 Cluj-Napoca, Romania

N. Herisanu and V. Marinca (eds.), *Acoustics and Vibration of Mechanical Structures—AVMS-2023*, Springer Proceedings in Physics 302,
https://doi.org/10.1007/978-3-031-48087-4_33

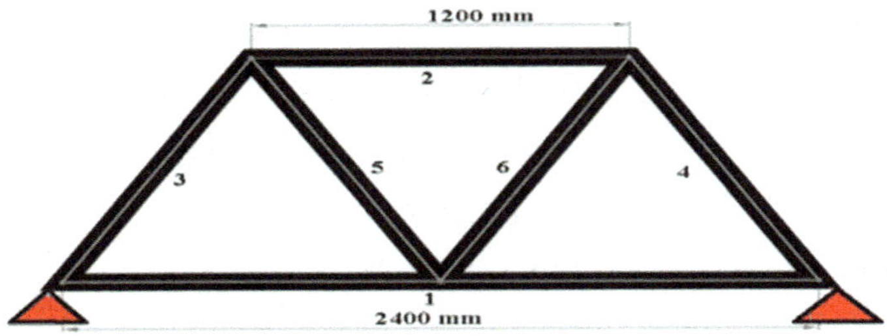

Fig. 33.1 Warren truss with six bar

impacts or shocks, which can be useful in some situations but can also increase the risk of damage or failure [4, 5].

In contrast, materials with lower moduli of elasticity are generally more flexible and deformable, leading to lower natural frequencies and higher levels of vibration. These materials may also absorb more energy during impacts or shocks, reducing the risk of damage or injury but potentially also reducing the material's effectiveness in certain applications [6–8].

In order to obtain the natural frequencies and compare them for the same structure in this study, a single-span Warren truss with six elements was subjected to a numerical modal analysis. However, the elasticity modulus of one of the elements, which were considered to be weak or damaged, was purposefully altered. Overall, the modulus of elasticity is an important factor to consider in understanding and predicting the dynamic behavior of materials, structures or systems [9, 10].

The modal analysis approach for a Warren truss (Fig. 33.1) (elements truss, beam and solids) from the point of view of dynamic behavior start from consideration the under the action of its own weight, the structural elements of the truss are subject to bending, compression and tension [11, 12].

In this study, structural steel, from the SolidWorks library, rectangular tubes with the cross section dimensions of $120 \times 80 \times 8$ mm were used.

The length of the base is 2400 mm and the elements of 1200 mm (Fig. 33.1).

The modulus of elasticity should generally be taken into account when designing truss structures as it can significantly affect their dynamic behavior, which in turn affects their performance and safety. To determine the natural frequencies and mode shapes of a truss, numerical modal analysis must be used to solve the structure's equations of motion. Finite element analysis software is commonly used to accomplish this. It discretizes the structure into a collection of elements connected by nodes, and then it solves for the eigenvalues and eigenvectors of the resulting system matrix [13, 14].

In the case of a truss, modal analysis can provide valuable information about the vibrational behavior of the structure and help identify potential design issues or failure modes [15, 16].

In order to simulate the reduction of the rigidity of the structure's elements, due to the occurrence of the fatigue for example, and to highlight the changes in the dynamic behavior of the structure, the modulus of elasticity was reduced by 25% for each element of the structure one after the other.

33.2 Numerical Modal Analyzes

In SolidWorks, modal analyzes the elements can be considered as: truss, beam, or solids. To visualize the vibration modes and to obtain the natural frequencies of the Warren truss, the modal analysis by finite element method was used.

The analyzed Warren truss (Fig. 33.1) has a single-span, hinged at the ends, on which only its own weight acts. The process involves meshing the structure intro smaller elements and using numerical methods to solve for the eigenvalues and eigenvectors of the truss.

33.3 Analyzed Cases

The following cases were analyzed:

- All the truss bars are made of structural steel with $E = 2.1 \times 10^{11}$ N/m^2. The natural frequencies f_U were determined;
- The natural frequencies f_D obtained by modifying each truss bar, numbered from 1 to 6 at Fig. 33.1, with $E = 1.6 \times 10^{11}$ N/m^2 were compared with the natural frequencies f_U obtained in the case of the structure is made with $E = 2.1 \times 10^{11}$ N/m^2. The relative frequencies shift ε [%] were calculated with relation (33.1):

$$\varepsilon = \frac{f_U - f_D}{f_U} \times 100 \tag{33.1}$$

- For the cases mentioned above, it were considered three types of structural elements: truss, beam and solid.

33.4 Results

For each analyzed case, the natural frequencies and the relative frequencies shift ε for first sixth vibration modes are presented in Tables 33.1, 33.2 and 33.3.

Comparison of the relative frequencies shift are illustrated in the Figs. 33.2, 33.3 and 33.4

Table 33.1 Natural frequencies values and relative frequency shift when elements are defined as truss

Modified element	f_1 [Hz]	f_2 [Hz]	f_3 [Hz]	f_4 [Hz]	f_5 [Hz]	f_6 [Hz]
–	240.58	321.66	563.93	630	809.5	936.76
1	218.92	316.97	547	601.08	804.77	929.25
ε_1 [%]	9.0032	1.45806	3.00215	4.59047	0.58431	0.80170
2	238.63	313.61	560.29	627.49	776.54	891.77
ε_2 [%]	0.8105	2.50264	0.64547	0.39841	4.07165	4.80272
3	229.63	319.19	527.84	587.87	790.85	905.48
ε_3 [%]	4.5515	0.76789	6.39973	6.68730	2.30389	3.33917
4	236.43	301.05	534.66	592.69	784.09	904.48
ε_4 [%]	1.7249	6.40739	5.19036	5.92222	3.13897	3.44592
5	236.66	310.24	537.33	604.62	757.78	892.87
ε_5 [%]	1.6293	3.55033	4.71690	4.02857	6.38912	4.68530
6	236.82	321.66	542.08	609.18	804.5	905.5
ε_6 [%]	1.5628	0	3.87459	3.30476	0.61766	3.33703

Table 33.2 Natural frequencies values and relative frequency shift when elements are defined as solid

Modified element	f_1 [Hz]	f_2 [Hz]	f_3 [Hz]	f_4 [Hz]	f_5 [Hz]	f_6 [Hz]
–	11.117	47.42	60.587	168.73	177.79	204.51
1	10.717	43.498	58.789	158.57	171.16	197.84
ε_1 [%]	3.59809	8.28624	2.96763	6.02145	3.72912	3.26145
2	11.091	46.933	58.926	164.5	175.14	201.43
ε_2 [%]	0.23388	1.043687	2.74151	2.50697	1.49052	1.50604
3	11.096	46.796	59.121	161.51	176.19	194.33
ε_3 [%]	0.18890	1.33255	2.41966	4.27903	0.89994	4.97775
4	10.16	46.788	58.642	165.61	172.72	196.96
ε_4 [%]	8.60844	1.34941	3.21025	1.84911	2.85168	3.69175
5	11.087	46.938	59.399	165.81	174.86	201.85
ε_5 [%]	0.26986	1.03314	1.96081	1.73058	1.64801	1.30067
6	11.087	46.938	59.399	165.81	174.86	201.85
ε_6 [%]	0.26986	1.03314	1.96081	1.73058	1.64801	1.30067

Table 33.3 Natural frequencies values and relative frequency shift when elements are defined as beam

Modified element	f_1 [Hz]	f_2 [Hz]	f_3 [Hz]	f_4 [Hz]	f_5 [Hz]	f_6 [Hz]
–	199.41	212.46	219.24	240.71	249.66	323.32
1	187.63	204.17	206.67	233.51	249.66	310.43
ε_{11} [%]	5.90743	3.90191	5.73344	2.99115	0	3.98676
2	194.95	208.59	219.24	234.3	238.83	314.1
ε_2 [%]	2.23660	1.82152	0	2.66296	4.33790	2.85166
3	187.56	205.68	213.24	234.18	239.75	302.67
ε_3 [%]	5.94253	3.19119	2.73673	2.71281	3.96940	6.38686
4	199.21	203.83	215.59	233.04	249.66	308.01
ε_4 [%]	0.10030	4.06194	1.66484	3.18641	0	4.73525
5	198.29	207.29	215.2	235.76	239.86	318.73
ε_5 [%]	0.56166	2.43340	1.84273	2.05642	3.92534	1.41965
6	193.63	209.95	215.58	239.86	240.36	318.49
ε_6 [%]	2.89855	1.18140	1.66940	0.35312	3.72507	1.49388

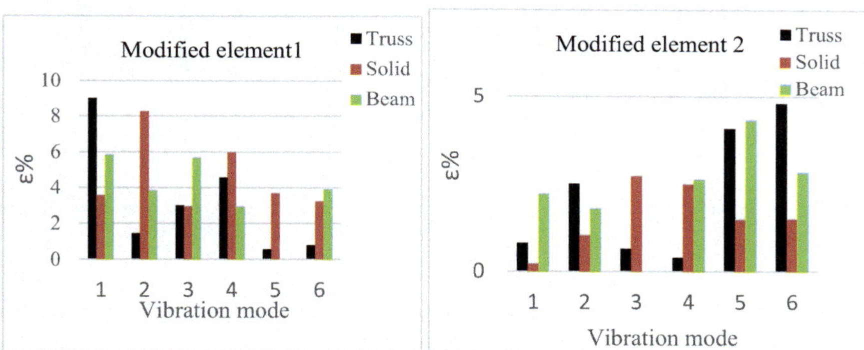

Fig. 33.2 Comparison of the relative frequency shift in case of elements 1–2 have E = 160,000 N/mm^2

33.5 Conclusions

In the paper are presented the deviations of the natural frequencies for the first 6 modes of vibration at a Warren truss when one element of the structure has the low elasticity modulus compared to the other elements of the structure.

The numerical modal analysis performed in SolidWorks was carried out in three variants, respectively, the elements of the structure were treated as: truss, beam and solid.

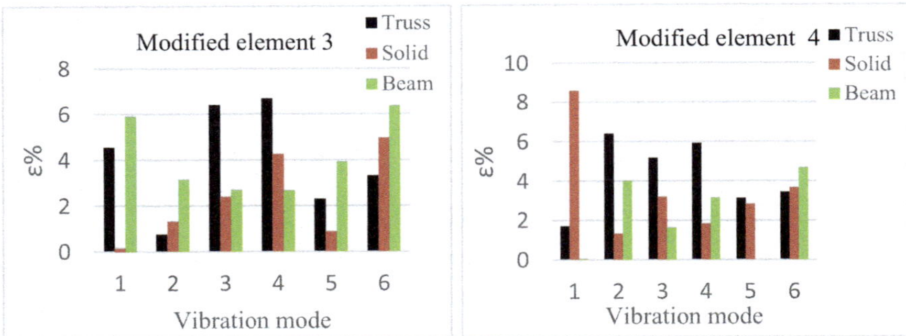

Fig. 33.3 Comparison of the relative frequency shift in case of elements 3–4 have E = 160,000 N/mm^2

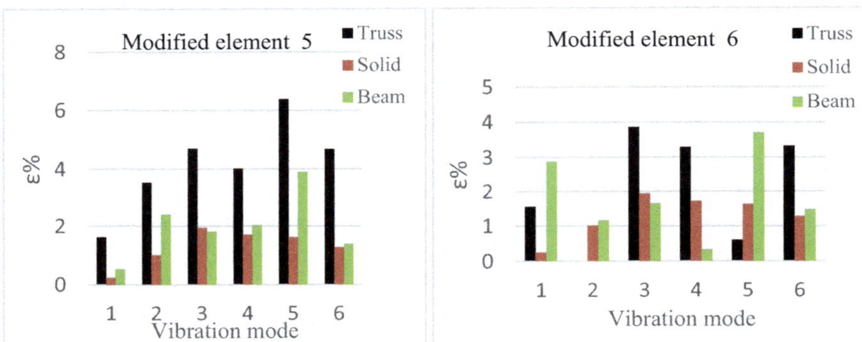

Fig. 33.4 Comparison of the relative frequency shift in case of elements 5–6 have E = 160,000 N/mm^2

The results of the natural frequencies and their deviations from the healthy beam for the first six vibration modes are shown in Tables 33.1, 33.2 and 33.3, and the evolution of the deviations for the three cases analyzed can be traced in Figs. 33.2, 33.3 and 33.4.

In the case of numerical modal analysis with elements assimilated as truss, the results obtained take into account only the behavior of the elements of the structure subjected to tension/compression under the action of their own weight. For this case, the SolidWorks program only offers the first six natural frequencies, respectively, modal shapes. The element with the highest weight on the dynamic behavior of the structure is the element:

- 1, at which the reduction of the elasticity module by 25% produces the maximum effect at mode 1 with the value 9%;
- 2, at which the reduction of the modulus of elasticity by 25% produces the maximum effect at modes 6 with the value 4.8%;

- 3, at which the reduction of the modulus of elasticity by 25% produces the maximum effect in mode 4 with the value 6.7% %;
- 4 and 5 at which the reduction of the modulus of elasticity by 25% produces the maximum effect in modes 2 and 5 with the value 6.4%;
- 6, at which the reduction of the elasticity module by 25% produces the maximum effect at mode 3 with the value 3.9%.

In the case of numerical modal analysis with the elements assimilated as beam, the results obtained consider only the behavior of the elements of the structure subjected to bending under the action of their own weight. There are no limitations of the program on the number of own frequencies obtained. The element with the highest weight on the dynamic behavior of the beam is the element:

- 1, at which the reduction of the elasticity module by 25% produces the maximum effect in mode 1 with the value 5.9%;
- 2, at which the reduction of the elasticity module by 25% produces the maximum effect in mode 5 with the value 4.3%;
- 3 and 4 at which the reduction of the modulus of elasticity by 25% produces the maximum effect in mode 6 with the value 6.3%, respectively 4.7%;
- 5 and 6, at which the reduction of the modulus of elasticity by 25% produces the maximum effect in mode 5 with the value 3.9% respectively 3.7%.

In the case of numerical modal analysis with the elements assimilated as solid, the program treats the model single part. There are no limitations of the program on the number of own frequencies obtained. The element with the highest weight on the dynamic behavior of the beam is the element:

- 1, at which the reduction of the elasticity module by 25% produces the maximum effect at mode 2 with the value 8.2%;
- 2, at which the reduction of the elasticity module by 25% produces the maximum effect at mode 3 with the value 2.7%;
- 3 and 4, at which the reduction of the elasticity module by 25% produces the maximum effect in mode 6 and 1 with the value 4.9% respectively 8.6%;
- 5 and 6, at which the reduction of the elasticity module by 25% produces the maximum effect in mode 3 with the value 2%.

The authors wanted to emphasize that in the dynamic behavior of structures by reducing the stiffness of an element of the structure, the modification of natural frequencies with the greatest weight is not reflected only in vibration modes 1 or 2, regardless of the type of element assimilated for numerical modal analysis.

References

1. Y.J. Yan, L. Cheng, Z.Y. Wu, L.H. Yam, Development in vibration based structural damage detection technique. Mech. Syst. Signal Proc. **21**, 2198–2211 (2007)
2. T.B. Loh, Y. Wu, S.H. Goh, K.H. Kong, K.L. Goh, J.J. Chong, https://www.mdpi.com/2079-9292/11/18/2826, A. Lahe, A. Braunbruck, A. Klauson, An exact solution of truss vibration problems. Proc. Est. Acad. Sci. **68**, 244–263 (2019)
3. Z.-I. Praisach, D.A. Pîrşan, The influence of the change of the longitudinal modulus of elasticity on the dynamic behavior of a Warren truss. Engineering **67**(1), 2022
4. S.L. Garrett, Elasticity of solids, in *Understanding Acoustics. Graduate Texts in Physics* (Springer, Cham, 2020). https://doi.org/10.1007/978-3-030-44787-8_4
5. G.R. Gillich, Z.I. Praisach, Robust method to identify damages in beams based on frequency shift analysis, in *Proceedings of SPIE The International Society for Optical Engineering*, vol. 8348. (San Diego, CA, 2012), p. 83481D. [Google Scholar]
6. D.A. Pîrşan, The influence of the Young modulus on two elements of a Warren truss on the dynamic behavior of the structure. J. Fac. Eng. Reşiţa Babeş-Bolyai Univ. Cluj-Napoca, **67**(1) (2022). https://doi.org/10.24193/subbeng.2022.1 Published Online: 2022–11–21 ISSN (Online): 2734-7680
7. P Avitabile, Experimental modal analysis: a simple non-mathematical presentation. Sound Vib. **35**, 20–31 (2001). [Web of Science ®], [Google Scholar]
8. Z.-I. Praisach, D.A. Pîrşan, The influence of the truss bar on the dynamic behavior of a Warren truss by changing the modulus of elasticity, in *60th International Conference on Vibroengineering in Resita* (Romania, 2022). 18 Nov 2022
9. S. Alampalli, G. Fu, E.W. Dillon, Signal versus noise in damage detection by experimental modal analysis. J. Struct. Eng. **123**, 237–245 (1997)
10. Praisach, Z.-I., D.A. Pîrşan, The influence of the change of the longitudinal modulus of elasticity on the dynamic behavior of a Warren truss. J. Fac. Eng. Reşiţa Babeş-Bolyai Univ. from Cluj-Napoca, 67(1), (2022). https://doi.org/10.24193/subbeng.2022.1 Published Online: 2022–11–21 ISSN (Online): 2734-7680
11. M.G. Amani, J.D. Riera, R.O. Curadelli, Identification of changes in the stiffness and damping matrices of linear structures through ambient vibrations. Struct. Control. Health Monit. **14**, 1155–1169 (2007)
12. H. Yasui, H. Marukawa, Y. Momomura şi T. Ohkuma, Analytical study on wind-induced vibration of power transmission towers. J. Wind Eng. Indus. Aerodyn. **83**, 431–441 (1999)
13. G. Li, H. Zhang, R. Wang, Z.-Q. Dong, Y. Ding-Hao, Seismic damage characteristics and evaluation of aboveground-underground coupled structures. Eng. Struct. (2023). https://doi.org/10.1016/j.engstruct.2023.115871,283,(115871)
14. W. Gao, N. Kessissoglou, Dynamic response analysis of stochastic truss structures under non-stationary random excitation using the random factor method. **196**(25–28)
15. A.S. Tzimas, K.A. Skalomenos, D.E. Beskos, A damage-controlled behavior factor (EC8) for seismic design of steel irregular space MRFs, ce/papers, **4**, 2–4, (2029–2035), (2021). https://doi.org/10.1002/cepa.1518
16. K.J. Bathe, *Finite Element Procedures* (2nd edition,), (Prentice Hall, Pearson Education, Massachusetts, USA, 2014)

Chapter 34
Identification of the Segments with Changed Density in Inhomogeneous Beams

Alexandra-Teodora Aman, Cristian Tufisi, Tiberiu Manescu, and Gilbert-Rainer Gillich

Abstract This paper presents an original algorithm for identifying the segments with changed density in inhomogeneous beams. The algorithm is based on the calculation of relative frequency shifts for a free-free beam using the squared normalized modal curvature and the strain energy method while considering damage coefficients in the affected segments with increased density. The relative frequency shifts are then used to train an artificial neural network that can detect and estimate the density of the affected segments, So, inhomogeneities in beams, for example as the result of roasting, are determined. The proposed method was tested and validated using a numerical simulation. We succeeded to find all segments with altered density, which demonstrates the effectiveness and accuracy of the method. The proposed approach can potentially be used in real-world applications to identify inhomogeneities in beams of various materials.

34.1 Introduction

Structures with inhomogeneities are often encountered in engineering practice. Inhomogeneity can have undesirable causes, such as the heterogeneous structure of the material due to faulty manufacturing or the exposure of structures to environmental factors (corrosion, moisture, etc.). Biological cellular and bone materials are another class of inhomogeneous materials. Sometimes, inhomogeneity is generated intentionally, to provide customized properties to the structures; among these properties,

A.-T. Aman · T. Manescu · G.-R. Gillich (✉)
Doctoral School of Engineering, Babes-Bolyai University, 400084 Cluj-Napoca, Romania
e-mail: gilbert.gillich@ubbcluj.ro

C. Tufisi · G.-R. Gillich
Department of Engineering Science, Babes-Bolyai University, 400084 Cluj-Napoca, Romania

T. Manescu
Department of Industrial Engineering, Aurel Vlaicu University, 310130 Arad, Romania

N. Herisanu and V. Marinca (eds.), *Acoustics and Vibration of Mechanical Structures—AVMS-2023*, Springer Proceedings in Physics 302,
https://doi.org/10.1007/978-3-031-48087-4_34

we can mention the deflection, the vibration behavior, etc. Inhomogeneity can be associated with large engineering structures, but also with nanostructures.

Researchers have developed several models for materials with spatially varying microstructure [1]. This kind of material needs special attention because it presents a spatial variation in specific weight and elastic properties. Other researchers focused on developing techniques to assess the moisture content in wood [2], or to assess corrosion in metallic structures [3, 4].

The free vibration of non-uniform axially functionally graded thick beams is studied in [5]; this study involves the Timoshenko beam theory. The same topic is approached in [6], where a precise integration solution for the static and dynamic responses of axially graded solid beams is found. On the other hand, paper [7] presents a study on the coupled thermic, elastic, and dynamic behavior of axially functionally graded nanobeams.

This paper's authors are unaware of any studies that use artificial intelligence to identify inhomogeneities in the structure. In this work, we present how it is possible to create a database that contains information about inhomogeneities in beams with free-free boundary conditions associated with the change in frequency response. The information is used to train a neural network that finally allows the identification of segments with modified density.

34.2 The Frequency Drop Due to Mass Increase

In this section, we propose a mathematical relationship that can be used to calculate the decrease in natural frequencies of a beam with free ends if it has a constant section but with variable specific mass. This allows the creation of a database that links the position of the segments with modified specific mass with the decrease in the natural frequencies of the bending vibration modes. The mathematical relationship is valid for all types of bending vibrations, and any type of boundary conditions, but next, we will address the problem of vibrations along the weak axis. Therefore, when we refer to bending vibrations, we refer to bending vibrations on the weak axis. The database is used for training an ANN model using MATLAB, that will be able to detect which segment or segments are affected by loss of mass. The method is presented in Fig. 34.1. By incorporating ANN into the process, the detection accuracy can be significantly improved as it learns complex relationships between input data and output responses. Training the ANN using a dataset of RFS values enables it to accurately classify new measurements and identify the presence and location of damage, even in the presence of environmental variations and measurement uncertainties. So far we know, there are no methods to identify inhomogeneities in structures that use artificial intelligence.

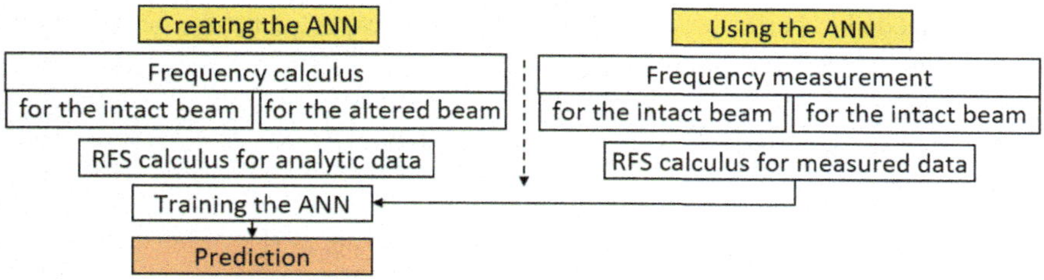

Fig. 34.1 Developed method: on the left side is the depicted the training phase, on the right side is depicted the data generation from measurements

34.2.1 Method Applied for Generating the Analytical Training Data

The well-known expression of the frequency of uniform beams is:

$$f_i = \frac{\lambda_i^2}{2\pi}\sqrt{\frac{EI}{\rho AL^4}} \tag{34.1}$$

where we used the notations: λ for the eigenvalue, E for the modulus of elasticity, I for the moment of inertia, ρ for the mass density, A for the cross-section area, and L for the beam length. The index i indicates the bending vibration mode number.

In the case of uniform beams, the specific mass $\overline{m} = \rho A$ is constant. Therefore, we can calculate the kinetic energy at a given point x on the beam using the mathematical relationship:

$$U_{Ki} = \frac{1}{2}\phi_{i\,\max}\int_0^{x_1}\left\{\frac{\overline{m}\left[4\overline{\phi}_i(x)\right]^2}{T_i^2}\right\}dx \tag{34.2}$$

In the above equation, we denoted $\phi_{i\,\max}(x)$ the maximum beam deflection achieved in mode i at location x. With T_i we denoted the period, that is the time necessary to perform four times the distance $\phi_{i\,\max}(x)$.

If we take into consideration a beam with an increased specific mass $\overline{m}_C$ on the segment x_1-x_2, the kinetic energy will be:

$$\tilde{U}_{Ki} = \frac{1}{2}\phi_{i\,\max}\left\{\int_0^{x_1}\frac{\overline{m}\left[4\overline{\phi}_i(x)\right]^2}{\tilde{T}_i^2}dx + \int_{x1}^{x_2}\frac{\overline{m}_C\left[4\overline{\phi}_i(x)\right]^2}{\tilde{T}_i^2}dx + \int_{x_2}^{L}\frac{\overline{m}\left[4\overline{\phi}_i(x)\right]^2}{\tilde{T}_i^2}dx\right\} \tag{34.3}$$

We can consider two beams that have the same kinetic energy and maximum displacement. In this case, the terms in the curly brackets in Eqs. (34.2) and (34.3) are equal. The mode shapes of the two beams are quite similar, but an increase in the time needed to perform one cycle of vibration is expected for the beam with the bigger mass. Hence, we obtain the equality:

$$\int_0^L \frac{\overline{m}\left[4\overline{\phi}_i(x)\right]^2}{T_i^2}dx = \int_0^{x_1} \frac{\overline{m}\left[4\overline{\phi}_i(x)\right]^2}{\tilde{T}_i^2}dx + \int_{x1}^{x_2} \frac{\overline{m}_C\left[4\overline{\phi}_i(x)\right]^2}{\tilde{T}_i^2}dx + \int_{x_2}^{L} \frac{\overline{m}\left[4\overline{\phi}_i(x)\right]^2}{\tilde{T}_i^2}dx \tag{34.4}$$

Replacing the period with the frequency and after the reduction of some terms, we obtain the equation:

$$f_i^2 \int_0^L \overline{m}\left[\overline{\phi}_i(x)\right]^2 dx = \tilde{f}_i^2 \int_0^{x_1} \overline{m}\left[\overline{\phi}_i(x)\right]^2 dx + \int_{x1}^{x_2} \overline{m}_C\left[\overline{\phi}_i(x)\right]^2 dx + \int_{x_2}^{L} \overline{m}\left[\overline{\phi}_i(x)\right]^2 dx \tag{34.5}$$

From Eq. (34.5), after reduction of $\overline{m}$, the frequency of the beam with a variable specific mass distribution becomes:

$$\tilde{f}_i = f_i\left(\frac{\int_0^L \left[\overline{\phi}_i(x)\right]^2 dx}{\int_0^{x_1}\left[\overline{\phi}_i(x)\right]^2 dx + \frac{\overline{m}_C}{\overline{m}}\int_{x1}^{x_2}\left[\overline{\phi}_i(x)\right]^2 dx + \int_{x_2}^{L}\left[\overline{\phi}_i(x)\right]^2 dx}\right)^{0.5} \tag{34.6}$$

Equation (34.6) can be used to calculate the frequencies of a beam with a beam that has three segments with different specific masses. Note that, the cross-section area and the stiffness are constant over the entire length of the beam.

In Fig. 34.2 we exemplify the mass participation for the second weak-axis bending vibration modes, highlighting the influence of the increased specific mass that is associated with the second sector of the beam. The increase in the area of the beam with increased mass, the red curve, is proportional to the local values of the squared normalized deflection and the mass ratio $r_C = \overline{m}_C/\overline{m}$.

If analyzing Fig. 34.2, we can trace the following conclusions:

- For a given vibration mode and length of a segment, the increase of the mass participation value depends on the position of the segment.
- For a given vibration mode and length of a segment, the increase of the mass participation value depends on the increase of the specific mass.
- The mass participation value is different for the same segment and mass increase for two different vibration modes.

It can be observed in Fig. 34.2 that a very low mass participation change is obtained for mode 1, while for mode 2 the mass participation alteration is significant.

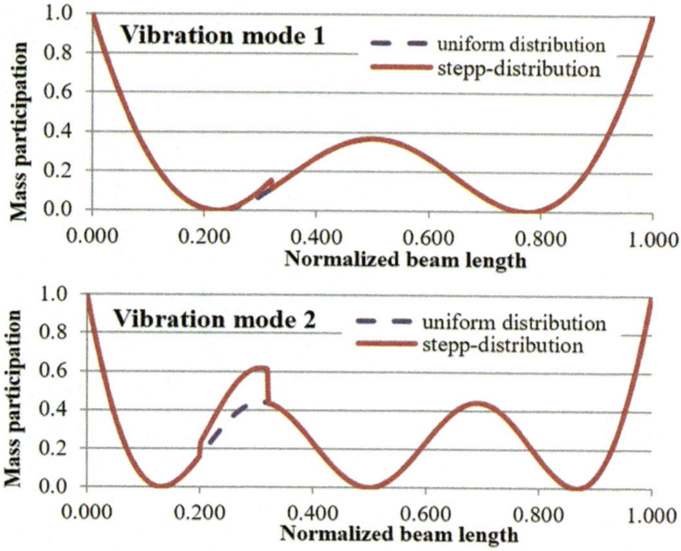

Fig. 34.2 Mass participation diagrams for the first two bending vibration modes

34.2.2 Training of the ANN

We propose to find inhomogeneities in beams. Since we can find the changes in the natural frequencies for any inhomogeneity scenario, we can create a database containing the Relative Frequency Shifts (RFS) as INPUT data and the specific mass of the segments as OUTPUT data. The RFS are calculated after a procedure presented in [8]. Detection of inhomogeneities becomes an inverse problem, from the response of the structure we determine the affected segments.

Equation (7) can be generalized, in the sense that we can divide the beam into any number of segments; in this case the denominator contains one term for each segment. In this study, we consider 20 segments of equal length, among which all have the same specific mass, except one or two segments which have another specific mass (may be different one from the other). The creation of scenarios is performed with Python, and the training was performed using a toolkit of MATLAB.

The number of considered vibration modes is six, which is proved to be reliable for detecting structural changes [9]. These constitute the six neurons in the Input layer. To this first layer, the ANN architecture contains two hidden layers with 40 neurons each, an output layer with 10 neurons that are the ten beam segments of the half-beam, and these indicate the output with ten predicted rigidities. Note that, due to symmetry, we consider just a half of the beam, thus the position of the segments with a different modulus of elasticity is indicated relative to the single free end. On the real beam, the mirrored segment from the other half of the beam can be the targeted one.

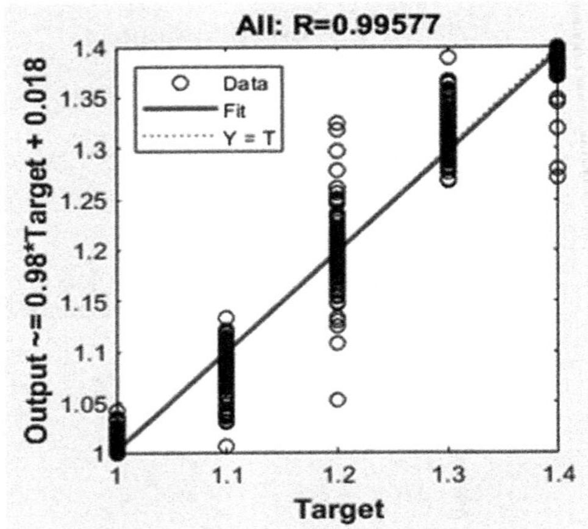

Fig. 34.3 ANN regression plot

The ANN has following characteristics:

- The network is a feedforward neural network with backpropagation;
- Bayesian regularization is used;
- The validation is made via the mean squared error.

From the scenarios in the database, we set 70% of the data for training, 15% for validation and 15% for testing. The data is selected randomly. During the training, regression plots are generated after each epoch. They represent the correlation between the predicted outputs (positions and modulus of elasticity) and the real one.

The training is performed as following. For each input set, weights are proposed in the training phase. Afterward, these are validated with another set of data. In the case of success, the weights remain definitive. It follows the test phase, in which the actual model must predict outputs for data extracted from a third set of data. The plot with the overlapped results (are presented in Fig. 34.3. The R^2 values bigger than 0.99 indicate an accurate model.

34.3 Numerical Validation of the Method

To prove the reliability of the method, we perform simulations for several scenarios involving finite element models (FEM). The test specimen is a prismatic steel beam, with the dimensions 1000 × 20 × 5 mm, and has a mass density of 7850 kg/m^3. The material has the modulus of elasticity $E = 2{\cdot}10^{11}$ Pa.

The beam was divided into 20 segments of 50 mm each. First, we found the first six frequencies of the homogeneous beam. Afterward, we increased the mass of one or two segments at a time, after the nine scenarios presented in Table 34.1.

Table 34.1 Beam model with an altered mass on several segments—created scenarios

Case	1	2	3	4	5	6	7	8	9
Segment number	5	1	1	2	3	1	7	8	11
Mass ratio r_C	1.4	1.1	1.1	1.1	1.1	1.2	1.3	1.1	1.1
Segment number	–	9	10	10	10	4	9	9	18
Mass ratio r_C	–	1.1	1.1	1.1	1.1	1.3	1.4	1.3	1.1

Table 34.2 Predictions for the mass alteration

Case	1	2	3	4	5	6	7	8	9
Segment number	5	1	1	2	3	1	7	8	11
Mass ratio r_C	1.376	1.084	1.097	1.068	1.099	1.179	1.162	1.093	1.097
Segment number	–	9	10	10	10	4	9	9	18
Mass ratio r_C	–	1.039	1.079	1.122	1.102	1.322	1.304	1.309	1.072

For each scenario, we derived again the natural frequencies. With these frequencies, we calculated the RFSs for all nine scenarios and tested the ANN. Comparing lines one and three in Table 34.1 with the same lines in Table 34.2, we observe the values coincide. This means, the segments with altered masses are correctly identified. Now, comparing the lines two and four in Tables 34.1 and 34.2, we observe a quite good fit. This means we estimated the mass ratios well. The results are promising, and prove the ability of the method to identify inhomogeneities in beams.

The stiffnesses of all segments were estimated with the corresponding accuracy. Besides the fact that the altered segments were detected, for the segments with $r_C = 1$ a value of the modulus of elasticity close to one was found.

34.4 Conclusions

The proposed method for detecting segments with different masses in inhomogeneous beams, which involves frequency measurement and machine learning techniques, can be successfully used for real systems. This is valid even if the real case does not correspond to a case used in training. The method requires no information about the material characteristics, the single parameter being the mass ratio r_C. Thus, it can be applied to structures of any material. Moreover, because the method uses just the mode shapes as information about the dynamic behavior, we can apply the method for any boundary conditions of the structure when the right mode shape function is used.

For simulating real-world cases with the objective of testing the ANN, we have generated for a free-free beam nine scenarios considering that one or two segments have altered masses. We were able finding in all analyzed cases the segment or the segments with different masses. This was possible even for a small mass change (r_C = 1.1). Also, we observed that nearly located segments with altered mass (example cases 7 and 8) are correctly identified. Thus, we conclude that the proposed method is suitable to evaluate real structures. Further research will be made on more complex structures and for different types of materials, respectively for functionally graded materials.

References

1. M. Sadd, *Elasticity—Cap. 14 Nonhomogeneous Elasticity*. 4th edn. (Academic Press, 2021)
2. M. Autengruber, M. Lukacevic, C. Gröstlinger, J. Eberhardsteiner, J. Füssl, Numerical assessment of wood moisture content-based assignments to service classes in EC 5 and a prediction concept for moisture-induced stresses solely using relative humidity data. Eng. Struct. **245**, 112849 (2021)
3. R.K. Landolfo, L. Cascini, F. Portioli, Modeling of metal structure corrosion damage: a state of the art report. Sustainability **2**(7), 2163–2175 (2010)
4. G.R. Gillich, Z.I. Praisach, V. Iancu, H. Furdui, I. Negru, Natural frequency changes due to severe corrosion in metallic structures. Strojniški Vestn. J. Mech. Eng. **61**(12), 721–730 (2015)
5. H. Lohar, A. Mitra, S. Sahoo, Free vibration of initially deflected axially functionally graded non-uniform timoshenko beams on elastic foundation. Rom. J. Acoust. Vib. **15**(2), 75–89 (2019)
6. S. Qi, P. Zhang, G. Zhang, J. Ren, C. Yang, Precise integration solutions for the static and dynamic responses of axially graded solid beams. Eng. Anal. Boundary Elem. **147**, 152–163 (2023)
7. K. Kiani, Thermo-elasto-dynamic analysis of axially functionally graded non-uniform nanobeams with surface energy. Int. J. Eng. Sci. **106**, 57–76 (2016)
8. G.R. Gillich, Z.I. Praisach, M.A. Wahab, N. Gillich, I.C. Mituletu, C. Nitescu, Free vibration of a perfectly clamped-free beam with stepwise eccentric distributed masses. Shock. Vib.Vib. **2012**, 2086274 (2012)
9. N. Gillich, C. Tufisi, C. Sacarea, C.V. Rusu, G.-R. Gillich, Z.-I. Praisach, M. Ardeljan, Beam damage assessment using natural frequency shift and machine learning. Sensors **22**(3), 1118 (2022)

Chapter 35
Experimental Modal Analysis Involving the Use of Artificial Intelligence to Accurately Evaluate the Natural Frequencies of Engineering Structures

Daniela Georgiana Burtea, Rusalin-Lucian Paun, Nicoleta Gillich, and Gilbert-Rainer Gillich

Abstract There are several well-known methods that can be used to improve the estimate of the frequency for short signals such as the response of a structure to an impulsive excitation. However, the accuracy of the estimates is insufficient, thus more advanced methods are requested. We propose a machine learning approach to estimate the frequency values between the spectral lines of the Discrete Fourier Transform (DFT). The method involves Artificial intelligence and uses the data obtained by applying the DFT to a sinusoidal signal with known frequency and amplitude and varying lengths in time. The peak amplitudes in the spectrum have a similar shape to that of the sinusoidal signal generated. To determine this amplitude, we preprocess the signal by zero padding. Then we find the frequency and amplitude for three points on the main lobe of the DFT, namely the maximizer and its two neighbors. These points, derived for each signal length, represent the INPUT data used to train an ANN. The OUTPUT data consists of the frequency of the generated signal. Tests performed with generated signals with a different frequency and with data achieved from measurements confirm the validity of the method.

35.1 Introduction

Modal analysis is a procedure to estimate the natural frequencies of structures and its mode shapes, to get information about the dynamic behavior of the structure [1]. If it is made in the design phase, numerical simulations on virtual models or experiments on physical models or prototypes can predict the resonant frequencies [2]. If these are

D. G. Burtea · R.-L. Paun · N. Gillich (✉) · G.-R. Gillich
Doctoral School of Engineering, Babes-Bolyai University, 400084 Cluj-Napoca, Romania
e-mail: nicoleta.gillich@ubbcluj.ro

G.-R. Gillich
Department of Engineering Science, Babes-Bolyai University, 400084 Cluj-Napoca, Romania

N. Herisanu and V. Marinca (eds.), *Acoustics and Vibration of Mechanical Structures—AVMS-2023*, Springer Proceedings in Physics 302,
https://doi.org/10.1007/978-3-031-48087-4_35

in certain ranges, the functioning of the engineering system can be harmed, thus the structure is redesigned. Modal analysis is performed also in the exploitation phase when adequate monitoring ensures observing abnormal functioning in real time [3]. The abnormal functioning can be produced by wear of mobile elements, changes of the mechanical or physical properties of the materials or by structural damage.

There are two types of modal analysis which can be performed on real structures: (i) experimental operational analysis [4], when the excitation is produced artificially by an operator; (ii) operational modal analysis [5], when the excitation comes from the environmental or operational loads.

The first step in estimating the natural frequencies is to acquire the vibration response of the structure. Then, the signal in the time domain is converted into the frequency domain by calculating the Discrete Fourier Transform (DFT). For short signals, which include a small number of cycles, the frequency resolution is rough. Therefore, special algorithms are requested for an accurate frequency estimation. A review of procedures applied to improve the frequency readability is given in [6]. However, despite the application of advanced algorithms that increase time and computational costs, the frequencies are still not read with adequate precision.

In this paper we propose a modal analysis equipment and methodology that lead in obtaining accurate frequency estimates. It consists in a special excitation procedure and a signal analysis that involves the calculus of the DFT and artificial intelligence (AI).

35.2 Standard Vibration Signal Acquisition and Processing

The most common algorithm to transform the signal from the time domain in the frequency domain is calculating its DFT. This approach is simple and effective if the signal contains numerous vibration cycles, which means it is long or it contains high frequencies. In the opposite cases, the estimations are inaccurate.

Impulsive excitation is the preferred excitation method in modal analysis, especially for small and medium structures. A typical DFT spectrum is presented in Fig. 35.1.

Apparently, the frequencies are well defined, and these can be found with high accuracy. But, if we look at an enlarged part of the spectrum (see Fig. 35.2), we will see that the amplitudes are displayed next to the spectral lines that have the distance Δf between them. This distance depends on the signal length t_S. It is known that between the signal length and the frequency resolution is the relation $\Delta f = 1/t_S$. Since the spectral lines depend on t_S. Thus, neither the frequencies nor the amplitudes displayed are the real ones.

Depending on the signal time length, the biggest amplitude A_k is found somewhere on main lobe of the *sinc* function represented with a blue line in Fig. 35.2. It is a matter of chance to find this amplitude at the tip of the *sinc* function, which indicates the real frequency. Usually between the spectral line k containing the maximizer A_k and the interline position $k + \delta$ on which we find the tip of the *sinc* function. The

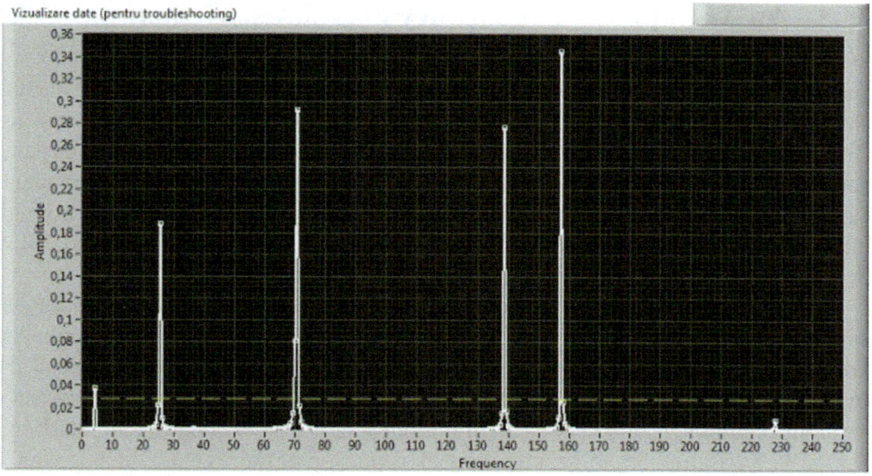

Fig. 35.1 The DFT spectrum for a vibration signal acquired after impulsive excitation

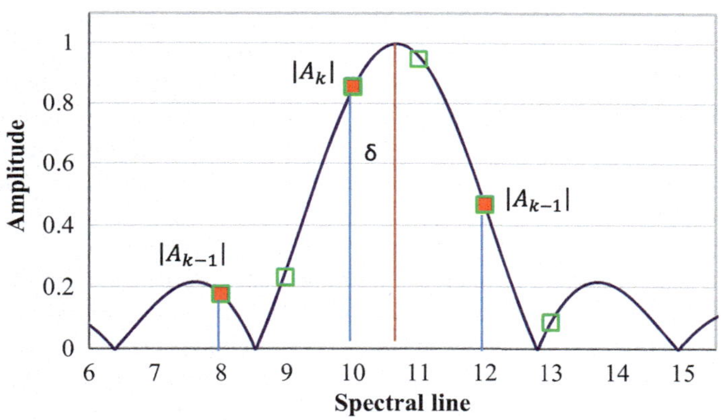

Fig. 35.2 The spectral lines, marked with blue, found in the standard DFT

interpolation methods try to find the correction term δ and, in consequence, the real frequency.

One of the simplest methods to increase the number of spectral lines is to add samples with zero amplitude to the original signal—the so-called zero-padding [7]. This ensures an increased number of spectral lines. If doubling the signal length t_S, the number of spectral lined becomes double. It is possible to get a smaller δ, hence a better frequency estimation. As mentioned, interpolation methods do not always succeed.

35.3 An AI-Supported Frequency Estimation Method

To improve the frequency estimates, we focus on two aspects: (i) the first regards obtaining a suitable vibration signa, (ii) the second regards the signal post-processing. Obtaining a signal with a relevant amplitude of one harmonic component, the targeted one, is important to avoid digital noise. If two closely located harmonic components have comparable amplitudes, the lateral lobes of one component produce a nonuniform increase of the amplitudes of the main lobe of the second component [8]. Thus, frequency estimation becomes even harder.

A method to avoid this effect is to apply a controlled excitation with a load that has the frequency close to the resonant frequency of the targeted harmonic component. This excitation can be done with acoustic pressure [9]. In this way we control the energy transfer to the structure. We will obtain a big amplitude for the targeted harmonic component and very small amplitudes for the other components, as shown in Fig. 35.3a. In Fig. 35.3b we present a short portion of the signal, which is almost sinusoidal and allows an effective frequency analysis. The DFT spectrum is presented in Fig. 35.4.

The second step is to improve the results obtained by involving the DFT. We observed the errors in frequency estimation are repetitive in respect to the signal length. This encouraged us to apply an AI technique to derive δ. To this aim, we generated sinusoidal signals with the known frequency of $f_r = 5$ Hz and amplitude $A = 1$ and performed zero-padding to double the signal length. This ensured us that we will find on the main lobe the maximizer A_k and its two neighbors A_{k-1} and A_{k+1}, respectively. In addition, we calculate the resulted frequency f_o and the resulted correction coefficient δ.

The original signal has 2149 samples, and the sampling frequency is $r = 1000$ Hz. Now we crop the signal by extracting two samples and calculate again the amplitudes and the frequency; the process is repeated until A_k is not anymore on the main lobe. This data, shown in Table 35.1, constitutes the Input data and the Output data for a ANN. The ANN contains an Input layer with three neurons (the amplitudes), three

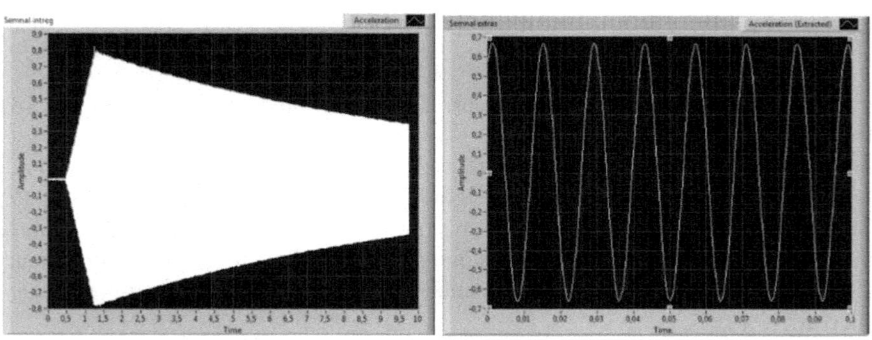

Fig. 35.3 Response signal achieved when a controlled-energy excitation is applied

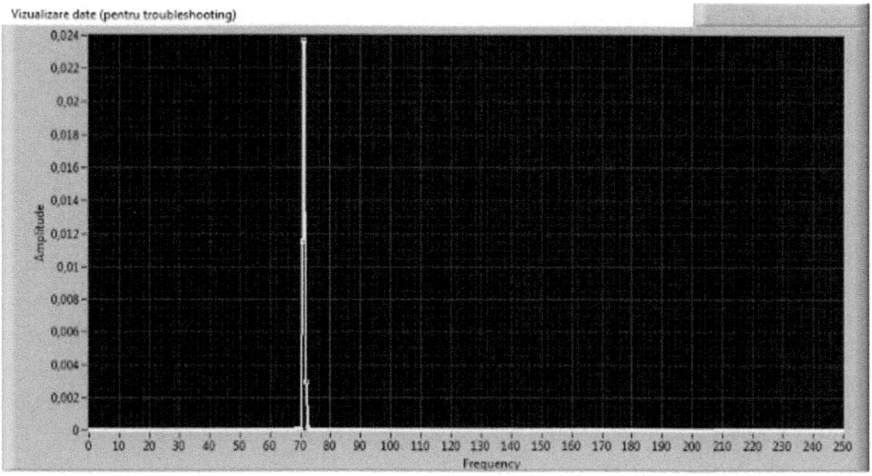

Fig. 35.4 The DFT spectrum for a vibration signal achieved for a controlled-energy excitation

Table 35.1 Examples of input and output data

Nr. of samples N	Input			Output δ
	$A_{k\text{-}1}$	$A_{k\text{-}1}$	$A_{k\text{-}1}$	
2149	0.34056317	1	0.99144261	1.485
2147	0.35362327	1	0.97444708	1.465
2145	0.36667106	1	0.95761995	1.445
2101	0.64594329	1	0.62834404	1.005
2099	0.65834327	1	0.61533858	0.985
2055	0.958326898	1	0.361511871	0.545
2053	0.97470059	1	0.350551739	0.525
2051	0.991461997	1	0.339541093	0.505

Hidden layers with ten neurons each, and one Output with one neuron (the correction coefficient). We consider:

- A Feed-forward backpropagation network
- The Hyperbolic Tangent as activation function
- The Mean Square Error for performance validation
- The Bayesian Regularization as training algorithm.

The data in Table 35.1 is normalized, the amplitudes in the DFT being divided to the maximum value of the series. This is made to generalize the ANN model, which is so able to predict the frequencies for signals with any amplitude. However, the amplitudes of the analyzed signals must be normalized as well.

Table 35.2 Test data and frequency results

Samples N	Input			Output δ_E	Estimated frequency f_E (Hz)
	$A_{k\text{-}1}$	$A_{k\text{-}1}$	$A_{k\text{-}1}$		
2043	0.37993	1	0.9409282	1.4289	5.000954
2031	0.45791	1	0.8440793	1.3056	5.000147
2017	0.54711	1	0.7382367	1.1621	4.999280
2005	0.62197	1	0.6542038	1.0437	4.999675
1991	0.70887	1	0.5642647	0.9054	5.000100
1979	0.78545	1	0.4934039	0.7864	5.000353
1965	0.88174	1	0.4157072	0.6417	4.999160
1953	0.97470	1	0.3505204	0.5397	5.003764

35.4 Results and Discussion

We tested the ANN against data calculated for a signal with the same sampling frequency $r = 1000$ and the amplitude $A = 2$. Table 35.2 present the number of samples, the normalized amplitudes in the DFT and the estimated correction coefficient δ_E. The latter parameter is the output of the trained network. With this parameter, and with the frequency displayed at the spectral line k, we find the frequency f_E. The equation used to calculate the estimated frequency is:

$$f_E = k\Delta f + \delta_E \tag{35.1}$$

One can observe that, for all cases, the error in estimating the frequencies is less than 0.003 Hz. These results ensure a higher accuracy compared with that achieved with other methods [6], which qualify the method as extremely reliable and useful in industrial applications.

35.5 Conclusions

The method proposed in this paper presumes a controlled excitation with an equipment ensuring a sinusoidal non-contact load. The resulted structural response is analyzed in the frequency domain; the DFT data is used as input in an ANN which can identify the real frequency at an inter-line position.

The tests performed to prove the reliability of the method showed that the results are highly accurate, even if the test signal has a different amplitude than the training signal. Being simple and reliable, the method can be used to perform modal analysis for lightweight metallic or composite structures that can be excited with sound pressure.

References

1. D.J. Ewins, *Modal Testing: Theory, Practice and Application*, 2nd edn. (Wiley, London, 2000)
2. Theoretical basis of modal analysis. Am. J. Mech. Eng. **1**(7), 173–117 (2013)
3. L. Zhang, Y. Wang, S. Zhang, G. Hao, The fault diagnosis of machine based on modal analysis, in *2009 International Conference on Measuring Technology and Mechatronics Automation* (IEEE Explore, Zhangjiajie, China, 2009), pp. 738–741
4. N.M.M. Maia, J.M.M. Silva, Modal analysis identification techniques. Phil. Trans. R. Soc. A **359**, 29–40 (2001)
5. C. Wang, J. Wang, T. Zhang, Operational modal analysis for slow linear time-varying structures based on moving window second order blind identification. Signal. Process. **133**(C), 169–186 (2017)
6. A.A. Minda, C.I. Barbinita, G.R. Gillich, A review of interpolation methods used for frequency estimation. Rom. J. Acoust. Vib. **17**(1), 21–26 (2020)
7. Y.A. Al-Jawhar, K. Nidzam Ramli, M.A. Taher, N.S.M. Shah, L. Audah, M.S. Ahmed, Zero-padding techniques in OFDM systems. Int. J. Electr. Eng. Inf. **10**(4), 704–725 (2018)
8. M. Tufoi, G.R. Gillich, I.C. Mituletu, C. Hatiegan, An analysis of the dynamic behavior of rectangular plates from a damage detection approach. Rom. J. Acoust. Vib. **12**(2), 146–150 (2015)
9. I.C. Mituletu, G.R. Gillich, N.M.M. Maia, A method for an accurate estimation of natural frequencies using swept-sine acoustic excitation. Mech. Syst. Signal Process. **116**, 693–709 (2019)

Chapter 36
Static and Dynamic Behavior of Composite Sandwich Bars with Crushed Corn Cob Core

Alexandru Ioan Rădoi, Marius Marinel Stănescu, Dumitru Bolcu, Cosmin Mihai Mirițoiu, Claudiu Nicolicescu, and Alexandru Bolcu

Abstract For this research we have built some composite sandwich bars in this way: the core is made from crushed corn cob and the samples faces are made from natural fabrics: cotton, flax and silk. The connection of these elements (core and the faces) is made by using epoxy resin Resoltech 1050 with its hardener Resoltech 1058 s. All the samples are molded at room temperature (20–23 °C) and there was waited 48 h for the resin polymerization with its hardener. From the plates obtained there were cut samples to be tested by three methods: tensile test and bending on universal testing machines (for the tensile test there was used an Instron machine and for the bending there was used an LBG Testing Equipment machine) and free vibrations by clamping the bars at one end and leaving the other end free. From the static tests there were determined the Young modulus values, breaking strength (both tensile and bending), elongation at break and maximum force. From the free vibrations, by applying Euler–Bernoulli theory, the next dynamic parameters were obtained: the eingenfrequency for the first eigenmode, damping factor, dynamic Young modulus, dynamic stiffness and loss factor. The general conclusion obtained from all the tests was that the samples reinforced with cotton had the best mechanical properties no matter the applied test (static or dynamic).

A. I. Rădoi · M. M. Stănescu · D. Bolcu · C. M. Mirițoiu (✉) · A. Bolcu
Faculty of Mechanics, University of Craiova, Calea București Street, No. 107, Craiova, Romania
e-mail: miritoiucosmin@yahoo.com

C. Nicolicescu
Faculty of Mechanics, University of Craiova, 1 Călugăreni Street, 220037 Drobeta-Turnu Severin, Romania

N. Herisanu and V. Marinca (eds.), *Acoustics and Vibration of Mechanical Structures—AVMS-2023*, Springer Proceedings in Physics 302,
https://doi.org/10.1007/978-3-031-48087-4_36

36.1 Introduction

The usage of agricultural waste in composite materials manufacturing has become more and more frequent in the latest researches. The tendency is to find other engineering applications for these materials and to avoid burning them in open field. One of the agricultural wastes that is obtained after the corn tillage is the corn cob. For example, according to [1], in Thailand, there is produced 4.6 million of tons/year resulting around 1.7 million of tons/year of corn cob. In the past years, it was used to heat up the houses in stoves. But after the appearance of gas and electricity plants, the usage of corn cob for home heating has been reduced.

In this research, there is tried to give another application for the corn cob: to be used as a reinforcement in building composite materials with usage in furniture industry.

In previous researches, there were determined the mechanical properties of the corn cob. For example, firsts tests of quasi-static radial-compression tests were made in [2]. There was found that the corn variety and harvest date affect the cob modulus of toughness. It was also found that the elastic modulus was affected by harvest date but not by variety.

In other researches the corn cob was used as a component part in building cement, mortars, clay brick or lightweight concrete [3–5]. There are known other researches where there are built composites with crushed corn cob. For example, in [6], there are built composites with lignosulphonate matrix and crushed corn cob as reinforcement. There were used corn cob particles with the size between 0.009 and 0.18 cm and all the created samples were tested to compression. Some mechanical characteristics were obtained: breaking strength around 19 MPa and the Young modulus around 310 MPa. In [7], the crushed corn cob and Kraft lignin were combined with polylactic acid matrix. The best mechanical properties at flexural strength were obtained for the samples with 12% corn cob and Kraft lignin. In [8] there are manufactured composites with matrix from polylactic acid (PLA) reinforced with crushed corn cob (CC) with the percentage between 10 un to 40%. The mechanical, thermal and morphological properties were studied for these samples. Another study of PLA composites filled with CC was also made in [9]. There were studied the tensile properties and microstructures for samples with percentages between 5 up to 60% CC particles. In [10, 11] there are studied samples with CC for water absorption and there were obtained dimensional variations of the matrix, formation of micro-cracks and low mechanical strength due to the CC hydrophilic nature.

For this research, some sandwich composites were built in this way: the core is made with crushed corn cob and the faces with natural fabrics (cotton, flax, silk). The connection between these elements is made by using epoxy resin Resoltech 1050 with its hardener Resoltech 1058 s.

36.2 Materials and Methods

As stated in the above section, for this research some composite sandwich plates were moulded, by using the lay-on-hands technique, in this way: the core is made from crushed corn cob and the faces are made with natural fibers (cotton, silk and flax). There were used 4 layers of natural fibers for each face. The connection between the faces and the core is made by using epoxy resin Resoltech 1050 type with Resoltech 1058 s as hardener. There was fulfilled the producer's recommendation: 100 parts resin with 33 parts hardener, by weight. There was waited for 48 h for the resin polymerization to be completed.

From the obtained plates, there were cut 15 samples for each type of composite to be tested for tensile according to the ASTM D3039 standard [12]. The samples had the dimensions 250 × 25 × 12 mm. For the bending test, from the samples plates, another 15 samples were cut. The tests was made according to the ASTM D790-17 standard [13]. The samples had the dimensions 150 × 12.7 × 12 mm. For the free vibrations experiment, there were used samples with the same dimensions like the ones used for tensile test. A general view with a sample for each type of studied material is given in Figs. 36.1, 36.2 and 36.3.

For the tensile test there was used an universal testing machine Instron 1000 HDX with the maximum force of 1000 kN. For the bending test there was used an LBG universal testing machine with the maximum force of 100 kN. For the free vibrations, there was used an accelerometer Bruel and Kjaer for the recordings. The accelerometer is connected to a signal conditioner Nexus type which was also connected to a data acquisition system Spider 8. The acquisition system was connected to a laptop, and the interface between them was made by using the software Catman Easy.

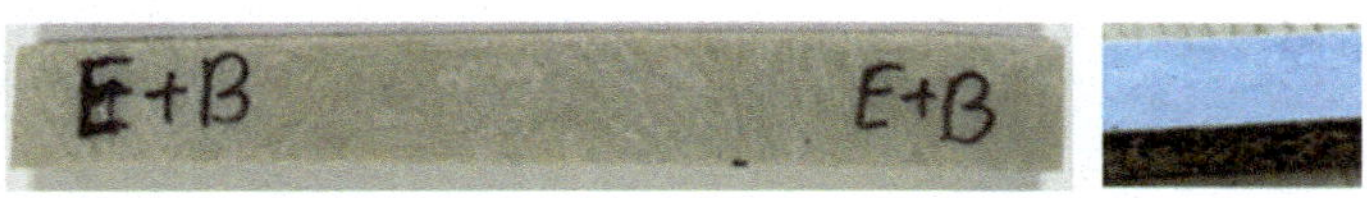

Fig. 36.1 An example with the sample reinforced with cotton, core made from crushed corn cobs and the matrix from epoxy resin

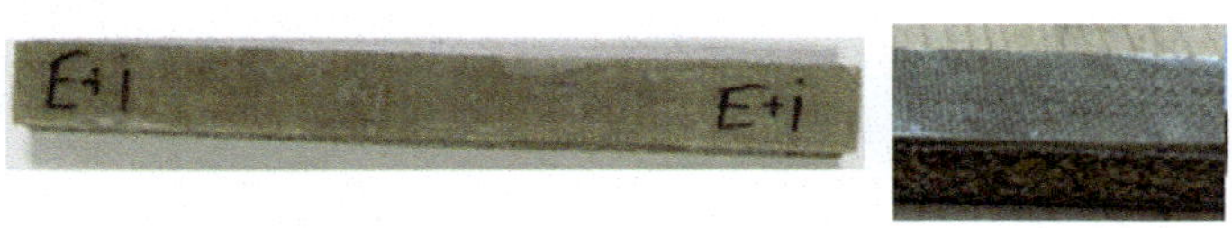

Fig. 36.2 An example with the sample reinforced with flax, core made from crushed corn cobs and the matrix from epoxy resin

Fig. 36.3 An example with the sample reinforced with silk, core made from crushed corn cobs and the matrix from epoxy resin

36.3 Results and Discussions

The results obtained from the tensile test (modulus of elasticity, breaking strength and elongation at break) are written in Table 36.1 and the ones from the bending test (maximum force at bending, breaking strength at bending and maximum traverse stroke) are written in Table 36.2.

Some stress–strain graphics, obtained from the tensile test, for representative samples are presented in Figs. 36.4, 36.5 and 36.6. Some force-traverse stroke graphics, obtained from the bending test, for representative samples are presented in Figs. 36.4, 36.5 and 36.6. There is named as representative sample, the specimen that has the mechanical properties, obtained from the destructive tests, close to the mean values written in Tables 36.1 and 36.2.

By analyzing the results from the Tables 36.1 and 36.2 and Figs. 36.4, 36.5, 36.6, 36.7, 36.8 and 36.9, it can be seen that the samples reinforced with cotton have the best mechanical properties when compared to the ones reinforced with flax and silk. The free vibrations were studied by using the next experimental setup: the bar was embedded at on end with the other one being free. On the free end an accelerometer was mounted and a force was applied to the bar. Then, the force was removed and the bar was let to freely vibrate. From the free vibrations there could be determined the

Table 36.1 All the results obtained from the tensile test

Sample	Modulus of elasticity E (MPa)	Breaking strength at tensile (MPa)	Elongation at break A (%)
Cotton	6650 ± 117	84 ± 1.21	8.08 ± 0.92
Flax	6290 ± 128	82 ± 2.29	5.5 ± 0.44
Silk	3560 ± 128	71 ± 1.82	10.4 ± 2.06

Table 36.2 All the results obtained from the bending test

Sample	Maximum force at bending F_{max} (kN)	Breaking strength at bending (MPa)	Traverse stroke (mm)
Cotton	13.33 ± 1.22	46.2 ± 1.08	10.76 ± 1.08
Flax	11.4 ± 0.92	39.8 ± 1.44	7.1 ± 0.32
Silk	9.25 ± 1.25	32 ± 1.48	12 ± 1.07

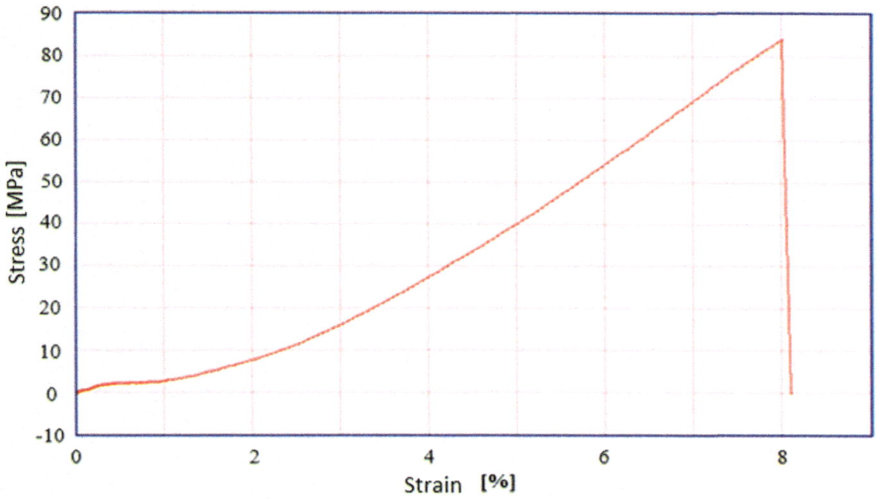

Fig. 36.4 Stress–strain graphic for a sample reinforced with cotton, obtained from the tensile test

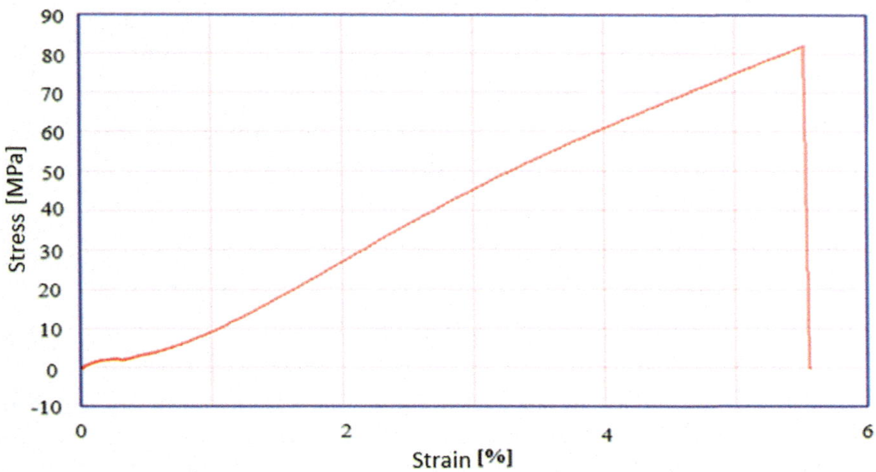

Fig. 36.5 Stress–strain graphic for a sample reinforced with flax, obtained from the tensile test

next mechanical parameters: the eigenfrequency of the first eigenmode (Eq. 36.1), the damping factor per unit mass (Eq. 36.2), the damping factor per unit length (Eq. 36.3), dynamic stiffness (Eq. 36.4), dynamic Young modulus (Eq. 36.5) and the loss factor (Eq. 36.6).

$$\nu = \mathrm{T}^{-1} \tag{36.1}$$

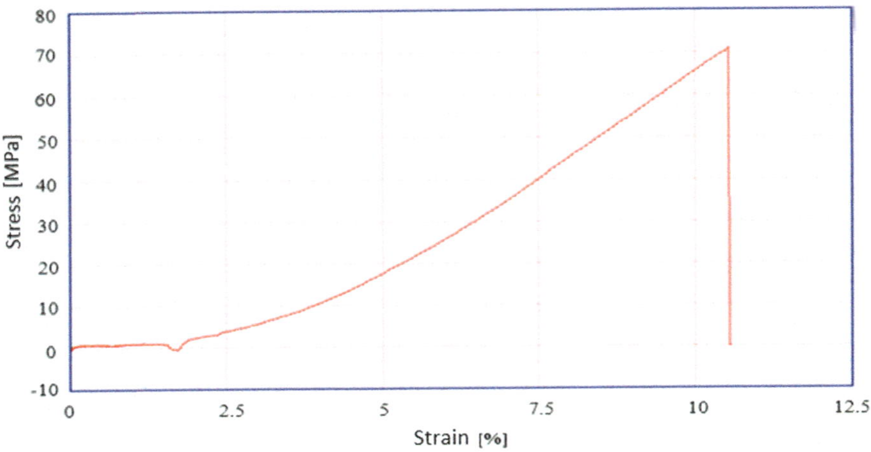

Fig. 36.6 Stress–strain graphic for a sample reinforced with silk, obtained from the tensile test

$$\mu = \Delta t^{-1} \cdot \ln \frac{v_1}{v_2} \tag{36.2}$$

$$C = 2 \cdot \mu \cdot \rho \cdot H \cdot B \tag{36.3}$$

$$EI = 39.5 \cdot \rho \cdot B \cdot H \cdot \left(\nu \cdot \frac{L^2}{\vartheta^2}\right)^2 \tag{36.4}$$

$$E = \rho \cdot \left(6.2 \cdot \nu \cdot \frac{L^2}{H}\right)^2 \tag{36.5}$$

$$\eta = 0.32 \cdot \frac{\mu}{\nu} \tag{36.6}$$

Equations (36.1–36.6) are obtained from the Euler–Bernoulli theory and were used in previous studies to determine the dynamic behavior of composites (see for example [14]). In the formulas (36.1–36.6) there was marked with η the loss factor, ν—the frequency of the first eigenmode, μ—the damping factor per unit mass, ρ—the material density, C—the damping factor per unit length, B, H the samples width and thickness, ϑ a constant equal cu 1.875 and it is obtained because the bar is clamped at one end and with the other end left free, Δt the time gap from two maximum values, v_1, v_2 the maximum values in the diagram chosen to determine the damping factor. The free vibrations were recorded for several samples free lengths: 100, 120, 140, 160 and 180 mm. In Figs. 36.10, 36.11 and 36.12 there are given the experimental recordings and the determination of the damping factor per unit mass and eigenfrequency, for each sample. The values correspond to the free length of 140 mm.

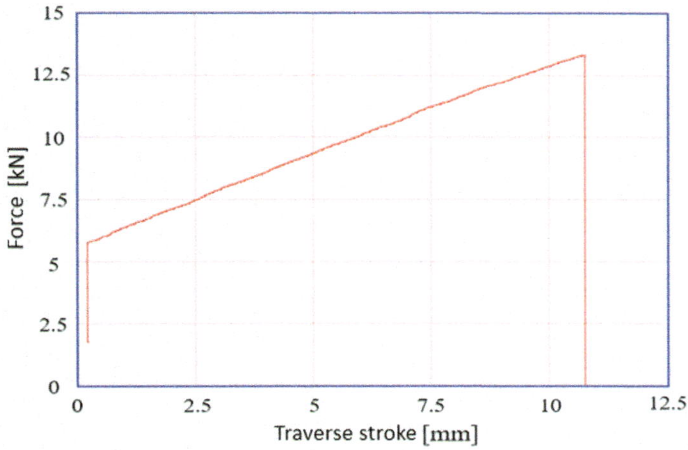

Fig. 36.7 Stress–strain graphic for a sample reinforced with cotton, obtained from the bending test

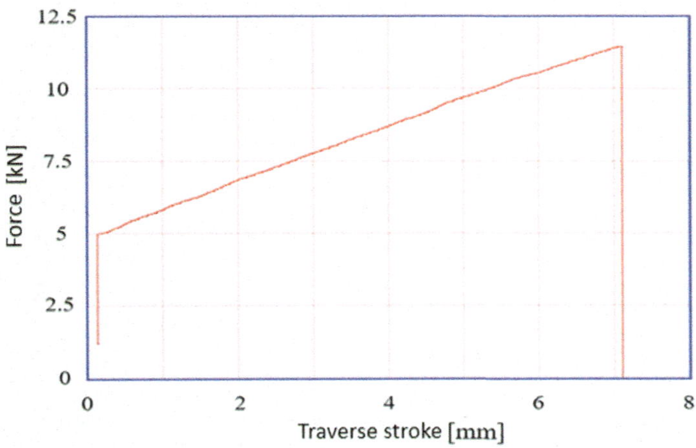

Fig. 36.8 Stress–strain graphic for a sample reinforced with flax, obtained from the bending test

From the Table 36.3 results, it can be seen that the highest dynamic Young modulus and dynamic stiffness is obtained for the samples reinforced with cotton. There can be also be seen that, no matter the reinforcer type (cotton, silk or flax), the damping factor per unit mass and the eigenfrequency decrease with the bar's free length increase. The dependence between these parameters and the samples free length is almost exponentially.

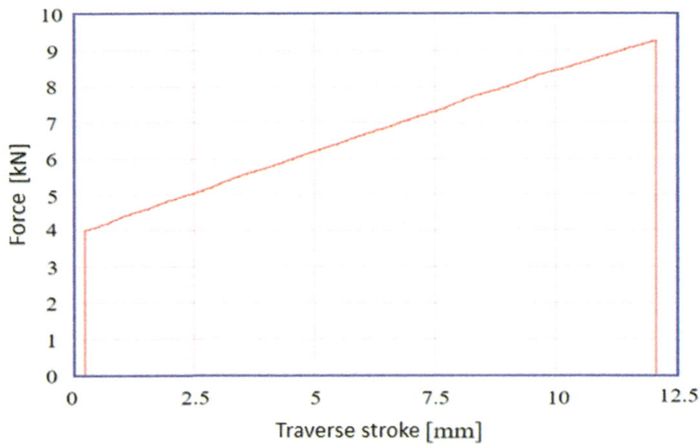

Fig. 36.9 Stress–strain graphic for a sample reinforced with silk, obtained from the bending test

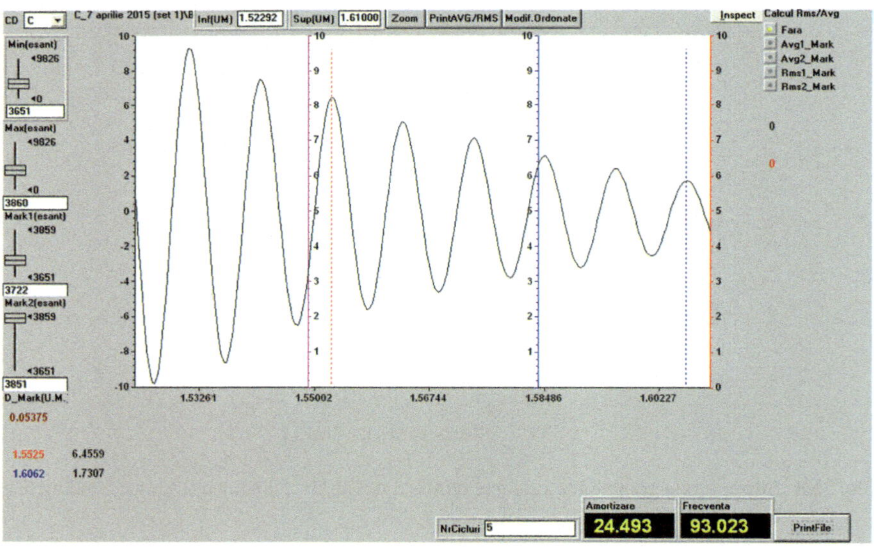

Fig. 36.10 The damping factor and eigenfrequency, for the samples reinforced with flax (140 mm—free length)

36.4 Conclusions

For this research, some composite sandwich bars with the core made from crushed corn cob and the faces from natural fabrics (cotton, silk, flax) are build. The connection between all these components is made with epoxy resin Resoltech 1050 with

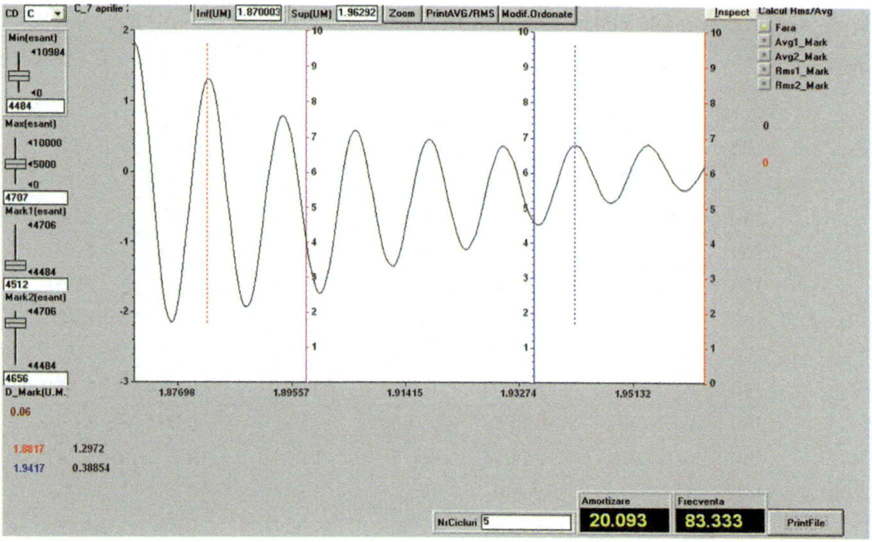

Fig. 36.11 The damping factor and eigenfrequency, for the samples reinforced with silk (140 mm—free length)

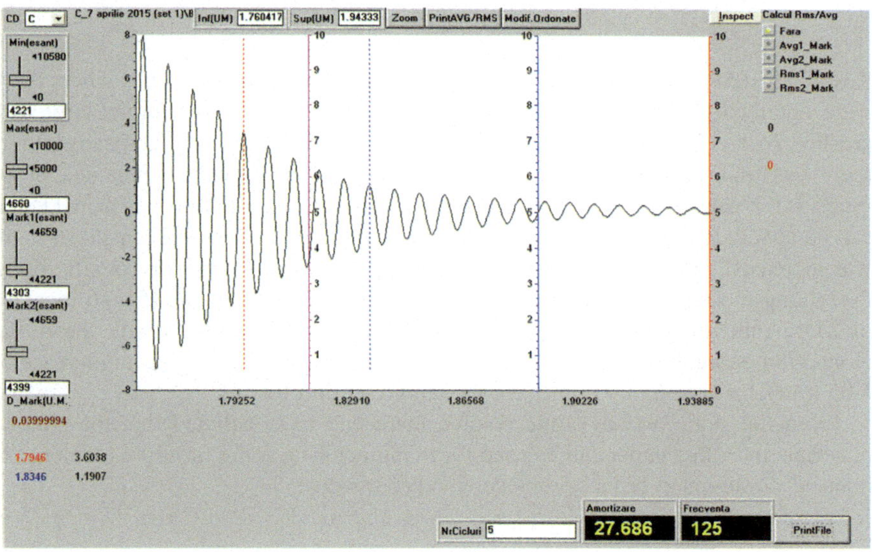

Fig. 36.12 The damping factor and eigenfrequency, for the samples reinforced with cotton (140 mm—free length)

Table 36.3 All the results obtained from the free vibrations

Type	Free length [mm]	μ [Ns/m/kg]	N [1/s]	E [MPa]	EI [Nm^2]	η	C [Ns/m/m]
Silk	100	40.02	161.7	789.7	1.546	0.079	14.80
	120	25.37	112.2	788.1	1.543	0.072	9.387
	140	20.72	83.33	805	1.576	0.079	7.666
	160	17.29	63.18	789.4	1.546	0.087	6.397
	180	14.49	49.24	768	1.504	0.094	5.361
Flax	100	47.25	181.4	993.9	2.052	0.083	18.42
	120	29.96	125.9	992.1	2.048	0.076	11.68
	140	24.49	93.02	1003	2.07	0.084	9.551
	160	20.63	70.88	993.5	2.051	0.093	8.046
	180	17.30	55.24	966.6	1.995	0.1	6.747
Cotton	100	52.41	243.9	1795	3.801	0.068	20.96
	120	37.09	169.5	1798	3.807	0.07	14.83
	140	27.68	125.0	1811	3.835	0.07	11.07
	160	20.85	95.61	1808	3.827	0.069	8.34
	180	13.52	75.37	1799	3.81	0.057	5.408

Resoltech 1058 s as hardener. This is a continuation of previous studies where there were obtained low mechanical values for composite samples reinforced only with crushed corn cob (see for example [15]). So, in order to remove this disadvantage, there was important to reinforce the bars with fibers to increase the mechanical properties. If the values obtained from this study are compared with the ones from [15], it can be concluded that, by inserting the faces with natural fibers, the tensile strength was increased 14 times for cotton, 13.6 times for flax and 11.8 times for silk. From the Young modulus point of view, it was obtained, compared to [15], an increase of 34.45 times for cotton, 32.6 times for flax and 18.45 times for silk. From the elongation at break point of view, it was obtained, compared to [15], an increase of 5.05 times for cotton, 3.43 times for flax and 6.5 times for silk.

From the static and dynamic results, there can be concluded that the studied materials from this paper can be used for manufacturing some furniture elements or house decorations in order to replace wood or sawdust.

References

1. S. Lawanwadeekul, A. Srisuwan, N. Phonphuak, P. Chindaprasirt, Enhancement of porosity and strength of clay brick fired at reduced temperature with the aid of corn cob and waste glass. Constr. Build. Mater. **369**, 1–8 (2023)

2. U.G.N. Anazodo, E.R. Norris, Effects of genetic and cultural practices on the mechanical properties of corn cobs. J. Agric. Engng. Res. **26**, 97–107 (1981)
3. D.A. Adesanya, A.A. Raheem, A study of the workability and compressive strength characteristics of corn cob ash blended cement concrete. Constr. Build. Mater. **23**, 311–317 (2009)
4. D.A. Adesanya, A.A. Raheem, A study of the permeability and acid attack of corn cob ash blended cements. Constr. Build. Mater. **24**, 403–409 (2010)
5. D.A. Adesanya, A.A. Raheem, Development of corn cob ash blended cement. Constr. Build. Mater. **23**, 347–352 (2009)
6. A. Tribot, C. Delattre, E. Badel, C.-G. Dussap, P. Michaud, H. de Baynast, Design of experiments for bio-based composites with lignosulfonates matrix and corn cob fibers. Ind. Crops Prod. **123**, 539–545 (2018)
7. H. de Baynast, A. Tribot, B. Niez, F. Audonnet, E. Badel, G. Cesar, C.-G. Dussap, E. Gastaldi, L. Massacrier, P. Michaud, C. Delattre, Effects of Kraft lignin and corn cob agro-residue on the properties of injected-moulded biocomposites. Ind. Crops Prod. **177**, 1–12 (2022)
8. K.S. Chun, S. Husseinsyah, Polylactic acid/corn cob eco-composites. J. Thermoplast. Compos. Mater. **27**, 1667–1678 (2014)
9. G. Faludi, G. Dora, K. Renner, J. Moczo, B. Pukanszky, Biocomposite from polylactic acid and lignocellulosic fibers: structure-property correlations. Carbohydr. Polym. **92**, 1767–1775 (2013)
10. Z.N. Azwa, B.F. Yousif, A.C. Manalo, W. Karunasena, A review on the degradability of polymeric composites based on natural fibres. Mater. Des. **47**, 424–442 (2013)
11. N.M. Stark, D.J. Gardner, Outdoor durability of wood-polymer composites, in *Wood-Polymer Composites* (Elsevier Ltd., 2008), pp. 142–165
12. ASTM D3039, Standard Test Method for Tensile Properties of Polymer Matrix Composite Materials, D3039/D3039M Standard Test Method for Tensile Properties of Polymer Matrix Composite Materials (astm.org). Last accessed 12 May 2023
13. ASTM D790-17, Standard Test Methods for Flexural Properties of Unreinforced and Reinforced Plastics and Electrical Insulating Materials, D790 Standard Test Methods for Flexural Properties of Unreinforced and Reinforced Plastics and Electrical Insulating Materials (astm.org). Last accessed 12 May 2023
14. C.M. Miriţoiu, M.M. Stănescu, D. Bolcu, Researches regarding the mechanical properties of a new hybrid vegetal resin. Materiale Plastice **57**, 37–45 (2020)
15. C.M. Miriţoiu, M.M. Stănescu, D. Bolcu, A.I. Rădoi, C. Nicolicescu, A. Diniţă, A study about some mechanical properties for composites reinforced with corn cob powder. Materiale Plastice **58**, 133–140 (2021)

Chapter 37
Resonance Frequency of ASA 3D Printed Models. Simulation Versus Physical Testing

Karoly Menyhardt, Remus Stefan Maruta, and Ramona Nagy

Abstract Focusing on the dynamic characteristics like resonant frequency, damping, and mode shapes impact the object's behavior under vibrations, this paper presents a comparison for resonance frequency of Acrylonitrile Styrene Acrylate (ASA) 3D printed cantilever beam based on results gathered from analytical calculus, finite element analysis and physical testing.

37.1 Introduction

Acrylonitrile Styrene Acrylate (ASA) is a thermoplastic polymer that is a copolymer of acrylonitrile, styrene, and acrylate. It is a relatively new material that is increasingly being used in various applications due to its desirable physical properties that overcome those of Acrylonitrile Butadiene Styrene (ABS) [1, 2]. UV Resistance, weather resistance, improved chemical resistance, reduced warping and improved printability, surface finish and appearance make ASA an appealing choice of material in 3D printing compared to ABS.

Most 3D printed parts have static applications, that are characterized by the object's mechanical properties, structural integrity, and load-bearing capabilities. Strength, stiffness, Young's modulus, fracture toughness, hardness, dimensional accuracy, load-bearing capacity—collectively determine the structural performance and suitability of a 3D printed object for various applications. From a dynamic point of view, the parameters are related to the object's mechanical properties and structural behavior: resonant frequency, damping, stiffness, mode shapes.

K. Menyhardt (✉) · R. S. Maruta · R. Nagy
Politehnica University Timisoara, Timisoara Timis, Romania
e-mail: karoly.menyhartdt@upt.ro

R. S. Maruta
e-mail: remus.maruta@upt.ro

R. Nagy
e-mail: ramona.nagy@upt.ro

N. Herisanu and V. Marinca (eds.), *Acoustics and Vibration of Mechanical Structures—AVMS-2023*, Springer Proceedings in Physics 302,
https://doi.org/10.1007/978-3-031-48087-4_37

Analytical calculations of resonant frequencies play a vital role in the design, while Finite Element Analysis (FEA) evaluation offer an insight into optimization of structures and components, helping to visualize the mode shapes in a time and cost-efficient way.

Nonetheless, physical testing is the most important tool in validating analytical or FEA analysis as it provides empirical evidence, validates models, identifies uncertainties, calibrates models, validates designs, builds confidence, and mitigates risks. It enhances the understanding and trust in the analytical and numerical tools used in engineering analysis, leading to more accurate and reliable predictions.

Thus, it's important to validate the simulation results against physical testing whenever possible, to ensure that the simulation accurately represents the behavior of the actual system [3].

In this paper we present the differences in resonant frequencies obtained through physical testing and simulation using FEM for a cantilever beam model created through 3D printing from ASA filament.

37.2 Material Characterization and Geometry

ASA has excellent weather resistance, which makes it ideal for outdoor applications. It can withstand exposure to UV radiation, heat, and moisture without degrading or losing its physical properties, has high impact resistance, which makes it ideal for applications where durability is important (Table 37.1). Relative ease of processing makes ASA an ideal candidate to be molded, extruded or thermoformed [4, 5].

The dimensions of a beam that are best suited for testing its resonance frequencies depend on various factors such as the material properties, the intended use of the beam, and the available testing equipment.

In general, the length of the beam is the most critical parameter for determining its resonance frequencies. The longer the bar, the lower its resonance frequencies. Therefore, if the intended use of the beam involves low-frequency vibrations, a longer beam should be used for testing. The cross-sectional dimensions of the beam also play a role in determining its resonance frequencies. Generally, beams with larger cross-sectional dimensions have lower resonance frequencies compared to beams with smaller cross-sectional dimensions.

Table 37.1 ASA versus ABS

	Tensile strength [MPa]	Flexural strength [MPa]	Impact strength [kJ/m^2]	Heat deflection [°C]	Density [g/cm^3]	UV resistance	Chemical resistance
ASA	50–75	70–100	10–25	80–100	1.05–1.1	Excellent	Good
ABS	40–50	50–80	15–35	70–95	1.03–1.07	Moderate	Moderate

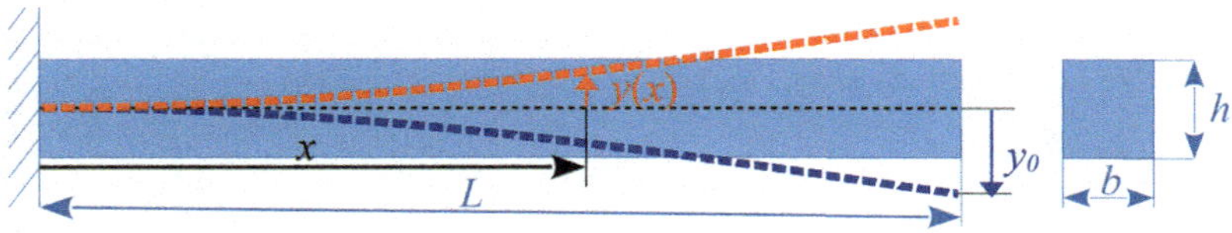

Fig. 37.1 Cantilever beam

In order to make the comparison, a well-established cantilever beam model was considered having L = 120 mm, b = 10 mm and h = 10 mm (Fig. 37.1).

37.3 Analytical Approach

Analytical calculations allow engineers and designers to optimize the design of a structure or component to avoid resonance at critical frequencies [6]. By predicting resonant frequencies analytically, potential issues such as excessive vibrations, fatigue failure, or reduced performance can be identified and addressed early in the design stage. The beam vibrates with maximum amplitude at its resonance frequency, and this vibration is characterized by the specific mode of vibration associated with that frequency.

Resonant frequency depends on the object's mass, stiffness, geometry and boundary conditions. Higher resonant frequencies indicate stiffer objects with faster vibration response. Stiffness is a measure of how resistant an object is to deformation under an applied force. It determines the object's response to dynamic loads and influences its natural frequencies.

If the bar is clamped at one end and free at the other end—cantilever beam, the bar will vibrate in its second mode of vibration, where there is one node (zero displacement) at the clamped end and one antinode (maximum displacement) at the free end [7, 8].

Assuming that the lateral displacement is harmonic [9], with the frequency $f = \frac{\omega}{2\pi}$, , synchronous at all points along the bar (Fig. 37.1):

$$y(x,t) = y_0(x)\cos\omega t,\ y_0(x) = y_0 \cdot \left(1 - \cos\frac{\pi x}{2L}\right) \tag{37.1}$$

where ω is the angular frequency and *L* is the length of the beam.

It is known that potential energy *Ep* is:

$$E_p = \int_V \frac{\sigma^2}{2E} dV \tag{37.2}$$

where E is the elastic modulus of the material (Young), σ—normal stress at bending given by Navier's relation is:

$$\sigma = \frac{M_{iz}}{I_z} \cdot y \tag{37.3}$$

and the elementary volume of the beam is dV.

The potential energy can be rewritten as

$$E_p = \frac{1}{2} \cdot \int_L EI_z \left(\frac{\partial^2 y}{\partial x^2} \right)^2 \cdot dx \tag{37.4}$$

where I_z is the geometric moment of inertia of the beam cross-section and can be calculated for a rectangular shape as:

$$I_z = \frac{bh^3}{12} \tag{37.5}$$

The maximum value is obtained for the maximum amplitude of the fixed beam (y_0), of constant rigidity EI_z.

$$E_{p_{max}} = \frac{1}{2} \cdot \int_L EI_z \left(\frac{\partial^2 y_0}{\partial x^2} \right)^2 \cdot dx = \frac{1}{2} \cdot EI_z \cdot \int_L \left(\frac{\partial^2 y_0}{\partial x^2} \right)^2 \cdot dx = \frac{1}{2} EI_z y_0^2 L \cdot \frac{\pi^4}{32L^4} \tag{37.6}$$

The kinetic energy E_c knowing the speed for a point on the beam is $v = \frac{\partial y}{\partial t}$ can be written:

$$E_c = \frac{1}{2} \cdot \int_M \left(\frac{\partial^2 y}{\partial x^2} \right)^2 \cdot dm = \frac{1}{2} \cdot \omega^2 \cdot \rho \cdot A \cdot \int_L y^2 \cdot dx \tag{37.7}$$

where M is the mass of the beam, A is the area of the transversal section and ρ is the density of the material.

The maximum kinetic energy of the beam is calculated for the maximum displacement y_0:

$$E_{C_{max}} = \frac{1}{2} \cdot \omega^2 \cdot \rho \cdot A \cdot \int_L y_0^2 \cdot dx = \frac{1}{2} \omega^2 \rho A y_0^2 L \cdot \left(\frac{3}{2} - \frac{4}{\pi} \right) \tag{37.8}$$

Equalizing the maximum potential energy with the maximum kinetic energy Rayleigh's formula is obtained for fundamental angular frequency:

Table 37.2 Analytically determined first resonant frequency based on elastic modulus

Theoretical E [MPa]	Frequecy [Hz]	Experimental E [MPa]	Frequency [Hz]
2020	157.9675	1420.4	132.2188

$$\omega^2 = \frac{EI_z \cdot \int_L \left(\frac{\partial^2 y_0}{\partial x^2}\right)^2 \cdot dx}{\rho \cdot A \cdot \int_L y_0^2 \cdot dx} = C \cdot \frac{EI_z}{\rho \cdot AL^4} = 13.42401 \frac{EI_z}{L^4 \rho A} \tag{37.9}$$

Thus, the natural frequency for the first mode of vibration will be determined as a function of a constant C which depends on the approximation $y(x, t)$ of the beam deformation (37.1) (Table 37.2).

This calculation assumes that the bar is perfectly homogeneous and has uniform properties throughout its length, and that it's vibrating in its first mode of vibration.

37.4 Finite Element Analysis

Mathematical modeling techniques such as finite element analysis (FEA) can also be used to determine the resonant frequencies of a vibrating beam. These techniques involve discretizing the bar into a finite number of small elements and solving the equations of motion to obtain the natural frequencies and associated mode shapes of vibration. To calculate the effective stiffness and effective mass, we need to consider the material properties and beam dimensions. ASA is a thermoplastic material, so we'll need to know its Young's modulus (E) and density (ρ).

Determining Young's elastic modulus requires measuring the stress and strain of a material. The minimum number of samples needed to determine Young's modulus depends on several factors such as the type of material, the level of accuracy required, and the variability of the material properties. As per ISO 527-2 the recommended testing a minimum is of five samples. We tested 5 (3D printed) specimens on a Zwick/Roell tensile test machine obtaining the results shown in Table 37.3. The samples were printed in horizontal position, $\boldsymbol{L} * \boldsymbol{b}$ parallel with the print bed.

Finite Element Method (FEM) software uses mathematical models to simulate the behavior of a structure or component under different conditions, including vibration. The accuracy of the simulation results depends on several factors, such as the accuracy of the model, the mesh size, and the material properties used in the simulation. The

Table 37.3 Measured elastic moduli

Specimen no.	1	2	3	4	5
E [MPa]	1427	1435	1454	1430	1456
Average value	1420.4		Median value	1435	

Table 37.4 ASA filament manufacturer mechanical characteristics

Tensile strength	Tensile modulus	Elongation at break	Specific gravity
47.5 MPa	2020 MPa	15%	1.11 g/cm^3

Table 37.5 FEM resonance frequencies

Mode	1st	2nd	3rd	4th	5th
Theoretical frequency—TF [Hz]	158.98	963.9	1605.5	2573.2	2960.36
Practical frequency—PF [Hz]	134.67	816.49	1359.65	2179.68	2507.6

FEA model used material properties provided by the datasheet of the manufacturer as shown in (Table 37.4) as well as those measured in-house [10].

Only the first frequency was considered for further study due to the fact that 3D printed objects are not usually used at high frequencies [11] (Table 37.5). Theoretical frequency (TF) was determined by FEA (Ansys2021R2) using the data provided by manufacturer while Practical frequency (PF) was determined using the data from the experimental measurement on tested specimens. The samples were fixed from 4 faces like in a vice, leaving the desired $\boldsymbol{L} = 120$ mm distance free. The generated mesh had a standard element of size of 2.29 mm, resulting 9757 nodes and 5932 elements.

37.5 Physical Testing of 3D Printed Beam

In some cases, the resonant frequencies obtained through physical testing and simulation may be similar, especially if the simulation model accurately represents the physical system and the input conditions are well-defined. However, there may also be differences due to simplifications or assumptions made in the simulation model or discrepancies in the material properties used in the simulation compared to the actual material properties of the tested sample.

When 3D printing a cantilever beam (or any other object) some of the most important factors include layer height, infill density, printing orientation, material properties and printing defects. The layer height determines the thickness of each printed layer and can affect the accuracy and surface finish of the part. Finer layer heights can produce smoother and more accurate parts, which can lead to more accurate resonant frequency results (Table 37.6).

For a 0.4 mm nozzle, the recommended layer height for most FDM (Fused Deposition Modeling) 3D printers is between 0.2 and 0.3 mm. This is because printing with a layer height greater than the diameter of the nozzle can cause issues such as under-extrusion, poor adhesion between layers, and a rough surface finish.

The infill density refers to the amount of material that is printed inside the part to provide structural support. Higher infill densities can result in stiffer and more

Table 37.6 Creality CR-10 max 3D printer properties

Nozzle diameter [mm]	0.4	Wall thickness [mm]	1.2
Layer height [mm]	0.2	Infill density [%]	100
Head temperature [°C]	250	Heat bed temperature [°C]	95
Orientation	Vertical (beam length parallel with the table)		

rigid parts, which can affect the resonant frequency of the beam. While a lower infill percentage results in a less stiff part with a lower resonant frequency. However, the specific effect of infill percentage on resonant frequency also depends on the geometry and material of the part [12].

The orientation of the beam during printing can affect the anisotropic behavior of the material and the strength of the part. For example, printing the beam in the vertical orientation (i.e., with the length of the beam parallel to the printing bed) can result in stronger and stiffer parts than printing in the horizontal orientation [13–15].

Printing defects such as warping, delamination, or voids can also affect the resonant frequency of the beam. These defects can introduce local stress concentrations, which can alter the stiffness and natural frequency of the beam.

In physical testing, the resonant frequency of a sample is typically determined by subjecting it to controlled vibration and measuring its response. The testing conditions, such as the method of vibration excitation, the boundary conditions, and the measurement system, can affect the resonant frequency obtained. The specimens were tested in 2 possible orientations as shown in Fig. 37.2, having the layers horizontally and vertically aligned.

Experimental testing involved exciting the crimped bar attached to an electrodynamic shaker with a known force/displacement and measuring the resulting vibrations using a monoaxial accelerometer. The resonant frequencies can then be determined by analyzing the frequency response of the bar/accelerometer and identifying the peaks corresponding to the natural frequencies of vibration (Table 37.7).

Generally, the more samples that are tested, the more statistically significant the results will be. However, testing more samples can also increase the time and cost of the testing. ISO 13373: "Vibration condition monitoring—Part 1: General procedures", ISO 10816: "Mechanical vibration—Evaluation of machine vibration by measurements on non-rotating parts" recommend testing a minimum of three samples

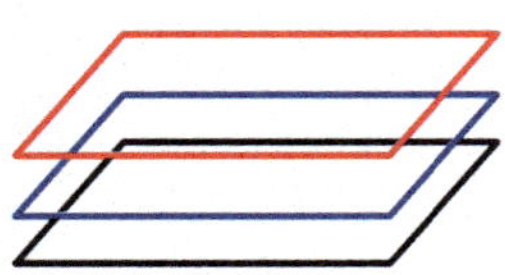

Orientation H - with layers parallel with the horizontal plane

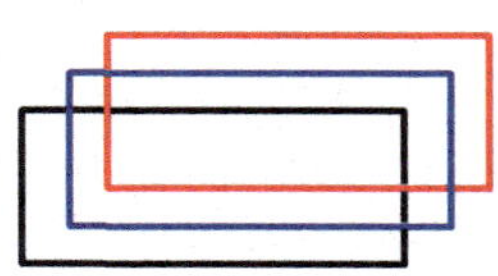

Orientation V - with layers parallel with the vertical plane

Fig. 37.2 Layers orientation

Table 37.7 Measured resonance frequencies

Specimen no.	1	2	3	4	5	6	7	8	9	Average value	Median value
Frequency for orientation H [Hz]	131	124	140	135	143	142	165	134	144	140	140
Frequency for orentation V [Hz]	144	121	154	145	159	150	159	156	159	150	154

to determine the resonant frequency of the beam, with the understanding that more samples may be necessary for better statistical accuracy.

37.6 Conclusions

FEA offers a cost-effective and time-efficient approach to analyze resonance frequencies compared to experimental testing. It eliminates the need for physical prototypes or extensive testing, which can be time-consuming and costly. FEA allows for virtual testing, enabling engineers to explore different design options and assess their resonance characteristics rapidly. In order to accurately determine the resonant frequencies in FEA, Young's elastic moduli were experimentally determined. The exact vibration modes and frequencies depend on the specific geometry and material properties of the beam.

Because FEA model was considered to be homogeneous, obviously there are some discrepancies with the physical model which is made from a matrix of nonhomogeneous wires (melted filament) fused in various proportions (Table 37.8).

It can be observed that the first natural frequency obtained in physical tests was 140 Hz if the piece was fixed in the same way as its growth mode on the 3D printer and if it was rotated by an angle of 90°, the frequency was154 Hz. In technical applications, this property is useful because if the operation of the equipment is at a given frequency, which cannot be changed, the growth mode of the components can

Table 37.8 Frequency comparison

Test type	First resonance frequency [Hz]
Analytical with theoretical elastic modulus	157.9675
Analytical with measured elastic modulus	132.2188
FEA resolved theoretical elastic modulus	158.98
FEA resolved with measured elastic modulus	134.67
Measured frequency for layer orientation H	140
Measured frequency for layer orientation V	154

influence the resonance frequencies of the system, thus the resonance phenomenon can be avoided more easily.

Physical testing plays a crucial role in validating analytical or FEA (Finite Element Analysis) for several reasons: experimental verification, model validation, identification of uncertainties, calibration of models, design validation.

Overall, it is also important to consider the printing characteristics and material properties when 3D printing any type of geometry for modal analysis and resonant frequency testing. Careful consideration of these factors can help to ensure accurate and reliable results.

References

1. All you need to know about ASA for 3D printing. https://www.3dnatives.com/en/asa-3d-printing-170820204/
2. ABS (Curbell Plastics): https://www.curbellplastics.com/materials/plastics/abs/
3. W. Young, et al., *Roark's Formulas for Stress and Strain*, 8th edn. (McGraw-Hill, 2011)
4. F.F. Yildirim et al., The effects of the weathering methods on the properties of the ABS, ASA and PMMA polymers. Polymer Test. (2022). https://doi.org/10.1016/j.polymertesting.2022.107484
5. R. Kishore, K. Mohan, Performance analysis of acrylonitrile styrene acrylate (ASA) polymer blends for automotive applications. Procedia Mater. Sci. **6**, 1833–1843 (2014)
6. S.P. Timoshenko, J.M. Gere, *Theory of Elastic Stability* (McGraw-Hill, 1961)
7. J.M. Gere, S.P. Timoshenko, *Mechanics of Materials* (PWS Publishing Co., 1997)
8. R.C. Hibbeler, *Mechanics of Materials* (Pearson, 2013)
9. M. Rades, *Vibratii Mecanice* (Printech, 2008)
10. Form Futura. https://c.cdnmp.net/490505258/custom/prod/1_fisa_tehnica_1245.pdf
11. P. Elomaa, J. Pekkanen, M. Kärki, Mechanical properties of 3D-printed PLA under tension and dynamic loading conditions. Int. J. Mech. Eng. Robot. Res. **8**(5), 773–778 (2019)
12. X. Tan et al., Experimental investigation on the effect of infill density on mechanical properties and vibrational characteristics of FDM-printed lattice structures. Materials **13**(18), 4007 (2020)
13. Q. Liang, Y.T. Chuan, Y.S. Wong, Effects of build orientation on tensile strength of 3D printed ABS parts fabricated by fused deposition modelling. Virtual Phys. Prototyping **12**(3), 199–206 (2017)
14. M. Sava et al., Characteristics of 3D printable bronze PLA-based filament composites for gaskets. Materials **14**(16), 4770 (2021)
15. D.I. Stoia et al., Properties and medical applications of biocompatible polyamide in additive manufacturing. Mater. Plast. (2021)

Chapter 38
Modal Analysis of Reinforced Concrete Bridges with Multiple Neoprene Bearings

Nicusor Dragan, Adrian Mihai Goanta, Maria Aurora Potirniche, Gigel-Florin Capatana, and Gina Diana Musca Anghelache

Abstract The study presents the modal analysis of a bridge made of reinforced concrete beams subjected to external dynamic actions from road traffic and/or seismic movements. The bridge is modeled as a 6DOF solid body with viscoelastic bearings that schematize the neoprene support devices. The case study presents the modal analysis of the bridge on the A3 Transilvania Highway Romania (between Târgu Mureș and Cluj Napoca, subsection 2B, Săvădisla, Km 29 + 602.75 ↔ Km 29 + 801.25), modeled as a solid body with neoprene bearings devices 300 × 500 × 81 mm, for which they were determined and analyzed the modal vibration modes. A series of experimental results obtained from "in-situ" dynamic tests are presented, using the virtual instrumentation technique; the experimental determinations were carried on Săvădisla bridge. A total numer of 3 × 16 tests were performed in order to determine the response of the bridge structure to dynamic traffic forces (shocks, vibrations). A truck with a total mass of 41 tons generated the shocks by passing over the standardized obstacles chemically fixed on the surface of the bridge; in total, 16 sets of measurements were performed, differentiated by the following parameters: v = 10–50 km/h, h = 40–80 mm, with/without traction.

38.1 The Bridge in Vălişoara on A3 Transilvania Highway

The viaduct on the A3 Transilvania highway between Tg. Mureș and Cluj Napoca is located next to Vălișoara village (Săvădisla commune, Cluj County), between km 29 + 602.75 and km 29 + 801.25 m, on subsection 2B (Fig. 38.1). The reinforced concrete structure of the viaduct (piers, abutments, viaduct deck) supports in each direction, a 12.00 m wide carriageway that includes two traffic lanes with a width of 3.75 m each, a parking lane with a width of 3.50 m and two safety lanes with a

N. Dragan (✉) · A. M. Goanta · M. A. Potirniche · G.-F. Capatana · G. D. M. Anghelache
Research Center for Mechanics of Machines and Technological Equipments, Engineering and Agronomy Faculty of Braila, Engineering Sciences and Management, Dunarea de Jos University of Galati, 810017 Braila, Romania
e-mail: nicusor.dragan@ugal.ro

N. Herisanu and V. Marinca (eds.), *Acoustics and Vibration of Mechanical Structures—AVMS-2023*, Springer Proceedings in Physics 302,
https://doi.org/10.1007/978-3-031-48087-4_38

width of 0.50 m each. The total width of the superstructure is 13.60 m and consists of the width of the carriageway to which are added the two spaces of 0.80 m for the placement of safety parapets, one on each side (Fig. 38.2). The carriageway has a unique transverse slope of 2.5% for both directions of traffic [1].

The infrastructure of the bridge consists of 2 abutments and 4 pillars for each direction of traffic. The bridge deck is made up of 4 × 5 prefabricated U-type beams, arranged at a distance of 3.32 m between the axes in the transverse plane and a 25 cm thick reinforced concrete slab being poured over the beams in order to rigidly connect them (Fig. 38.2). The beams are made of pre-compressed reinforced concrete of class

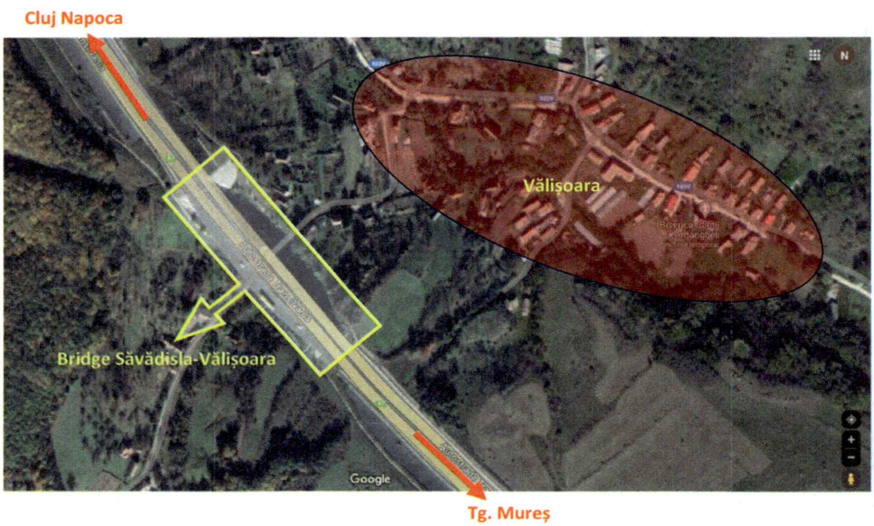

Fig. 38.1 Position of the bridge on the A3 Transilvania highway [Google maps], (Săvădisla-Vălișoara, subsection 2B, K m 29 + 602.75 ↔ K m 29 + 801.25)

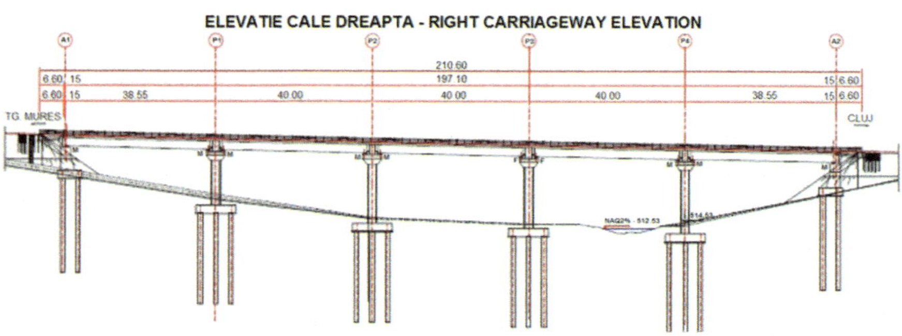

Fig. 38.2 The structure of the bridge on the A3 Transylvania highway (lateral view). Reprinted from [2] with permissions. © Romanian Acoustic Society, all rights reserved

C35/45, and the overplate of reinforced concrete of class C25/30. The prestressed reinforced U beam reinforcements are made of full-section concrete-steel (bars) and cable steel strands, pre-tensioned or post-tensioned. The bridge has 5 equal openings of 40 m. The support of the viaduct deck on the infrastructure elements, abutments and piers, was carried out by Freyssinet neoprene bearing apparatus 300 × 500 × 81 mm [1].

38.2 3D Modeling of the Bridge Deck

The geometric modeling of the deck of the Vălișoara bridge was performed in Siemens PLM-NX 7 on the basis of the technical execution documentation provided by the bridge designer IPTANA Bucharest Romania [1]. Figure 38.3 shows the simplified 3D model axonometric view of a single reinforced concrete U-beam used for the bridge; the bridge was built of 20 U-beams connected on the over-deck roadway (25 cm thick reinforced concrete rigid plate/slab and 5 cm thick asphalt layer).

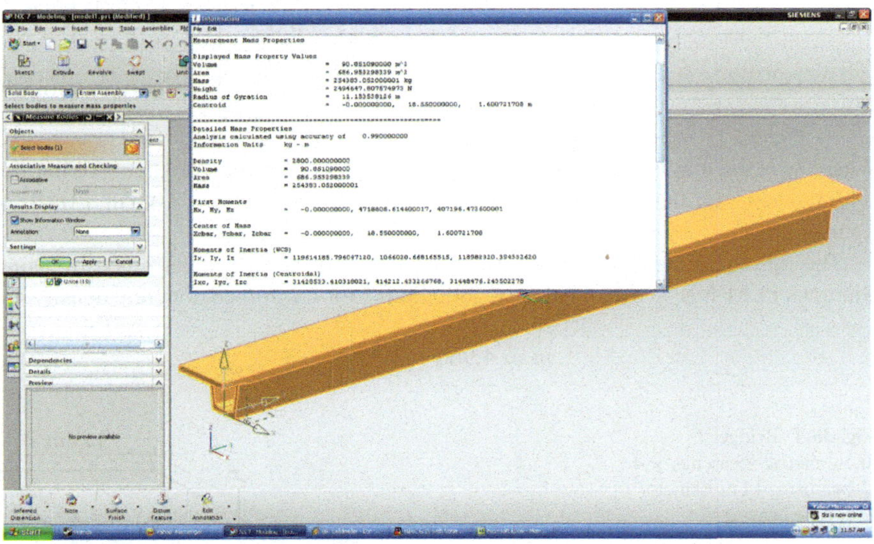

Fig. 38.3 Siemens NX 7 3D U-beam modeling–calculation of inertial characteristics

38.3 Inertial Characteristics of the Bridge

Figure 38.4 shows the simplified model from for the bridge modeled as a 6DOF rigid body made of 20 reinforced concrete beams, individually supported on 4 identical neoprene support devices [3–5]. The 20 U-beams are grouped in 4 rows (Cx axis) of 5 arches each (Cy axis), their solidarization being done constructively by a concrete slab (plus a 5 cm thick asphalt overlay, without functionality for beams connection).

In the central system axis Cxyz, where $\begin{bmatrix} x & y & z \end{bmatrix}$ are the translation displacements of the mass center C and $\begin{bmatrix} \Phi_x & \Phi_y & \Phi_z \end{bmatrix}$ are the rotation angles of the bridge, the vector of general coordinates can be written as $\{q\} = \begin{bmatrix} x \; y \; z & \Phi_x & \Phi_y & \Phi_z \end{bmatrix}^T$. Due to the structural symmetry of the bridge (on the vertical axis Cz), the inertia/mass matrix of the bridge is diagonal and can be written as follows [6–8].

$$[M] = DIAG\left(m, m, m, J_x, J_y, J_z\right) = \begin{bmatrix} \begin{bmatrix} m & 0 & 0 \\ 0 & m & 0 \\ 0 & 0 & m \end{bmatrix} & \begin{bmatrix} 0\,0\,0 \\ 0\,0\,0 \\ 0\,0\,0 \end{bmatrix} \\ \begin{bmatrix} 0\,0\,0 \\ 0\,0\,0 \\ 0\,0\,0 \end{bmatrix} & \begin{bmatrix} J_x & 0 & 0 \\ 0 & J_y & 0 \\ 0 & 0 & J_z \end{bmatrix} \end{bmatrix}$$

$$= \begin{bmatrix} [M_{11}] & [M_{12}] \\ [M_{21}] & [M_{22}] \end{bmatrix}, \qquad (38.1)$$

where m is the mass of the bridge and $\left(J_x, J_y, J_z\right)$ are the mass moments of inertia, with the M.U. for the submatrice: $[M_{11}] \rightarrow$ kg, $[M_{12}] \equiv [M_{12}]^T \rightarrow$ kg m, $[M_{22}] \rightarrow$ kg m^2,

The mass and the mass moments of inertia of the bridge, determined by using Siemens PLM-NX 7, with a value of 2600 kg/m^3 for reinforced concrete density are

$$m = 4.960 \times 10^6 [\text{kg}]$$

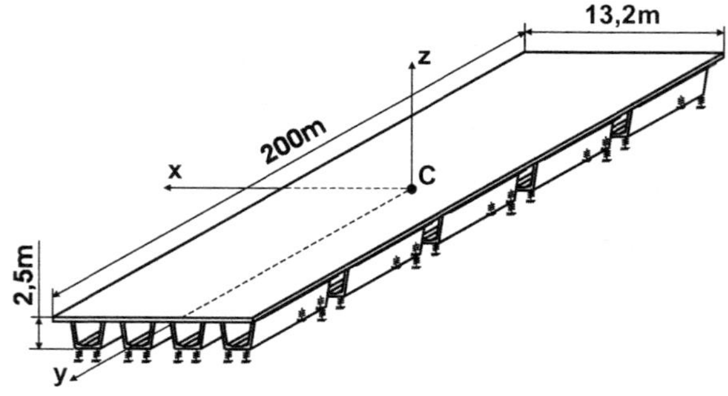

Fig. 38.4 Bridge axonometric−5 arches × 4 U-beams × 4 Freyssinet neoprene bearingsn. Reprinted from [2] with permissions. © Romanian Acoustic Society, all rights reserved

$$J_x = 16475 \times 10^6 [\text{kg m}^2]$$

$$J_y = 75.663 \times 10^6 [\text{kg m}^2]$$

$$J_z = 16542 \times 10^6 [\text{kg m}^2],$$

the mass matrix becoming:

$$[M] = 10^6 \times \text{DIAG}(4.960, 4.960, 4.960, 16475, 75.663, 16542). \tag{38.2}$$

38.4 Elasticity Matrix of the Bridge

The bridge modeled as 6DOF solid body has 80 identical triortogonal elastic bearings with stiffness $k_x = 3.15 \times 10^6 [N/m], k_y = 3.15 \times 10^6 [N/m], k_z = 650 \times 10^6 [N/m]$. Considering the model of the bridge as a 6DOF solid body with symmetry related to the central vertical axis Cz, the elasticity matrix has the general the form [8–10].

$$[k] = \begin{bmatrix} \Sigma k_{ix} & 0 & 0 & 0 & \Sigma z_i k_{ix} & 0 \\ 0 & \Sigma k_{iy} & 0 & -\Sigma z_i k_{iy} & 0 & 0 \\ 0 & 0 & \Sigma k_{iz} & 0 & 0 & 0 \\ 0 & -\Sigma z_i k_{iy} & 0 & \Sigma\left(k_{iy} z_i^2 + k_{iz} y_i^2\right) & 0 & 0 \\ \Sigma z_i k_{ix} & 0 & 0 & 0 & \Sigma\left(k_{iz} x_i^2 + k_{ix} z_i^2\right) & 0 \\ 0 & 0 & 0 & 0 & 0 & \Sigma\left(k_{iz} x_i^2 + k_{ix} z_i^2\right) \end{bmatrix} \tag{38.3}$$

where $B(x_i, y_i, z_i) i = \overline{1, 80}$ are the coordinates of the bridge bearings points connection in the central axis system Cxyz and the axial triortogonal stiffness are $\left(k_{ix}, k_{iy}, k_{iz}\right) = \left(k_x, k_y, k_z\right) = (3.15, 3.15, 650) \times 10^6 N/m$.

Taking into consideration the axial stiffness of the bearings and the coordinates of the bridge bearings, the elasticity matrix of the 6DOF mechanical system becomes

$$[K] = 10^6 \begin{bmatrix} 252 & 0 & 0 & 0 & -367 & 0 \\ 0 & 252 & 0 & 367 & 0 & 0 \\ 0 & 0 & 52000 & 0 & 0 & 0 \\ 0 & 367 & 0 & 183342000 & 0 & 0 \\ -367 & 0 & 0 & 0 & 724113 & 0 \\ 0 & 0 & 0 & 0 & 0 & 892009 \end{bmatrix} \tag{38.4}$$

where the M.U. of the elasticity coefficients are: for submatrix $[K_{11}]$, yellow highlight–*N/m*; for submatrices $[K_{12}] \equiv [K_{21}]^T$, red highlight–*N*; for submatrix $[K_{22}]$, green highlight–*Nm*

38.5 Dynamic Matrix of the Bridge. Decoupled Subsystems with Coupled Movements

Taking into consideration the mass matrix done by (38.2) and the elasticity matrix done by (38.4), the dynamic matrix of the bridge modeled as a 6DOF solid body with 80 elastic bearing becomes [8]:

$$[D] = [M]^{-1} \times [K] = \begin{bmatrix} 50.806 & 0 & 0 & 0 & -73.992 & 0 \\ 0 & 50.806 & 0 & 73.992 & 0 & 0 \\ 0 & 0 & 10483.871 & 0 & 0 & 0 \\ 0 & 0.022 & 0 & 11128.498 & 0 & 0 \\ -4.850 & 0 & 0 & 0 & 9570.239 & 0 \\ 0 & 0 & 0 & 0 & 0 & 53.924 \end{bmatrix} [s^{-2}] \tag{38.5}$$

From the analysis of the dynamic matrix structure (38.5) of the 6DOF mechanical system results that its movements are decoupled into 4 subsystems with coupled movements, the constitutive equations of free vibration being as follows [2–5]:

- subsystem (x, Φ_y)–swaying + roll rotation, green highlight

$$\begin{Bmatrix} \ddot{x} \\ \ddot{\Phi}_y \end{Bmatrix} + \begin{bmatrix} 50.806 & -73.992 \\ -4.850 & 9570.239 \end{bmatrix} \begin{Bmatrix} x \\ \Phi_y \end{Bmatrix} = \begin{Bmatrix} 0 \\ 0 \end{Bmatrix} \begin{bmatrix} \mathrm{ms}^{-2} \\ \mathrm{s}^{-2} \end{bmatrix} \tag{38.6}$$

- subsystem (y, Φ_x)–surging + pitch rotation, yellow highlight

$$\begin{Bmatrix} \ddot{y} \\ \ddot{\Phi}_x \end{Bmatrix} + \begin{bmatrix} 50.806 & 73.992 \\ 0.022 & 11128.498 \end{bmatrix} \begin{Bmatrix} y \\ \Phi_x \end{Bmatrix} = \begin{Bmatrix} 0 \\ 0 \end{Bmatrix} \begin{bmatrix} \mathrm{ms}^{-2} \\ \mathrm{s}^{-2} \end{bmatrix} \tag{38.7}$$

- decupled movement (z)–vertical translation, red highlight

$$\ddot{z} + 10483.871z = 0\left[\mathrm{ms}^{-2}\right] \tag{38.8}$$

- decupled movement (Φ_z)–yaw rotation, blue highlight

$$\ddot{\Phi}_z + 53.924\Phi_z = 0\left[\mathrm{ms}^{-2}\right]. \tag{38.9}$$

38.6 Modal Analysis of the Bridge with Decoupled Movements

Taking into consideration the dynamic matrix $[D]$ of a linear nDOF mechanical system, the eigenvalues $\Lambda_i\ i = \overline{1, n}$, the eigenpulsations $p_i\ i = \overline{1, n}$, the eigenfrequencies $p_i\ i = \overline{1, n}$ and the eigenvectors $\{v_i\}\ i = \overline{1, n}$ can be determined by algebraic calculus as follows [8, 11–14]

$$det([D] - \Lambda[I_n]) = 0 \ \rightarrow \Lambda_i\ i = \overline{1, n} \ \rightarrow p_i = \sqrt{\Lambda_i} \ \rightarrow f_i = \frac{p_i}{2\pi} \tag{38.10}$$

$$\Lambda_i\ i = \overline{1, n} : [D]\{v_i\} = \Lambda_i\{v_i\} \ \rightarrow ([D] - \Lambda_i[I_n])\{v_i\} = \{0_n\} \tag{38.11}$$

where $[I_n]$ is the n order identity matrix and $\{0_n\}$ is the n order null vector.

Table 38.1 shows the results of the modal analysis of the movements of the decoupled subsystems due to the structural symmetries of the 6DOF model of the mechanical system of the bridge.

38.7 Experimental Dynamic Tests of the Bridge

The dynamic tests were performed on the construction site of the A3 Transilvania highway, at the viaduct located at km 29 + 602.75 ↔ 29 + 801.25 (in Săvădisla, between Târgu Mureş and Cluj) by a team from the acoustics and vibrations laboratory of the Institute of research for construction equipment and technologies ICECON S.A. Bucharest; the dynamic impulsive load were generated by a truck with four axles and a mass of 41 tons passing over a standardized obstacle with height of 4 cm (acc. STAS 12,504-86), fixed with chemical anchors on the surface of the bridge, Fig. 38.5; the speeds of passing over the obstacle were 10 km/h, 20 km/h, 30 km/h and 50 km/h, and the tests were carried out in two different driving situations of the drive axle (no. 2): with and without traction. The acceleration signals were acquired on three channels (axis x, axis y, axis z) by a USB port four-channel acquisition board (National Instruments NI 9233) connected to a PC workstation [15, 16]. The experimental data were processed, analyzed and visualized by virtual instrumentation in the LabView®. The three acceleration signals were analyzed in time domain and in frequency domain (FFT analysis). Figure 38.6 shows the time domain signal and Fig. 38.7 shows the frequency domain of vertical acceleration of the central point of the bridge [17, 18].

Table 38.2 shows the values for modal frequencies and the amplitudes of vertical acceleration in the case of dynamic test with traction and crossing speed $v = 10$ km/h.

Table 38.1 Modal analysis of the bridge. eigenfrequencies and eigenvectors

Subsystem movements	Eigenvalues (s^{-2})	Eigenpulsation (rad/s)	Eigenfrequency (Hz)	Eigenvector
(x, Φ_y)	50.7683	7.125	1.134	$\{v_1\} = \begin{Bmatrix} 1962.777 \\ 1 \end{Bmatrix} \begin{pmatrix} m/rad \\ - \end{pmatrix}$
	9570.2767	98.828	15.570	$\{v_2\} = \begin{Bmatrix} -0.00773 \\ 1 \end{Bmatrix} \begin{pmatrix} m/rad \\ - \end{pmatrix}$
(y, Φ_x)	50.8059	7.128	1.134	$\{v_3\} = \begin{Bmatrix} -503531.46 \\ 1 \end{Bmatrix} \begin{pmatrix} m/rad \\ - \end{pmatrix}$
	11,128.4981	105.492	16.790	$\{v_4\} = \begin{Bmatrix} 0.006679 \\ 1 \end{Bmatrix} \begin{pmatrix} m/rad \\ - \end{pmatrix}$
(z)	10,483.87	102.391	16.296	–
(Φ_z)	52.922	7.343	1.169	–

Fig. 38.5 Dynamic testing of the bridge (41t truck; obstacle on the road h = 40 mm)

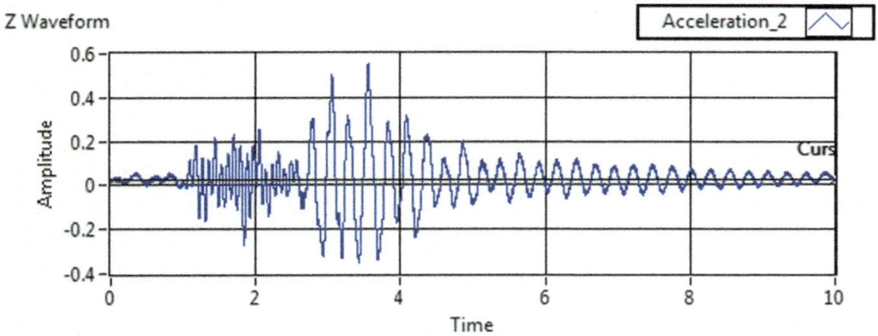

Fig. 38.6 Time domain vertical acceleration ($v = 10$ km/h, with traction)

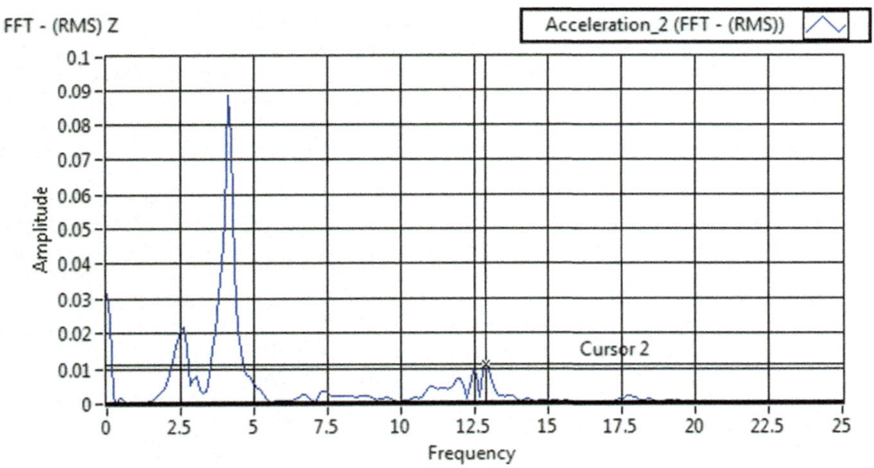

Fig. 38.7 Frequency domain vertical acceleration ($v = 10$ km/h, with traction)

Table 38.2 Modal frequencies and amplitudes of vertical acceleration of the bridge ($v = 10$ km/h)

Mode	1	2	3	4	5
Frequency (Hz)	2.618	*4.145*	12.001	12.440	12.869
Acc. ampl. (ms^{-2})	0.02182	*0.08893*	0.00697	0.00946	0.01104

38.8 Conclusions

A 6DOF solid body with multiple viscous-elastic bearings and structural symmetries can be modeled as dynamic decoupled subsystems with coupled movements and, therefore the constitutive equations being easier to study analytically [19, 20].

The extreme values from Table 38.1 of the eigenvectors show that the movements the subsystems (x, Φ_y) and (y, Φ_x) are very weakly coupled, the movements being quasi-decoupled [21]. The values in Table 38.2 show that the dynamic response of the bridge deck is a modal response, the main frequency of vertical response being $f = 4.145$ Hz (yellow highlight) [22, 23].

The global modal damping characteristics of the bridge (δ—logarithmic decrement, ζ—damping ratio) can be determined from the time domain and frequency analysis of the acceleration signals. The relevance of the values determined by spectral analysis using FFT depends of the length "in time" of the analyzed signal.

References

1. S.C. Iptana, S.A. Bucharest, Technical documentation of viaduct Săvădisla-Vălișoara, Km 29+602.75↔Km 29+801.25 (A3 Transilvania Highway, subsection 2B)
2. N. Dragan, I. Dzemyda, The innovative concept of dynamic analysis for the movements of the viaduct modeled as solid body with elastic bearings. Rom. J. Acoust. Vib. **8**(1), 13–20 (2011)
3. L. Sakalauskas, N. Dragan, O. Vasile, Studies concerning the optimization of the modal analysis of the Bechtel's viaducts—calculus of natural frequencies and eigenvalues. Rom. J. Acoust. Vib. **8**(1), 21–26 (2011)
4. R. Manolescu, G. Bașno, A. Potarniche, N. Dragan, Dynamic analysis of the elastic beared solid body with structural symmetries. Case study—decoupled vibrations of reinforced concrete parts, in *Proceedings of the Scientific Communications Session "SIMEC 2010"* (MatrixRom, Bucharest, 2011), pp. 69–74. ISSN 1842-8045
5. P. Bratu, N. Dragan, O. Vasile, The analysis of rigid body vibratory movements of the viaduct on neoprene support devices. in *Proceedings of the XIIIth National Congress of Roads and Bridges vol. II*, Poiana Brașov 15–17.09.2010, (Bucharest, 2010), pp. 185–194
6. P. Bratu, Vibrations of elastic systems. Tehnica, (Bucharest, 2000)
7. P. Bratu, Elastic structures analysis—Static and dynamic action behavior. Impuls, (Bucharest, 2011)
8. N. Dragan, Dynamics of inertial vibrating conveyors. Impuls, (Bucharest, 2003)
9. N. Dragan, B. Lificiu, M. Cantaragiu, Dynamic modeling of elastically supported solid body with structural symmetries. Case study—the modal calculation of the bridge built of U type reinforced concrete beams. in *Proceedings of the XIVth National symposium of construction equipment "SINUC 2010"*, ISSN 978-973-100-144-9, (Bucharest, 2010)
10. N. Dragan, A. Potarniche, Experimental analysis of the dynamics of reinforced concrete bridges subjected to traffic actions. in *Proceedings of the XIVth National symposium of construction equipment "SINUC 2010"*, (Bucharest, 2010). ISSN 978-973-100-144-9
11. D. Inman, *Vibration with Control* (John Wiley and Sons Ltd., New Jersey, 2006)
12. S. Rao, *Mechanical Vibrations*, 4th edn. (Pearson Education Inc., New Jersey, 2004)
13. P. Bratu, Dynamical behavior of the structures subjected to mechanical shocks, Research Trends in mechanics vol. III. Romanian Academy, (Bucharest, 2009)
14. N. Dragan, Dynamic analysis of reinforced concrete bridges–Modal calculus (eigenvalues, eigenfrequencies). Synth. Theor. Appl. Mech. **9**(4), 245–254, ISSN 2068-6331, MatrixRom, (Bucharest, 2018)
15. Gh. Panfiloiu, N. Dragan, Dynamic analysis of the vibrations of the mechanical elastic systems using virtual instrumentation. Rom. J. Acoust. Vib. **8**(1), 27–32 (2011)
16. Gh. Panfiloiu, N. Dragan, New systems for experimental analysis of the dynamics of machines and equipment–Virtual instrumentation. In Proceedings of the XIVth National symposium of construction equipment SINUC 2010, ISSN 978-973-100-1449, (MatrixRom, Bucharest, 2010)

17. P. Bratu, N. Dragan, in *Dynamic modelling of the response of the viaduct subjected on traffic actions— modal calculus*. in *the 11-th International Congress on Automotive and Transport Engineering CONAT 2010 Proceedings—Volume III "Automotive Vehicles and Environment"*, (Brasov, 2010), pp. 185–192, 16-17 Dec. ISSN 2069-0401
18. P. Bratu, O. Vasile, C. Zaharia, Experimental studies of a bridge made from reinforced concrete beams–Bridge of Transilvania highway. in *the Proceedings of 35th Danubia Adria Symposium on Advances in Experimental Mechanics*, (Sinaia, 2018), pp. 187–188
19. P. Bratu, N. Dragan, in *Dynamic analysis of machine-structure interaction based on the equivalent model of rigid to viscoelastic connections*. The Annals of "Dunarea de Jos" University of Galati, Fascicle XIV Mechanical Engineering, (IV), (Galati, 1997), pp. 9–14. ISSN 1224-5615
20. P. Bratu, N. Dragan, in *The analysis of uncoupled movements of the solid-body with elastic bearings model*. The Annals of "Dunarea de Jos" University of Galati, Fascicle XIV Mechanical Engineering, (IV), (Galati 1997), pp. 15–20. ISSN 1224-5615
21. V. Ovidiu, Shock evaluation on viaducts for heavy vehicles crossing over obstacles, in *20th International Congress on Sound and Vibration 2013-ICSV 20*, vol. 1, ed. by M. Crocker, B. Paosawatyanyong (Curran Associates Inc., Red Hook NY, 2018), pp.576–581
22. B. Polidor, V. Ovidiu, Establishing the effect of seismic shock action for a viaduct with viscoelastic isolators, in *20th International Congress on Sound and Vibration 2013 - ICSV 20*, vol. 2, ed. by M. Crocker, B. Paosawatyanyong (Curran Associates Inc., Red Hook NY, 2018), pp.1340–1345
23. A.M. Potirniche, O. Vasile, G.F. Capatana, Modal analysis of a mechanical system modeled as a 6 degrees-of-freedom solid body with elastic bearings and structural symmetries. Rom. J. Acoust. Vib. **19**(1), 36–40 (2022)

Correction to: Three-Dimensional Model of Wind Turbine NREL-Monopile 5 MW for Vibration Characterisation

Marcela R. Machado and Maciej Dutkiewicz

Correction to: Chapter 4 in: N. Herisanu and V. Marinca (eds.), *Acoustics and Vibration of Mechanical Structures—AVMS-2023*, Springer Proceedings in Physics 302, https://doi.org/10.1007/978-3-031-48087-4_4

The original version of the book was published without the acknowledgement section in Chapter 4 which has now been included. The book and the correction chapter have been updated with the changes.

The updated version of this chapter can be found at
https://doi.org/10.1007/978-3-031-48087-4_4

N. Herisanu and V. Marinca (eds.), *Acoustics and Vibration of Mechanical Structures—AVMS-2023*, Springer Proceedings in Physics 302,
https://doi.org/10.1007/978-3-031-48087-4_39

The manufacturer's authorised representative in the EU is Springer Nature Customer Service Centre GmbH, Europaplatz 3, 69115 Heidelberg, Germany. If you have any concerns regarding our products, please contact ProductSafety@springernature.com

Printed and bound by CPI Group (UK) Ltd, Croydon, CR0 4YY
01/10/2026
02225866-0015